INSTRUCTOR'S GUIDE FOR

BIOLOGY

CONCEPTS & CONNECTIONS

Second Edition

Campbell • Mitchell • Reece

Eugene J. Fenster
Longview Community College

Fred M. Rhoades
Western Washington University

An imprint of Addison Wesley Longman, Inc.

Menlo Park, California • Reading, Massachusetts
New York • Harlow, England • Don Mills, Ontario
Sydney • Mexico City • Madrid • Amsterdam

Sponsoring Editor: Lisa Moller
Associate Editor: Evelyn Dahlgren
Production Editor: Larry Olsen
Copy Editor: Brian Jones
Cover Designer: Yvo Riezebos
Cover Photo: Copyright © Dwight Kuhn

Copyright © 1997 by The Benjamin/Cummings Publishing Company, Inc.
All rights reserved. No part of this publication may be reproduced, stored in a database or retrieval system, or transmitted in any form or by any means, electronic, mechanical, photocopying, recording, or any other media or embodiments now known or hereafter to become known, without the prior written permission of the publisher. Printed in the United States of America. Published simultaneously in Canada.

ISBN 0-8053-2025-3

1 2 3 4 5 6 7 8 9 10—VG—00 99 98 97 96

The Benjamin/Cummings Publishing Company, Inc.
2725 Sand Hill Road
Menlo Park, California 94025

Contents

	How To Use This Book	iv
Chapter 1	Introduction: The Scientific Study of Life	1
Chapter 2	The Chemical Basis of Life	7
Chapter 3	The Molecules of Cells	14
Chapter 4	A Tour of the Cell	23
Chapter 5	The Working Cell	31
Chapter 6	How Cells Harvest Chemical Energy	39
Chapter 7	Photosynthesis: Using Light to Make Food	47
Chapter 8	The Cellular Basis of Reproduction and Inheritance	55
Chapter 9	Patterns of Inheritance	65
Chapter 10	Molecular Biology of the Gene	75
Chapter 11	The Control of Gene Expression	85
Chapter 12	DNA Technology	93
Chapter 13	The Human Genome	103
Chapter 14	How Populations Evolve	114
Chapter 15	The Origin of Species	126
Chapter 16	Tracing Evolutionary History	132
Chapter 17	The Origin and Evolution of Microbial Life: Prokaryotes and Protists	141
Chapter 18	Plants, Fungi, and the Colonization of Land	155
Chapter 19	The Evolution of Animal Diversity	166
Chapter 20	Unifying Concepts of Animal Structure and Function	180
Chapter 21	Nutrition and Digestion	189
Chapter 22	Respiration: The Exchange of Gases	200
Chapter 23	Circulation	207
Chapter 24	The Immune System	216
Chapter 25	Control of the Internal Environment	225
Chapter 26	Chemical Regulation	232
Chapter 27	Reproduction and Embryonic Development	240
Chapter 28	Nervous Systems	253
Chapter 29	The Senses	266
Chapter 30	How Animals Move	275
Chapter 31	Plant Structure, Reproduction, and Development	283
Chapter 32	Plant Nutrition and Transport	294
Chapter 33	Control Systems in Plants	302
Chapter 34	The Biosphere: An Introduction to Earth's Diverse Environments	309
Chapter 35	Population Dynamics	321
Chapter 36	Communities and Ecosystems	329
Chapter 37	Behavioral Adaptations to the Environment	342
Chapter 38	Human Evolution and Its Ecological Impact	354
Appendix A	Organizing a Course Syllabus	366
Appendix B	General Resources and References	368
Appendix C	Addresses of Resource Suppliers	373

How To Use This Book

Modular Format

Teaching introductory biology to students who are not science majors can be one of the most rewarding occupations and also one of the most challenging. Biologists are diverse in their ways of looking at living organisms. How do we organize these diverse approaches into a whole that nonscience majors can grasp and find interesting?

In *Biology: Concepts and Connections*, Second Edition, each chapter consists of a series of concept modules, each taking up no more than two facing pages. Each module's heading announces a focal concept. The module's text and illustrations support the concept with explanation and evidence. Each module is integrated into the context of the chapter's topic, and to the concepts covered by preceding and following modules. It is the hope of the authors that such an approach will help students organize the details of biology around an organized conceptual framework. In addition, the overall organization of the text is based on the interconnecting contexts of several broad themes: evolution, ecology, human biology, interdisciplinary approaches, and biological science as a process.

The structure of the chapters provides several ways for students to organize their study and evaluate their assimilation of the information. In your introductory lecture, encourage your students to learn the ways to use this structure. Be sure that you read the text authors' preface to further your understanding of the guiding principles behind the organization of *Biology: Concepts and Connections*.

Organization of This Instructor's Guide

As you set out to organize a syllabus for an introductory course in biology, you will have your own approach as both a teacher and a biologist. There are many factors to consider. Your department may provide a general curricular design that you must follow. You may want to emphasize the theme of biology as a scientific way of knowing. You will realize that most of the challenges facing humankind have at least some biological bases. However, some special topics may be currently "in the news," and you will want to cover them in greater detail. A goal to consider is that of providing your students with the opportunity to become an "educated voter." Learning the material is important, but of even greater importance is learning to learn and learning to think critically. Your course will likely include laboratory or discussion sections, where some of the course content will best be presented. Above all, you must constantly balance the breadth and depth of coverage of each topic.

As citizens and students, your audience will approach biology with a broad range of prior experience and diverse viewpoints. In addition to the challenges of the material and your students, other factors will likely affect your course design. Classes presented over one quarter or one semester must focus on some, but not all, topics. You may be lecturing to 25 students, or to 500. Some laboratories and discussion sections may be staffed by inexperienced graduate (or undergraduate) students.

We hope that the material in this *Instructor's Guide* will help you develop your course in the context of these factors. We have made an effort to provide both seasoned and first-time lecturers with ideas for teaching biology to nonscience majors using this innovative text. The modular format of *Biology: Concepts and Connections* should make it much easier to start your course organization. The chapters in the *Instructor's Guide* follow the chapters of the text and are organized as discussed below. Suggested ways to structure and organize the overall course content—from the perspective of cell first, diversity first, or ecology first, and for periods of various lengths—are presented in Appendix A.

Each chapter of this guide is divided into five sections. In the **Approach** section, we have tried to consider both the natural development of the material, as it relates to student interests and biases, and the particular pedagogy as developed in the text. In some cases, the latter organization may not be the same as the pedagogical development of the material in other texts you have used previously. Reviewing the Approach section should provide you with some ways to check your own biases and to begin to organize your thoughts parallel to those of the authors. Also included here are hints on making material more accessible and interesting.

The **Chapter Objectives** section lists the main teaching and learning goals you should consider setting for a chapter. Included are the main topics, concepts, and terms your lectures should cover. This section is the first place to look for an overview of the content of a chapter, without reading through the individual module headings or lecture outline.

The **Lecture Outline** section presents a hierarchically arranged outline of the module contents. Module headings appear as primary and secondary outline entries, in the order they appear in the chapter. Important subsidiary details appear as secondary or tertiary entries. To help structure the hierarchy, additional primary entries have been added to group broader topics. Within the Lecture Outline you will find extensive cross-referencing to other modules. Also check the Approach section for a discussion of any unusual pedagogical arrangement of the material.

The opening essays are of particular importance to the structure of the text. They ground the theme of each chapter firmly in an organismal context. This material should play the same role in your lectures. In the lecture outlines, the content of the opening essay has been incorporated in the introductory section for each chapter.

The close integration of illustrations and text is another important feature of this text and should be used in your lectures. Pertinent illustrations from the text are referenced in the lecture outlines at points where they would first be used in a lecture. These are available as overhead transparencies and masters from the supplements package, or as other illustrations that are not included in that package (see the conventions below).

Most chapters contain more material than can be covered in a single lecture. In this guide, no attempt has been made to construct outlines that take a standard amount of time. You will have to gauge how much to cover considering your own time constraints. By using the hierarchy of topics at the primary and secondary levels in the outline, you should be able to adapt the outline to your own coverage, including, omitting, or reordering modules as you see fit. Check Appendix A for suggestions on how to reorganize selected chapters if you choose a diversity-first or ecosystem-first overall course structure.

The **Class Activities** section includes ideas for activities to spark lecture. Most of the materials mentioned here should be available from local environments or stores without much trouble or expense. As indicated, some of the suggested activities will work best in smaller classes. Others work well even in lecture settings of several hundred students, or they can be adapted to your particular curriculum (incorporated into discussion sections, laboratory introductions, etc.).

The **Resources and References** section includes books, journal articles, and materials from biological and multimedia distributors. Because biology deals with dynamic processes, video and computer media with animated demonstrations are particularly valuable in presenting many of the concepts covered in a general biology course. There are numerous video and computer resources, and more become available with every new catalog that comes along. Such resources vary in quality in terms of content, mode of delivery, and appropriateness to non-science majors. The recommendations in this section are based on our personal experience with a resource, favorable reviews, and the reputation of the author or publisher. In order to be effective in using a multimedia resource, you must spend time reviewing the materials and preparing your presentation in advance.

At the end of the first chapter of each unit (chapters 1, 2, 8, 14, 17, 20, 31, and 34) you will find a list of **Internet Resources** for that unit. This is not meant to be an exhaustive list of sites. Each of these sources has links to other sites that will be useful. Each listing includes the title of the web page and its URL (address). Web addresses change frequently, and something that can be accessed one day may be unavailable the next day and again available the following day. If you cannot access a particular site, check for up-to-date links on the **Biology: Concepts and Connections,** Second Edition, web page (http://www.aw.com/bc/sci/bio/).

Appendix B includes general resources suggested for the text as a whole. Appendix C lists the addresses of distributors of video and computer resources. Most of these distributors have a preview policy that allows you to decide whether the product is suitable for your use.

Outline Conventions

Terms used in the Lecture Outline sections refer to modules in the following ways:

Module: A parenthetical reference to the module number in the text. Material in the lecture outline up to the next such module reference outlines the content of this module.

Preview: Material in the module referenced is covered later in the text in greater detail. The details in the present module are a taste of things to come.

Review: Material in the module or illustration referenced has been covered in a previous module. Usually, the information or illustration was particularly important or is topically or pedagogically relevant.

NOTE: These are side comments mentioning details not explicitly presented in the text.

References to text illustrations, numbered according to the module number, are made in the lecture outline at points where they should be introduced to students during lecture. Some of these illustrations are provided in the supplements package; others are not.

To distinguish the various types of figures, the following conventions are used:

Acetate: Color overhead transparency, available in the supplements package.

Master: Black-line master, available in the supplements package. These masters can be duplicated to provide student copies or to provide you with additional black-line transparencies for overhead projection.

Chapter-opening photo: This refers to the illustration accompanying the introductory module in each chapter. These are not included in the supplements package.

§ ∞ §

Know your audience.
Keep track of what your audience has understood of what you say to them.
Teaching less can be teaching more.

1
INTRODUCTION: THE SCIENTIFIC STUDY OF LIFE

APPROACH

The introductory lectures are critical in setting the tone for the remainder of the course. To appeal to the diverse interests and experiences of nonscience majors, it is helpful to present yourself first as a human being separate from the realm of science, and then to talk in general about your interests in biology (research and teaching).

Each instructor will have to deal with administrative details regarding the length of the course; lecture, discussion, and laboratory organization; the availability of outside resources (tutorial center, computer and audiovisual resources, etc.); and the grading system. Students appreciate knowing from the start exactly what will be required of them to complete your course. This is also the appropriate time for comments about attendance, how to read the text and otherwise prepare for lecture, and how to make up requirements when unexpected events such as illness cause a student to miss an assignment.

Since the modular approach to this text is different from the approach taken in many science texts, spend some time suggesting how students might best approach the material. You may choose not to cover all the modules in all the chapters in one course, so develop a consistent technique of assigning material that is required, optional, or to be omitted. Consult Appendix A at the end of this guide for suggestions on how to rearrange the content to fit your topic sequence, style, and course length.

This chapter weaves two introductory aspects together: common themes in biology and characteristics of life. Considerable detail is presented. Stress that this chapter is both an introduction and a conceptual preview of the important themes of the course. Take care not to intimidate students with an overwhelming amount of detail in the introductory lectures on these topics. Suggest they read the module headings for the main messages and the details of the text for a general overview, not total recall.

We hear a lot about the endangered rain forest these days. The emphasis is warranted because rain forest ecosystems are more diverse and are losing species faster than any others on Earth. However, be sure to link comments made about rain forests in this chapter to problems encountered in endangered ecosystems in other parts of the world. Most regions in this country have similar stories to tell regarding loss of habitat and species.

CHAPTER OBJECTIVES

Introduce the overall course content and modular approach taken by the text. Introduce and review the core themes of biology, taking care not to present too much detail too soon.

Review the universal characteristics of living things.

Name and explain the core themes in biology (evolution, hierarchy of organization, structure and function, science as a way of knowing) and the characteristics of life.

Review the process of science, pointing out its format, its strengths, and the areas of human concern to which it can and cannot be applied.

Describe the five-kingdom system of classifying the diverse organisms of the world.

Relate the study of biology to a variety of more general human concerns such as medicine, conservation, nutrition, overpopulation, and so on.

LECTURE OUTLINE

I. **Introduction to class.**

    A. Introductory comments about the science of biology.

    B. Self-introduction.

    C. Administrative details.

    D. Introduction to class content and course organization.

    E. Core themes in biology.

        1. Biology is a broad collection of subdisciplines. How do biologists organize their subject matter?

        2. Can we uncover some common concepts that underlie all biological study?

        NOTE: The core themes are mentioned as such in the outline below. As a preface to your review of the themes and characteristics of life, encourage students to recall the organisms they are familiar with from their own environments.

    F. To begin, we start our look at biology by focusing on a region of the world we rarely experience directly—the rain forest (*Opening Essay*).

    Preview: Rain forests and their ecological importance are also discussed in Chapters 34 and 38.

II. **Life's levels of organization define the scope of biology** *(Module 1.1)*.

    A. The levels at which life is organized are (from largest to smallest):

        1. Ecosystem.

        2. Community.

        3. Population.

        4. Organism.

        5. Organ system.

        6. Organ.

        7. Tissue.

        8. Cell.

        9. Molecule.

    NOTE: Distinguish between unicellular and multicellular organisms.

    B. These levels represent a hierarchy because each level is built of parts at successively lower levels of organization.

    C. Researchers tend to focus individual studies on one level. Later, the results can be organized into a broader picture of higher levels.

    Preview: The dynamics of higher levels of organization are discussed in Chapters 35 (populations) and 36 (communities and ecosystems).

III. **Scientists pose and test hypotheses to answer questions about nature** *(Module 1.2)*.

    A. Science can be viewed as a five-step process.

1. Observations—from others or results of earlier tests.
2. Questions—about unclear aspects of the observations: How? Why? When?
3. Hypotheses—tentative explanations of a phenomenon phrased in such a way as to be testable.
4. Predictions—logical, testable outcomes of the hypotheses developed by the use of deductive reasoning. Predictions take the form of *if (statement of hypothesis) is true, then (predictions)*.
5. Tests—to determine if the predictions are supported (fail to falsify) or falsified.

    Note: The process of science will not "prove" a hypothesis true.

    Results—used as evidence to support or falsify the hypotheses and usually become new observations in another cycle of investigation.

    NOTE: Use of the process of science is a common underlying theme in biology.

    Preview: In addition to Module 1.3, other good examples of the application of the process of science to a problem can be found in, but are not limited to, Modules 35.6 (evolution shapes life histories), 36.18 (studies at Hubbard Brook), and 37.1 (Tinbergen's studies of digger wasp nesting behavior).

IV. The process of science: A case study in a tropical rain forest (*Module 1.3*).

　A. Roberto Roca's studies of the oilbird's feeding habits can be used to illustrate the process of science.

　　1. Preliminary observations led to questions.

　　2. Questions led to the development of hypotheses by inference and analogy.

　　3. Hypotheses were tested by observation and experimentation.

　　　NOTE: Experiments require a control to provide a basis of comparison, and must be repeatable.

　　4. Science is an adaptable process and is tentative.

　　　NOTE: Repeated observation and experimentation may result in a challenge to a hypothesis. Such challenges may be successful; thus science is self-correcting.

　B. Science does not deal with hypotheses that are not testable (i.e., questions that are not concerned with the natural world).

V. Biological diversity can be arranged into five kingdoms (*Module 1.4*).

　A. Making sense of the diverse life forms in the world, or even in a complex environment such as a tropical rain forest, can be overwhelming. To organize and balance our view, we need a system for categorizing these living things.

　B. The five-kingdom system of classification enables biologists to group organisms according to their forms and functions.

　C. The five kingdoms.

　　NOTE: Keep it brief; remember, we'll return to this later.

1. Monera: Small single-celled organisms and the only cells without nuclei.
2. Protista: Protozoans and algae are single-celled organisms.
3. Plantae: Plants are complex in form and are photosynthetic (autotrophs).
4. Fungi: Molds, yeasts, and mushrooms. Decompose and absorb nutrients from dead organic matter.
5. Animalia: Animals are complex in form and eat other organisms (heterotrophs).

VI. Unity in diversity: All forms of life have common features *(Module 1.5)*.

A. Life is diverse, but there are common themes that all living things exhibit.

Preview: When viewed under a microscope, all cells look very similar (Chapter 4).

1. Cells of all species contain many of the same molecules. For example, all cells have DNA, and the genetic information in DNA is coded in the same way in all cells.
2. The information in DNA provides the information for the properties that distinguish life from nonlife: growth, development, energy, response to external stimuli, reproduction.

B. Review the five kingdoms with some illustrations, prefacing the review with this question: What characteristics can you think of that are true for all living things? (Figures 1.5A–E, Acetates for Chapters 17–19, or your own 35-mm slides)

C. You might want to focus on one group of organisms to emphasize the point that at each level of biological organization, there is diversity and similarity.

VII. Evolution explains the unity and diversity of life *(Module 1.6)*.

Preview: This Module previews Unit III, especially Chapter 14.

A. Theories are much broader in scope than hypotheses and have much greater explanatory power.

B. Darwin showed that evolution can explain the diversity of life and underlying commonalties of life's diversity.

C. Natural selection, as proposed by Charles Darwin in *The Origin of Species*, is an important mechanism by which evolution proceeds.

D. In particular, evolution explains why organisms are adapted to their environments. A pangolin is a particularly good example of an animal with many visible adaptations to a particular environment (Figure 1.6C).

VIII. Living organisms and their environments form interconnecting webs *(Module 1.7)*.

A. Living things do not exist in isolation from their environment. An organism's environment includes both living and nonliving components.

B. Plants, as well as certain monerans and protistans, make food from carbon dioxide, water, and the energy of sunlight by the process of photosynthesis.

C. All animals are ultimately dependent upon autotrophs (e.g., plants) for food.

D. Decomposers (such as bacteria, fungi, and small animals) of dead organisms are important for the recycling of complex dead organic matter into mineral nutrients that plants can use.

Preview: Such cycles can be broken down into simpler form, dealing with, say, just one chemical. A knowledge of nutrient cycles, discussed in Modules 36.14–36.18, is important for understanding ecology.

E. Relationships among the living and nonliving components of an ecosystem can be illustrated as a web (Master 1.7A and 1.7C).

NOTE: Point out that these figures greatly simplify the many complex interactions among the organisms represented and their environment, and that the arrows represent the direction of flow of energy and nutrients.

Preview: This material previews the discussion of tropic levels and webs found in Modules 36.8–36.12; also see Module 7.14.

IX. **Biology is connected to our lives in many ways (*Module 1.8*).**

A. Birth and death; human population; nutrition, exercise, and dieting; medical concerns of all kinds; agriculture, including forestry; biodiversity and endangered species; pollution and environmental changes due to global warming.

Preview: These environmental concerns are discussed in more detail in Modules 38.13–38.16; also see Module 7.14.

CLASS ACTIVITIES

1. Illustrate how the process of science is applied in everyday life, even if only on a subconscious level. For example, ask the students what they do if, when they flip a light switch the light does not go on, or what they do if their car will not start. Have them relate their activities to the process of science.

2. Focus student attention on biological diversity by using props. Bring some living things from each kingdom to lecture. Or use figures from the text (Figures 1.4A–E, Acetates for Chapters 17–19) or your own 35-mm slides to show as many diverse forms of living things as possible. Encourage students to reflect on the organisms with which they have direct experience in their own environments.

3. Ask the students to characterize life (*Module 1.5*). List on the board the characteristics of life the students think of. Focus the discussion by having them contrast living things with nonliving things. Challenge the students to disprove that, for example, a car is alive (gasoline is "metabolized," acceleration when stepping on the gas pedal is a response to a stimulus, etc.).

4. In small lecture or discussion sections, have students bring in current newspaper clippings on what they perceive to be a biological issue. (Virtually every newspaper every day will have something on medicine, ecology, biodiversity, pollution, nutrition, and countless other biological topics.) During lecture, have the class categorize each article according to its biological theme. Facilitate this activity so that the end result is to show students that biology is indeed a broad topic, and that student approaches and interests are equally broad.

RESOURCES AND REFERENCES

Apple Multimedia Lab. *Life Story*. Sunburst, 1993. A CD-ROM (optional videodisc), multimedia version of the story of the discovery of the structure of DNA. It includes clips from the BBC dramatization of James Watson's book *The Double Helix*, about Watson and Crick's discovery of DNA structure. Supporting material includes interviews with the real people, molecular animations, historical photographs, background material about the controversies surrounding how the scientific process operates, and how the story of the discovery was told. For instance, an interview with

Anne Sayre reveals her feelings on the inadequate portrayal of Rosalind Franklin's role in the story. Macintosh version only.

Eames, Office of Charles and Ray. *Powers of Ten*. Pyramid Films, 1977. A classic, 9.5-minute film focusing on human scale and scales much larger and much smaller. Available in 16-mm film and video formats.

Halle, F. "A Raft Atop the Rain Forest." *National Geographic*, October 1990. More about the research described in Chapter 1's opening essay.

Kimball, Robert. *Botanical Gardens*. Sunburst, Wings for Learning, 1993. This Macintosh software is a problem-solving program in which students experiment with the growth of different plants in varying environmental conditions.

Kosinski, Robert J. *Fish Farm. A Simulation of Commercial Aquaculture*. Redwood City, CA: Benjamin/Cummings, 1993. A computer software simulation that can be used to introduce concepts and activities involved in scientific study. The simulation uses the process of science, experiment planning, data acquisition, and data presentation. It models a fish farm, allowing the student to modify the species of fish used, pond size, aeration, food nutrient content, and food delivery. Macintosh, Apple II, and IBM PC versions. Available as part of the instructor's support package for this text.

Krajick, K. "The Secret Life of Backyard Trees." *Discover*, November 1995. Forest canopy research in North America.

Mayr, E. *The Growth of Biological Thought: Diversity, Evolution, and Inheritance.* Cambridge, MA: Harvard University Press, 1982. A classic book by one of the greatest evolutionary biologists of the century. (The concept of emergent properties is discussed on pp. 63-64.)

Moffett, M.W. *The High Frontier: Exploring the Tropical Rainforest Canopy*. Cambridge, MA: Harvard University Press, 1993. An illustrated account of current research in the tropics.

Moore, J.A. *Science as a Way of Knowing: The Foundations of Modern Biology*. Cambridge, MA: Harvard University Press, 1993.

Morrison, Philip, Phylis Morrison, and the Office of Charles and Ray Eames. *Powers of Ten*. New York: W. H. Freeman, 1982. A book about the relative size of things in the universe and the effect of adding another zero, based on the movie *Powers of Ten* by the Office of Charles and Ray Eames. A wealth of information about the scale of living realms, physical-chemical realms (down to 10^{-16} meters), and astronomical realms (up to 10^{25} meters).

Nova. *Do Scientists Cheat?* New York: WGBH, 1988. A 1-hour video giving a general presentation of the recent concerns about just how the scientific process works in the U.S. Available from Coronet/MTI Film and Video.

Stolzenburg, W. "Winged Saviors of the Forest." *Nature Conservancy*, March/April 1991. Roberto Roca's study of oilbirds in Venezuela.

INTERNET RESOURCES FOR INTRODUCTION

What killed the dinosaurs?

http://www.connix.com/~harry/k-t.htm

This web site might be useful to demonstrate that science is a process that involves debate.

2
THE CHEMICAL BASIS OF LIFE

APPROACH

Most nonscience majors will be somewhat dismayed at having to begin a biology course from a chemical perspective. More than most other subjects in biology, simple chemistry is review for the teacher but preview for the student. Time taken at this stage will keep students from being intimidated and help ease their way through tougher chemistry to come.

In order not to lose students to the abstractions inherent in this subject, it is important to develop this topic logically, building down from the tangible and then back up the hierarchical levels. The introductory module introduces this approach, and Module 2.1 follows up nicely. You need to provide concrete examples of tangible, organic objects and their chemical makeup.

Although the terminology is relatively simple, it may confuse students who haven't had a previous course in chemistry. It will help to identify students without prior chemistry and offer your individual guidance at this point.

It is important to present and explain the various conventions of illustrating molecules (as used in the text) in the course of introducing the concepts. Module 2.8 is a good place to start. You might modify Master 2.8 with additions of electron-cloud and space-filling models. Or, once you have introduced the basic terms and covalent bonding, use Masters 3.1 and 3.4C to review and introduce different molecular conventions.

Chemistry is dynamic. Using animated sequences that show molecular structures, bonding, and reactions may help make the overall process more understandable and preview the shorthand references to atoms and molecules.

CHAPTER OBJECTIVES

Review the concepts introduced in Chapter 1: hierarchy of organization and the close relationship between function and form. Note at what level the property of being alive emerges.

Identify the advantages and disadvantages of taking a reductionist approach, and discuss starting a course in biology by studying chemistry.

Distinguish between the terms *element* and *compound, atom* and *molecule, atom* and *isotope,* and *atom* and *ion*.

Describe the dominant atoms present in living organisms, and give some reasons why they are the most common.

Distinguish between physical (weight, charge, radioactivity) and chemical (reactivity) properties of matter. Note at what level the property of chemistry emerges.

Distinguish among the three types of bonds, their relative strengths and functions, and the methods by which they are diagrammed.

Describe the special properties water exhibits that make the molecule indispensable to living systems. Explain how these properties are related to hydrogen bonding.

Define pH, and explain why it is important that pH remains within natural limits.

LECTURE OUTLINE

I. Introduction.

   A. Taking a reductionist approach (*Opening Essay*).

   1. *Review:* Major hierarchical levels of organization (no more than four illustrated levels here, from the human viewpoint): the "supervisible" (larger than we can directly perceive = ecosystem), directly perceivable (organismal), microscopic (cellular), and submicroscopic (molecular) scales (Master 1.1; follow with Master 2.1A or Figure 2.1).

   2. *Review:* Special properties from Module 1.1 and the characteristics of life from Module 1.5.

   3. Discuss how a reductionist approach to the study of life clarifies our understanding of all higher levels of organization. Only by reducing organisms to smaller parts have biologists been able to understand the complex structures and functions of life.

   4. A full understanding of the organism level, for example, requires that we put the parts back together so that the unique properties of the higher levels of organization are expressed.

   5. Properties of life begin to emerge at the molecular level discussed in this and the next chapters, and are fully emergent at the cellular level in Chapter 4.

   B. Biological function starts at the chemical level (*Module 2.1*; Figure 2.1, Master 2.1A).

   1. The matter in all living things is made of separate chemicals. In plant leaves, chlorophyll is just one of these chemicals.

   2. At every level, structure and function are strongly interrelated. Chlorophyll is a complex chemical (molecule) composed of smaller parts (atoms), bonded together in a way that allows the molecule to capture ("catch") sunlight.

   3. Notice how the structure of each larger level surrounding the chlorophyll molecule adds function, which builds the overall function of the plant leaf.

   4. *Preview:* We will return to this example in the chapters on photosynthesis and plant function (Chapters 7, 31, and 32).

II. The hierarchical levels of matter.

   A. To understand chemical structure and function in life, we must start small and see how the structures at each level are combined into each higher level.

   Review: This hierarchical organization is first discussed in Module 1.1.

   B. Life requires about 25 chemical elements (*Module 2.2*; Table 2.2).

   1. Elements are the basic chemical units that cannot be broken apart by chemical processes.

   2. There are 92 naturally occurring, chemically individual atoms. (Be sure to mention that "atom" is the name for elemental unit.)

3. Of the 25 elements required by living organisms, four (oxygen, carbon, hydrogen, and nitrogen) are particularly abundant, making up 96% of the human substance.

C. Elements can combine to form compounds (*Module 2.3*).

1. Compounds contain two or more atoms in a fixed ratio.
2. Different combinations of atoms determine unique properties of each compound.

    NOTE: At this point in your discussion, it might be useful to explain that our knowledge about chemistry is the result of several hundred years' worth of direct and indirect observations on the behavior of matter. Also explain that conventions used to show atoms and molecules convey various types of information. They don't show atoms and molecules the way they really are.

D. Atoms consist of protons, neutrons, and electrons (*Module 2.4; Master 2.4A*).

1. Protons and neutrons occupy the central region of an atom (the atomic nucleus), and electrons occupy the region surrounding this central area.
2. Neutrons have a mass of 1 and a charge of 0; protons have a mass of 1 and a charge of +1; electrons have an effective mass of 0 and a charge of –1.
3. Atoms always have the same number of electrons as protons. This atomic number defines each element's unique properties.

    NOTE: Refer to a periodic table of the elements during your discussion for a wealth of interesting information on atomic structure.

4. The number of protons plus the number of neutrons determines each element's atomic weight.
5. Atoms with the same atomic number but different atomic weight (differing numbers of neutrons) are isotopes.
6. If an isotope is unstable, it changes to a stable form by releasing radioactivity. (Master 2.4B illustrates isotopes of carbon.)

E. Radioactive isotopes can help or harm us (*Module 2.5*).

1. Isotopes of the same element behave the same way chemically. Therefore, organisms are indiscriminate as to the isotopes they use. This fact may result in harm, or it can be the basis of beneficial medical techniques.
2. Radioactive isotopes can function as "markers" for their nonradioactive counterparts. Biologists use such markers in research, to study the fate of elements in living systems, and in diagnostic medical procedures (see Module 7.3 for an example).
3. If the radioactivity is high and/or in large quantity, it can harm organisms by damaging large molecules and causing abnormal chemistry in cells. Radioactivity can come from a source outside the organism. But particularly dangerous are radioactive isotopes of biologically important elements like iodine, or those that behave chemically like biologically important elements—such as cesium, which acts like calcium and is concentrated in bone.

F. Electron arrangement determines the chemical properties of an atom (*Module 2.6*).

The Chemical Basis of Life

1. The properties of elements emerge from their atoms' subatomic parts. The chemical individuality of an atom depends on the number of electrons in the atom's outer shell (Acetate 2.6).

2. Atoms react to form molecules. Reactions involve sharing or transferring outer electrons.

3. *Preview:* Molecules play many roles in living organisms: energy storage and a means to transfer stored energy for doing work, structural components, control of activity, communication between cells and/or whole organisms, and a means of information storage and retrieval (Chapter 3).

III. There are three major ways—types of bonds—in which elements are joined to build compounds.

NOTE: In each of the following bond types, two rules are satisfied: (1) the resulting compound is electrically neutral; (2) outer electron orbits are filled. The total energy available in a system determines which type will form and be maintained. Give one or two examples of each.

A. Ionic bonds are attractions between ions of opposite charge (*Module 2.7*).

1. Ions are atoms that have lost or gained electrons from or to their outer shell.

2. When electrons are lost, atoms become positive ions.

3. When electrons are gained, atoms become negative ions.

4. Two or more ions are associated with one another to form a molecule bonded ionically. A good example is sodium chloride (Master 2.7A, B).

B. Covalent bonds involve electron sharing (*Module 2.8*).

1. Each atom in a covalently bonded molecule shares electrons in its outer shell with the other atom(s) in the molecule.

2. Examples: H_2, O_2, CH_4, H_2O (Master 2.8).

3. Other larger examples: Masters 3.1 and 3.4C.

Preview: The importance of the ability of carbon to form up to four covalent bonds is discussed in Module 3.1.

C. Hydrogen bonds are the third type and are discussed below.

IV. Water is a characteristic of all life.

A. Water is a polar molecule (*Module 2.9*).

1. In nonpolar molecules (H_2, O_2, CH_4), covalent bonds are balanced because the electrons (and their negative charges) are equally shared by the atoms in the molecule.

2. In polar molecules (H_2O), the electrons around the atoms are not equally shared because the different atoms have different attractions for electrons. This results in one part of the molecule being slightly positive (usually where the hydrogen atoms are) and another part being slightly negative.

B. Overview: Water's polarity leads to hydrogen bonding and other unusual properties (*Module 2.10*).

1. The attraction of these slight positive and negative charges between different molecules (or different parts of the same molecule) results in a weak hydrogen bond (Master 2.9, Acetate 2.10A).
2. Hydrogen bonding occurs in other biologically important compounds.

    Preview: For example, the secondary structure of a polypeptide (Module 3.16).

C. Hydrogen bonds make liquid water cohesive (*Module 2.11*).
   1. Cohesion is the tendency of water molecules to stick together.
   2. Cohesion between water molecules allows them to form drops and be transported through the tissues of plants.
   3. Surface tension results from the cohesion of water molecules to each other so strongly that, like a trampoline, they will support a small weight such as an aquatic insect.

       Preview: Transpiration (Module 32.3) is an example of how living systems take advantage of, and is an illustration of the importance of, these characteristics of water.

D. Water's hydrogen bonds moderate temperature (*Module 2.12*).
   1. Breaking hydrogen bonds causes the temperature of water to rise more slowly when heated than the temperature of nonpolar liquids.
   2. This also causes water to vaporize less quickly.
   3. Forming hydrogen bonds causes the temperature of water to lower more slowly when cooled.

E. Ice is less dense than liquid water (*Module 2.13; Acetate 2.13*).
   1. Hydrogen bonds in ice result in an extremely stable, three-dimensional structure.
   2. A given volume of ice has fewer water molecules than an equal volume of liquid water and is therefore less dense.

F. Water is a versatile solvent (*Module 2.14; Acetate 2.14*).
   1. A solution is a homogeneous mixture of a liquid solvent and one or more solutes (solid or liquid compounds that dissolve in the solvent).
   2. Because water is polar, it readily forms solutions with a wide variety of other covalently bonded, polar compounds and with the charged ions of ionically bonded compounds.

V. Another property of compounds important to living things is pH (potential hydrogen).

A. pH expresses the tendency of water to ionize, dissociating into OH^- and H^+ (actually, H_3O^+).

B. The chemistry of life is sensitive to acidic and basic conditions (*Module 2.15; Master 2.15*).
   1. Biological pH ranges from 1 (stomach acid) to 9 (seawater).
   2. A pH of 7 = neutral; lower = acid; higher = basic.
   3. Biological fluids contain buffers, substances that resist changes in pH by reacting with, and neutralizing, H^+ or OH^- ions.

The Chemical Basis of Life

C. Acid precipitation threatens the environment (*Module 2.16*).

1. Compounds of sulfur and nitrogen are part of air pollutants released from the combustion of fuels. These compounds react with atmospheric water to form acidic compounds.

2. Low, acidic pH associated with acid rain can be harmful to organisms adapted to neutral pH.

3. Acidic conditions definitely harm aquatic environments and likely cause various imbalances in terrestrial environments such as forests.

   Preview: Module 36.18 discusses the impact of acid precipitation on a deciduous forest ecosystem (the Hubbard Brook studies).

4. The capacity to withstand changes in pH (buffering) is naturally a characteristic of some areas (for example, limestone buffers acid rain).

VI. Chemical reactions rearrange matter (*Module 2.17*; Master 2.17A, B).

A. Contrast ionic and covalent reactions. Solution in water is not a reaction per se.

B. Common reactions between covalently bonded molecules.

1. *Preview:* Use Acetate 3.3A, B, which show hydrolysis and synthesis reactions. Use Acetates 6.8 and 6.9 or Master 6.10 to demonstrate the roles of reactions in energy transfer and metabolism.

   NOTE: Only brief coverage is required here because this topic is a major focus of Chapter 3.

CLASS ACTIVITIES

1. Create a molecular hierarchy illustration, similar to Figure 2.1, using an animal example. Tracing the structural and functional levels of hemoglobin to red blood cells and the circulatory system is an interesting example because both hemoglobin and chlorophyll share the porphyrin ring structure.

2. Use samples of elemental sodium and chlorine and table salt to discuss bonding, elemental characteristics, and chemical changes during reactions. Other common household substances, with which most students have direct experience, can be used to demonstrate properties of elements and molecules and the reactions they undergo.

3. Emphasize the unique properties of water: its moderating influence on body temperature can be illustrated by comparing rather moderate coastal environments and their rather narrow temperature ranges with inland environments and their relatively wider temperature ranges. Ask what might have been the impact on the evolution of life if ice were not less dense than water. What would happen to life currently in lakes if ice sank? Ask why things get wet (water is sticky because of its polar nature). Pour a beaker full of water until it nearly overflows the brim; ask why it does not overflow. Ask why, if hydrogen bonds are weak, there can be a column of water several hundred feet high, as in Sequoias.

RESOURCES AND REFERENCES

Alberts, B.D. Bray, J. Lewis, M. Raff, K. Roberts, and J.D. Watson. *Molecular Biology of the Cell*, 3rd ed. New York: Garland, 1994. A comprehensive cell biology textbook for advanced students; clearly written and illustrated.

Atkins, P.W. *Molecules*. New York: Scientific American Library, 1987. A vividly illustrated tour of the world of molecules.

Becker, W.M., J.B. Reece, and M.F. Poenie. *The World of the Cell*, 3rd ed. Redwood City, CA: Benjamin/Cummings, 1996. A student-oriented text for undergraduates.

Gould, R. *Going Sour: The Science and Politics of Acid Rain*. Boston: Birkauser, 1985.

Graedel, Thomas E., and Paul J. Crutzen. "The Changing Atmosphere." *Scientific American*, March 1989. A good discussion of the chemical reactions in the atmosphere that produce acid rain.

Mathews, C.K., and K.E. van Holde. *Biochemistry*, 2nd ed. Redwood City, CA: Benjamin Cummings, 1996. A general biochemistry text, very well illustrated.

McLeod, Christopher, and Robert Lewis. *Downwind/Downstream*. Oley, PA: Bulldog Films, 1988. A film about the ecological, economic, and political impact of acid rain and snow, including mining and population pressures, in sensitive, high-valley environments in Colorado.

Mohnen, V.A. "The Challenge of Acid Rain." *Scientific American*, August 1988. The causes of the acid precipitation problem and possible solutions.

"The Molecules of Life." *Scientific American*, September 1985. An entire issue on molecular biology, including many sources for illustrations of atoms and molecules not found in the text.

National Geographic Society. *Medical Technology*. 1986. A videotape that explores modern advances in medical diagnostic technology, including imaging systems that use radioisotopes or radioactivity.

Olson, Arthur J., and David S. Goodsell. "Visualizing Biological Molecules." *Scientific American*, November 1992. Computer-generated images of biochemicals.

Stephens, Sharon. "Lapp Life After Chernobyl." *Natural History*, December 1987. Food chains through lichens bring radioactive elements into animals.

University of Maryland. *The World of Chemistry. 8. Chemical Bonds*. Washington, DC: Corporation for Public Broadcasting, 1990. The Annenberg/CPB Project. Part of a series of videotapes (also on videodisc) exploring the foundations of chemistry. This one explains the differences between covalent and ionic bonds using models and examples from nature.

Videodiscovery. *BioSci II Videodisk*. Seattle, WA: Videodiscovery, 1990. Images of molecules, molecular bonds, and much more.

INTERNET RESOURCES FOR UNIT I

CELLS alive!

http://www.comet.chv.va.us/quill

Immunity, *H. pylori* and ulcers, cytoskeleton, cell death. Video clips and 3-D animations.

History of the Light Microscope

http://www.duke.edu/~tj/hist/hist_mic.html

Internet Chemistry

http://naio.kcc.hawaii.edu/chemistry/redox_title.html

Some useful aspects, especially for photosynthesis and cell metabolism.

Photosynthesis Center

http://photoscience.la.asu.edu/photosyn/default.html

Look at the listings under Educational Resources.

3

THE MOLECULES OF CELLS

APPROACH

Understanding polymer structure is at the heart of understanding how cells work chemically. This chapter reviews and amplifies the material introduced in Chapter 2 and previews how larger molecules will fit into the cellular framework discussed in Chapter 4.

Although students are still being introduced to topics they cannot perceive directly (molecular bonding, three-dimensional representations of complex molecules, reactions between molecules, etc.), they will have had direct experience with some of the material (nutritional information; Modules 21.13–21.20) about carbohydrates, fats, and proteins; the debate about anabolic steroid use; diet; and disease).

Now is the time to make sure students can interpret chemical symbols properly, and to increase their understanding of chemical bonding and reactions. Take time to go through the step-by-step illustration of all the atoms, bonding processes, bond types, and symbolization, using the example of transthyretin. Some new conventions in the representation of protein structure are introduced here.

To a certain extent, student interest in the nutritional aspect of macromolecules should be put aside until later, after cells, tissues, and organs and metabolic processes have been introduced. The real focus of this chapter is on the chemical properties of macromolecules, not their nutritional roles.

Emergent properties are particularly strong at this level. However, it may be premature to go into much detail on exact macromolecular function until their function can be placed in the context of cellular and organismal structures and genetic processes.

Illustrations in this chapter can be augmented with concrete examples and figures from throughout the text. The chapters on cell structure (Chapter 4), cellular work (Chapters 5, 6, and 7), and molecular genetics (Chapter 10), and the units on animal (Chapters 20
....0–30) and plant (Chapters 31–33) form and function contain many good examples, some of which are included in the lecture outline below.

CHAPTER OBJECTIVES

Review elemental chemistry, focusing on the structure and properties of carbon.

Review the types of bonds, focusing on the buildup of levels of structure, particularly in carbohydrates and proteins.

Stress the relationship between molecular function and molecular form, and the role of bonding, usually at several levels, in creating molecular form.

Compare molecular reactions that build and degrade macromolecules, and distinguish between the terms *monomer* and *polymer* and *dehydration synthesis* and *hydrolysis*.

Introduce the idea of functional group. It is not important for students to remember the molecular details of the functional groups in Figure 3.2. But it is important for them to understand that certain groupings of atoms, with certain repeating chemical properties, occur commonly in the macromolecules of life.

Introduce and distinguish among the general classes of carbon-based molecules important in living systems (carbohydrates, lipids, proteins, and nucleic acids).

Preview the important roles each of these molecular types plays in organisms.

LECTURE OUTLINE

I. Introduction.

A. Life's beauty and properties begin to emerge at the macromolecular level (*Opening Essay*).

1. The central topic of this chapter is how smaller molecular units are assembled into larger ones.

2. Figure 3.0 (the close-up of spiderweb proteins) is a lovely starting example of macromolecular structure and function.

B. Life's molecular diversity is based on the properties of carbon (*Module 3.1*; Master 3.1).

1. Organic compounds contain at least one carbon atom.

2. Because there are 4 electrons in the outer shell of carbon, carbon has a strong tendency to fill the shell to 8 by covalently bonding to other atoms, particularly hydrogen, oxygen, and nitrogen. The 4 electrons in the outermost shell of carbon allow it to form complex structures (e.g., long, branched chains; ring structures). This is a major reason why carbon is the structural backbone of organic compounds.

Review: Covalent bonds (Module 2.8).

NOTE: At this point you might want to ask the class if any of them has ever seen the episode of the original *Star Trek* in which the Horta, a silicon-based life form, appeared; ask if silicon-based life makes chemical sense.

3. Point out the double bond in Master 3.1, explaining that it represents 4 shared electrons.

NOTE: Although the topic is not introduced here, you might mention that triple bonds (which occur in molecular nitrogen, among other places) are represented by three lines (6 shared electrons).

4. The way bonding occurs among atoms in molecules determines an overall shape.

5. Isomers are molecules with the same numbers of each atom but with different structural arrangements of the atoms.

C. Functional groups help determine the properties of organic compounds (*Module 3.2*).

1. These functional groups are generally attached to a carbon backbone of different macromolecules, and they exhibit consistent chemical properties.

2. They behave almost as if they were individual atoms.

3. Go over Table 3.2, discussing a few of the examples.

4. All of these functional groups are polar. Therefore, most of the molecules they are found on are polar.

II. Macromolecules are built and destroyed by chemical reactions.

A. Cells make a huge number of large molecules from a small set of small molecules (*Module 3.3*).

1. Monomer—the fundamental molecular unit.

2. Polymer—a macromolecule made by linking many of the same kind of fundamental unit.

B. Types of reactions involved (note that water is involved in both; Acetate 3.3A, B).

1. Dehydration synthesis—molecules synthesized by loss of a water molecule between reacting monomers; the most common way organic polymers are synthesized.

2. Hydrolysis—literally, "breaking apart with water"; the most common way organic polymers are degraded.

C. The study of molecular reactions in living systems is a broad topic that will be a theme throughout the course.

1. The reactions reviewed in this module are ones involved in the formation of molecular structures introduced in the remaining modules.

2. Life's chemical reactions occur in particular intracellular and extracellular environments and in controlled ways.

3. *Preview:* Chapter 4 will discuss the cellular framework on which and in which molecular reactions occur.

4. *Preview:* Chapters 5–7 introduce metabolism, cellular reactions involving energy uptake, storage, and release.

III. Carbohydrates.

A. Monosaccharides are the simplest carbohydrates (*Module 3.4*).

Note: The word "carbohydrate" indicates these compounds are made of carbon (*carbo*, C) and water (*hydrate*, H_2O). This is reiterated in the general formula $(CH_2O)_n$ for monosaccharides.

1. Show examples of glucose and fructose (Master 3.4B).

2. In solution, many monosaccharides form ring-shaped molecules (Master 3.4C).

3. *Preview:* The basic roles of simple sugars are as fuel to do work, as raw material for carbon backbones, and as the monomers from which disaccharides and polysaccharides are synthesized.

B. Cells link single sugars to form disaccharides (*Module 3.5*; Master 3.5).

NOTE: This is an example of dehydration synthesis (Module 3.3).

C. "How sweet it is . . ." (*Module 3.6*).

1. Humans are born preferring sweet-tasting foods and consume, on average in the U.S., 125 pounds each person per year.

NOTE: There seem to be important biological reasons for this predilection from the standpoint of both animals (the importance of recognizing and consuming energy sources) and many plants (enticing animals to help disperse their seeds inside sweet fruits).

2. The development and use of high-fructose corn syrup as the dominant sweetener in the United States is an interesting example of how technological change impinges on human biology and culture.

3. *Preview:* The discussion in this module continues with additional material on nutrition in Chapter 21.

D. Polysaccharides are long chains of sugar units (*Module 3.7;* Acetate 3.7).
   1. Using glucose as the monomer, different organisms build several different polymers: plant starch, animal starch (glycogen), and cellulose.

      NOTE: Hydrogens and functional groups are not shown on the acetate.

   2. Each of these molecules is synthesized by dehydration synthesis, but there are subtle differences in the covalent bonds that lead to different overall structures and functions.

      Review: Covalent bonds (Module 2.8).

   3. Plant starch has one kind of bond between monomers and is a long, relatively unbranched, coiled polymer. Plant starch is used for long-term energy storage only in plants. Animals can hydrolyze this polymer to obtain glucose.

   4. Glycogen has the same kind of bond between monomers, but it has relatively more side branches. Glycogen also is used for long-term energy storage only in animals. Animals can hydrolyze this polymer to obtain glucose.

   5. Cellulose has a different kind of bond between monomers, forming linear polymers that are cross-linked with other linear chains. Cellulose is the principal structural molecule in the cell walls of plants and algae. Animals cannot hydrolyze this polymer to obtain glucose (only certain bacteria, protozoans, and fungi can).

IV. Lipids.

A. Lipids include fats, which are mostly energy-storage molecules (*Module 3.8*).
   1. In lipids, carbon and hydrogen predominate; there is very little oxygen. General molecular formula: $(CH_2)_n$.
   2. Diverse types of lipids have different roles, but all are more or less hydrophobic.
   3. Fats are polymers of fatty acids and glycerol, formed by dehydration synthesis reactions (Acetate 3.8B, C).
   4. Saturated fats have no double bonds between carbons (the carbons are "saturated" with hydrogen atoms). The molecular backbones are flexible and tend to ball up into tight globules. Saturated fats like butter and lard are solid at room temperature.
   5. Most plant fats are unsaturated, whereas animals are richer in saturated fats.
   6. Unsaturated fats include many double bonds between carbons. This causes the molecules to be less flexible and they do not pack into solid globules. Unsaturated fats like olive oil and corn oil are liquid at room temperature.

      NOTE: By "hydrogenating" unsaturated oils, the double bonds are removed and the molecules become more solid at room temperature. These structurally modified fats are as detrimental as their naturally saturated counterparts in leading to atherosclerotic plaques.

B. Phospholipids, waxes, and steroids are lipids with a variety of functions (*Module 3.9*).

1. Phospholipids are a major component of cell membranes (Acetate 5.11A, B).

2. Waxes are effective hydrophobic coatings formed by many organisms (insects, plants, even humans) to ward off water. They consist of a fatty acid linked to an alcohol.

3. Steroids are lipids with backbones bent into rings. Cholesterol is an important steroid formed by animals (Master 3.9; notice that the diagram omits carbons and hydrogens at each intersection in the rings and just shows the backbone shape). Among other things, cholesterol functions in the digestion of fats and as starting material for the synthesis of female and male sex hormones.

4. *Preview:* The structural roles of phospholipid-containing membranes are introduced in Chapter 4, on cell structure; their molecular structure and function are discussed in Chapter 5, with other topics relating to cellular work. Like fats, they are polymers of fatty acids and glycerol, but include a PO_4 group in place of one fatty acid. This gives them the unique property of having a hydrophobic "tail" and a hydrophilic "head."

C. Anabolic steroids make big bodies and big problems (*Module 3.10*).

1. These are variants of the male hormone testosterone, which, among other roles, causes the buildup of muscle and bone mass during puberty in men.

NOTE: The term *anabolic* means "not from metabolism," that is, synthetic. Since college-age body builders may be tempted to use steroids, you might want to point out some of the medical problems such use would lead to, including testicular atrophy, liver cancer, breast development in males, masculinization of females, and antisocial behavior.

V. Proteins.

A. Proteins are essential to the structures and activities of life (*Module 3.11*). The general roles played by proteins include:

1. Structural (hair, cell cytoskeleton).

2. Contractile (as part of muscle and other motile cells, produce movement; Acetate 30.8).

3. Storage (sources of amino acids, such as egg white).

4. Defense (antibodies, membrane proteins; Master 5.13A–C).

5. Transport (hemoglobin, Master 22.10B; membrane proteins).

6. Signaling (hormones, membrane proteins).

7. Catalyst (enzymes, both free and membrane-bounded; Acetate 5.6).

B. Proteins are made from just 20 kinds of amino acids (*Module 3.12*).

1. Amino acids are characterized by each having an alpha ("central") carbon atom covalently bonded to one hydrogen, one amino group, one carboxyl group, and one other chemical group (symbolized by R in Master 3.12A).

Review: Covalent bonds (Module 2.8).

2. Each naturally occurring amino acid has one of 20 chemical groups (Master 3.12B), which give the amino acid particular properties.

C. Amino acids can be linked by peptide bonds (*Module 3.13*; Acetate 3.13).
   1. Using amino acids as monomers, organisms build polymers (polypeptide = proteins) by dehydration synthesis, forming peptide bonds between each former monomer.
   2. Protein peptide bonds can be broken down by hydrolysis, to release free amino acids.

   NOTE: Add a reverse arrow to Acetate 3.13 and label it "Hydrolysis."

D. Overview: A protein's specific shape determines its function (*Module 3.14*).
   1. Long polypeptide chains include numerous and various amino acids.
   2. The final structure of a protein, and thus its potential role, depend on the way these long, linear molecules fold up.
   3. Each collection of amino acids folds in a different way under natural conditions (Master 3.14A; compare with Figure 3.14B).
   4. Changes in heat, pH, saltiness, and so on, can cause proteins to unravel (denature).
   5. The four levels of structure are shown in the protein transthyretin in Acetate 3.15–3.18.

   NOTE: At each level in the diagrams, details are hidden to show the essential structure added at that level.

E. A protein's primary structure is its amino acid sequence (*Module 3.15*; Acetate 3.15–3.18A).
   1. Transthyretin is found in blood and is important in the transport of a thyroid hormone and vitamin A.
   2. Three-letter abbreviations represent amino acids; each amino acid is in a precise order in the chain.
   3. In transthyretin, there are four polypeptide chains, each with 127 amino acids.
   4. Changes in the primary structure of a protein can affect its overall structure and thus affect its (ability to) function.

F. Secondary structure is polypeptide coiling or folding produced by hydrogen bonding (*Module 3.16*; Acetate 3.15–3.18B).
   1. Hydrogen bonds occur between -NH and -C=O groups of amino acids in sequence along each polypeptide chain.

   Review: Hydrogen bonds (Module 2.10).

   2. Depending on where the groups are relative to one another, the secondary structure takes the shape of an alpha helix or a pleated sheet.
   3. The R groups do not play a role in secondary structure and are not diagrammed.

   NOTE: There are recently devised conventions in diagramming the secondary structures of proteins using cylinders, flat arrows, and lines to represent helical regions, beta pleated sheets, and nonhydrogen-bonded regions, respectively.

G. Tertiary structure is the overall shape of a polypeptide (*Module 3.17*; Acetate 3.15–3.18C).

1. Tertiary structure results from the clustering of hydrophobic and hydrophilic R groups and the bonding (hydrogen and ionic) between certain R groups along the coils and pleats.
2. In transthyretin, the tertiary shape is essentially globular.

H. Quaternary structure is the relationship among multiple polypeptides of a protein (*Module 3.18*; Acetate 3.15–3.18D).

1. Many (but not all) proteins consist of more than one primary chain.
2. Transthyretin consists of four chains, each identical. Other proteins might have all chains different or be additionally complexed with other atoms or molecules.

NOTE: Quaternary bonding is largely by hydrogen bonds (Module 2.10).

VI. Talking About Science: Linus Pauling contributed to our understanding of the chemistry of life (*Module 3.19*).

A. Dr. Pauling felt that there is value in reductively studying the chemistry of biology to answer questions about whole organisms.

B. He was the first to describe the coiled and pleated-sheet secondary structure of protein and the first to describe the structure of hemoglobin and the abnormal form found in the red blood cells of those with sickle-cell anemia.

C. Later in his life, Pauling was most noted for his work on the role of vitamin C in the maintenance of health.

D. Pauling also had a lifelong interest in the biology of aging.

E. As are many scientists, Pauling was politically active, being an advocate for a ban on the testing of nuclear weapons.

VII. Nucleic acids are information-rich polymers of nucleotides (*Module 3.20*).

A. Nucleotides are complex molecules composed of three functional parts (Acetate 3.20A, B).

1. Phosphate group.
2. Five-carbon sugar (deoxyribose in DNA; ribose in RNA).
3. Nitrogenous base. There are five basic types: A, T, G, C in DNA and A, U, G, C in RNA (Acetate 10.2B, C).

NOTE: DNA nucleotide sequences encode the information required for production of the primary structure of proteins; such sequences are called genes (Modules 10.7 and 10.8).

B. Nucleotide monomers join by dehydration synthesis between the sugar parts of each to form polynucleotides with a linear structure of their sugar-phosphate backbones (Acetates 3.20A, B; Acetate 10.2A).

C. Hydrogen bonding between nitrogenous bases causes the final structure of the nucleic acid.

Preview: The mechanisms by which these structures determine gene expression are discussed in Chapters 10 and 11.

1. In DNA, two linear chains are held together in a double helix (Acetate 3.20C).

2. In RNA, one linear chain may be wrapped around itself in places, forming an rRNA or tRNA molecule (Acetate 10.11A, B) or remain unbonded (mRNA; Acetate 10.13A).

CLASS ACTIVITIES

1. To show that the polysaccharide starch is composed of smaller sugars, pass out grains of wheat, and have students chew on them for 5–10 minutes. As they chew, explain how the hydrolysis of plant starch produces disaccharide maltose, which is noticeably sweet. Also point out that this only occurs in the presence of the right environment, including the salivary enzyme amylase and the proper pH. The exact chemical role of this enzyme need not be introduced at this point. In larger classes, you might want to restrict this activity to discussion groups.

2. Use three-dimensional, space-filling, or ball-and-stick models or stereo-pair illustrations of macromolecules to illustrate many of the concepts introduced in this chapter. Such concepts as monomer structure and the role of hydrogen bonding in maintaining structure are clarified. This works best with smaller classes. Be sure to give students a chance to manipulate the model at the end of lecture.

RESOURCES AND REFERENCES

Asimov, I. *The World of Carbon*, 2nd ed. New York: Macmillan, 1962. A primer on organic chemistry by one of America's most popular science writers.

BioLearning. *Nucleic Acids*. Jericho, NY: Videodiscovery, 1981. Tutorial software that compares DNA and RNA, emphasizing how hydrogen bonds form between pyrimidines and purines. Available from Carolina Biological Supply Company.

BioLearning. *Proteins*. Jericho, NY: Videodiscovery, 1981. Tutorial software that enables the user to build different amino acids, changing the R groups and constructing dipeptides and polypeptides. Also demonstrated are the various higher levels of structure in proteins and the processes of denaturation and hydrolysis of proteins. Available from Carolina Biological Supply Company.

Dushesne, L.C., and D.W. Larson. "Cellulose and the Evolution of Plant Life." *BioScience*, April 1989. The chemistry and natural history of the most abundant organic molecule in the biosphere.

Flannery, M.C. "Collagen: Complex and Crucial." *The American Biology Teacher*, November/December 1990. The structure and function of our body's most abundant organic molecule.

Horgan, John. "Profile: Linus Pauling. Stubbornly Ahead of His Time." *Scientific American*, March 1993. The latest short profile, with a recent photograph.

Richards, Frederic M. "The Protein Folding Problem." *Scientific American*, January 1991. A discussion of the problems in determining biologically active shapes of proteins from their primary amino acid structure alone.

Skolnick, Jeffrey, and Andrzej Kolinski. "Simulations of the Folding of a Globular Protein." *Science*, November 23, 1990. Includes a stereo-pair color figure of apoplatocyanin.

Stirling, Charles. *Chirality—Handedness: Through the Looking Glass. In the Hands of Giants*. Princeton, NJ: Films for the Humanities and Sciences, 1993. One of a series of lectures on human symmetry, this videotape discusses the structure and building of carbohydrates and proteins.

University of Maryland. *The World of Chemistry. 23. Proteins: Structure and Function*. Washington, DC: Corporation for Public Broadcasting, 1990. The Annenberg/CPB

Project. Part of a series of videotapes (also on videodisc) exploring the foundations of chemistry.

Videodiscovery. *Atoms to Anatomy*. Seattle, WA: Videodiscovery, 1993. Contains state-of-the-art visual images for teaching human anatomy and physiology (or introductory biology). High-resolution images and animations of molecules. Single- or double-sided CAV videodiscs and Macintosh HyperCard software available.

Zheng, Chong, Chung F. Wong, J. Andrew McCammon, and Peter G. Wolynes. "Quantum Simulation of Ferrocytochrome C." *Nature*, Vol. 223, 1988. Simulations of molecular interactions have been based on laws of classical mechanics. For large biomolecules, quantum effects such as electron tunneling may also play a role.

4

A Tour of the Cell

APPROACH

The material in this chapter is a key link between previous chapters on basic chemistry and much of the rest of the material in the text. A clear understanding of cell structure will provide students with a consistent framework for organizing the details to come: molecular processes in the next three chapters, genetics (Unit II), evolution and the diversity of life (Units III and IV), and the system-level studies of animals and plants (Units V and VI).

Considerable detail is included on the functions of the cellular parts discussed. Virtually all of these functions will be covered again later in the text, so it is important that students understand two main purposes of this chapter: an introduction to subcellular structures, and the tight relationship between structure and function at this level. It is unrealistic to expect students to remember specific functional details at this point. Focus on the *basic* functions of the organelles (Table 4.20).

Spend time during lecture viewing 35-mm transparencies, movies, or video clips of cellular components to "flesh out" the figures, diagrams, and discussion in the text.

Although most students will have observed cells, they may not have a good idea of what the total cellular environment is like. Spend some time describing the gel-like physical environment of the cell from the perspective of human scale. It is also important to impress students with the dynamic nature of cellular structures, because cells are anything but static, as illustrations might suggest. There are several good resources that can help you portray the dynamic activities of many of the organelles.

Membranes play a major part in most organelle structures, but membranes are not discussed in detail until the next chapter. Preview some of the material on membrane structure from Chapter 5 as you begin to review the structure of membrane-bounded organelles. This will help students interpret illustrations of the organelles, especially the electron micrographs.

CHAPTER OBJECTIVES

Portray cells as membrane-bounded containers of mixtures of molecules, with each cell's various life-maintaining processes set off in differentiated regions.

Contrast the structural differences between prokaryotic and eukaryotic cells. Point out that eukaryotic cells are larger and able to do more complex things because they efficiently partition functions within themselves by means of membranes.

Preview the unique structures found in prokaryotic cells, deferring detailed discussion of these until later chapters.

Contrast the structural differences between plant and animal cells.

Preview the structure of the eukaryotic nucleus, mitochondrion, and chloroplast, deferring detailed discussion of the various inner structures until later chapters.

Emphasize the complex structural and functional interconnections among the organelles of the endomembrane system.

Introduce the skeletal and surface structures of cells, which provide support, protection, and movement in—and communication with—the extracellular environment.

LECTURE OUTLINE

I. Introduction.

A. *Review:* All organisms are composed of cells (Module 1.5).

B. At a scale just below what humans can directly perceive, cells are invisible, but with a microscope, we can plainly see that all living things contain cells.

C. Cells have complex internal organization (*Opening Essay*).

D. Use a complex, single-celled organism (the protozoan *Trichodina*) to introduce the basic cellular framework (Master 4.0).

1. Outer surface structures face the environment (hooks and cilia).

2. The plasma membrane physically and chemically separates the outside environment from the cell's internal environment.

3. Cytoplasm includes the semifluid medium and organelles (recognizable, smaller structures) between the plasma membrane and the nucleus.

4. The nucleus contains genes in the form of DNA.

II. Microscopes provide windows to the world of the cell (*Module 4.1*).

A. Images formed by microscopes represent the object "under" the microscope.

B. Two aspects of microscope images are important.

1. Magnification: the number of times larger the image appears than the object actually is.

2. Resolution: clarity of the image.

C. Images are formed in different ways by three types of microscopes that produced the images in the text. Each of these microscopes has advantages relative to the others, and a range of scales at which it functions best (Master 4.2).

1. Light microscopes (LM) bend the light coming through an object. The bent light rays form larger images in the viewer's eyes. Well-resolved LM images are limited to 1000–2000 times larger than life size. The LM is particularly good for looking at living cells and tissues (Master 4.1A).

2. Scanning electron microscopes (SEM) compose images on a TV screen, from electrons that bounce off the surfaces of the object. SEM images are usually about 10,000–20,000 times larger than life size. The SEM is particularly good for showing organismal and cellular surfaces under high magnification (Figure 4.1B).

3. Transmission electron microscopes (TEM) compose images on camera film, from electrons that have traveled through very thin slices of the object and have been bent by magnetic lenses. TEM images are usually about 100,000–200,000 times larger than life size. The TEM is particularly useful for showing the internal structures of cells (Figure 4.1C).

III. Cell size.

A. Biologists use smaller units of measurement when discussing microscopic scales (micrometers, μm, at LM scales; nanometers, nm, at EM scales).

B. Cell sizes vary with their function (*Module 4.2*; Master 4.2).

1. *Review:* the scales of life (compare with Master 1.1).

2. Nerve cells are very long, to communicate between different parts of an animal's body.

3. Bird eggs are very large, mostly composed of food reserves.

4. Blood cells are very small, to allow them to flow through blood vessels, and to provide a large surface area (from many cells) for efficient gas exchange (Master 4.3).

C. Natural laws limit cell size (*Module 4.3*).

1. Large cells have a smaller ratio of surface area to volume than small cells (Master 4.3).

2. This fact imposes the upper limit on cell size (actually, cell volume) because materials have to flow across the surface to get to the inside. Larger cells require correspondingly greater surface area, which they do not have.

3. The small size of cells is limited by the total size of all the molecules required for cellular activity (DNA, ribosomes, life-process-governing proteins, etc.).

IV. Cell organization: Two very different types of cells have evolved.

A. Prokaryotic cells are small and structurally simple (*Module 4.4; Acetate 4.4*).

1. Usually relatively small (0.5–10 μm in width).

2. Lack a nucleus: DNA is in direct contact with cytoplasm and is coiled into a nucleoid region.

3. Cytoplasm includes ribosomes (protein factories) suspended in a semifluid.

4. Otherwise composed of a bounding plasma membrane, complex outer cell wall (a rigid container, often with a sticky outer coat), pili, and, sometimes, flagella.

Preview: The classification and evolution of the prokaryotes are discussed in Modules 16.4, 17.8, and 17.20.

B. Eukaryotic cells are partitioned into functional compartments (*Module 4.5*).

1. Usually relatively larger (10–100 μm or more) in width.

2. Internally complex, with organelles of two types: membranous and nonmembranous.

3. Membranous organelles found in eukaryotic cells include the nucleus, endoplasmic reticulum, Golgi apparatus, mitochondria, lysosomes, and peroxisomes.

4. Nonmembranous organelles found in eukaryotic cells include ribosomes, microtubules, centrioles, flagella, and the cytoskeleton.

5. Animal cells are bounded by the plasma membrane alone, often have flagella, and lack a cell wall (Acetate 4.5A).

6. Plant cells are bounded by both a plasma membrane and a rigid cell wall (Acetate 4.5B). In addition, plant cells usually have a central vacuole and chloroplasts, lack centrioles, and usually lack lysosomes and flagella.

7. Cells of eukaryotes in other kingdoms vary in structure and components (protists: Master 4.1A; Figures 4.13B, 17.22A–D, 17.25A, B; fungi: Figure 18.17D).

8. *Preview:* Membranes play an important role in defining many cellular structures. Introduce the phospholipid bilayer and the protein mosaic model of membrane structure, reminding students that a thorough discussion of the structure and function of membranes will come in Chapter 5 (Acetate 5.12).

V. Eukaryotic cell organization.

A. The nucleus is the cell's genetic control center (*Module 4.6*; Acetate 4.6).

1. The nuclear envelope is a double membrane, perforated with pores through which material can pass into and out of the nucleus, which separates this organelle from the cytoplasm.

2. DNA can be seen as strands of chromatin dispersed inside the nucleus.

3. During cell reproduction, chromatin coils up into structures called chromosomes.

4. The nucleolus, also within the nucleus, is composed of chromatin, RNA, and protein. The function of nucleoli is the manufacture of ribosomes.

B. Overview: Many cell organelles are related through the endomembrane system (*Module 4.7*).

1. An extensive system of membranous organelles work together in the synthesis, storage, and export of molecules (Acetates 4.11B and 4.14).

2. Each of these organelles is bounded by a single membrane. Some are in the form of flattened sacs, some are rounded sacs, and some are tube-shaped.

3. Rough endoplasmic reticulum makes membrane and proteins (*Module 4.8*; Master 4.8A). Rough ER is composed of flattened sacs that often extend throughout the entire cytoplasm. Ribosomes on rough ER make proteins, some of which are incorporated into the membrane; other proteins are packaged in membranous sacs that bud off the rough ER (Master 4.8B).

4. Smooth endoplasmic reticulum has many functions (*Module 4.9*; Master 4.9). One job of smooth ER is to synthesize lipids. In other forms of smooth ER, enzymes help process materials as they are transported from one place to another. An example of this function is the detoxification of drugs by smooth ER in liver cells. Other functions of smooth ER include the storage of calcium ions that are required for muscle contraction.

NOTE: You may need to define enzymes as proteins functioning as biological catalysts (Module 5.5).

5. The Golgi apparatus finishes, sorts, and ships cell products (*Module 4.10*; Master 4.10). Transport vesicles from the ER fuse on one end of a Golgi stack to form flattened sacs. These sacs move through the stack like a pile of pancakes added at one end and eaten from the other. Molecular processing occurs in the sacs as they move through the Golgi. At the far end, modified molecules are released in transport vesicles.

6. Lysosomes digest the cell's food and wastes (*Module 4.11*; Acetate 4.11B). They are one kind of vesicle produced at the far end of the Golgi. Within these vesicles are hydrolytic enzymes that break down the contents of other vesicles with which they fuse (Acetate 3.3B).

7. Abnormal lysosomes cause fatal diseases (*Module 4.12*). A series of rare diseases result from the lack of certain hydrolytic enzymes from

lysosomes. In Pompe's disease, lysosomes lack glycogen-digesting enzyme. In Tay-Sachs disease, lysosomes lack lipid-digesting enzymes.

8. Vacuoles function in general cell maintenance (*Module 4.13*). *Vacuole* is the general term given to other membrane-bounded sacs. Plants have central vacuoles that function in storage, play roles in plant cell growth, and may function as large lysosomes (Figure 4.13A). Contractile vacuoles in cells of freshwater protists (both protozoans and algae) function in water balance (Figure 4.13B).

9. A review of the endomembrane system (*Module 4.14*; Acetate 4.14).

NOTE: Discuss the structural connections between the various organelles in this system. The red arrows show the functional connections.

C. Chloroplasts convert solar energy to chemical energy (*Module 4.15*; Acetate 4.15).

1. Found in most cells of plants and in cells of photosynthetic protists (algae).

2. Double-membrane-bounded.

3. Site of photosynthesis.

4. The structure of the organelle fits its function. As we will see, the capturing of light and electron energizing occur on the grana, and chemical reactions that form food-storage molecules occur in the stroma.

5. *Preview:* Photosynthesis is covered in detail in Chapter 7, and the origin of chloroplasts is discussed in Module 17.20.

D. Mitochondria harvest chemical energy from food (*Module 4.16*; Acetate 4.16).

1. Found in all cells of eukaryotes, except a few anaerobic protozoans.

2. Double-membrane-bounded.

3. Site of cellular respiration.

4. The structure of the organelle fits its function. As we will see, the ATP-generating electron transport system is embedded in the inner membrane (cristae), and chemical reactions occur in compartments between membranes.

5. *Preview:* Cellular respiration is covered in detail in Chapter 6, and the origin of mitochondria is discussed in Module 17.20.

E. Cells have an internal skeleton (*Module 4.17*; Figure 4.17A, Acetate 4.17B).

1. The organelles discussed up to this point, particularly the endomembrane system, provide cells with some support.

2. The cytoskeleton adds to this support.

3. It is composed of a three-dimensional meshwork of proteins arranged in three types of linear structures: solid rods composed of globular proteins (microfilaments), ropelike strands of fibrous proteins (intermediate filaments), and hollow tubes composed of globular proteins (microtubules).

4. Microtubules also provide anchors for organelles, function as conveyer belts along which particles and organelles move through the cell, and are the basis of ciliary and flagellar movement.

F. Cilia and flagella move when microtubules bend (*Module 4.18*).

1. Although the terms *cilium* and *flagellum* refer to similar structures, the structures were named when their internal similarities were not appreciated. Cilia are short, numerous, and usually complexly organized. Flagella are longer, fewer, and less complexly organized.

2. In both cases, these nonmembranous organelles are minute, tubular extensions of the plasma membrane that surround a complex arrangement of microtubules (Acetate 4.18A).

3. Cilia and flagella function to move whole cells or to move materials across or into cells.

4. The underlying structure consists of nine microtubule doublets arranged in a cylinder around a central pair of microtubules. At the base within the cell body (basal body), the structure is slightly different (Acetate 4.18A).

5. Various types of whipping movements of a whole flagellum or cilium occur when the microtubule doublets move relative to neighboring doublets. The connecting dynein arms apply the force (Acetate 4.18B).

6. *Preview:* Basal bodies are in the cytoplasm below these external extensions. They are identical in cross section to centrioles, which function in cell division (Acetate 4.5A, Modules 8.7 and 8.8, Master 8.7).

G. Cell surfaces protect, support, and join cells (*Module 4.19*).

1. Prokaryotic cells of bacteria and eukaryotic cells of many protists function independently of one another and relate directly to the outside environment.

2. In multicellular plants, cell walls protect and support individual cells and join neighboring cells into interconnected and coordinated groups (tissues) (Acetate 4.19A).

Preview: Plant cells and tissues (Module 31.5).

3. Plant cell walls are multilayered and are composed of various mixtures of polysaccharides and proteins. The dominant polysaccharide in plant cells is cellulose.

4. Plasmodesmata are channels through the cell walls connecting the cytoplasm of adjacent plant cells.

5. In multicellular animals, cells are often covered with sticky layers of polysaccharides and proteins, which protect the cells but do not provide much support (Acetate 4.19B).

6. In animal tissues, cells are joined by several types of junctions. Tight junctions provide leak-proof barriers; anchoring junctions join cells to each other but allow passage of materials along the spaces between cells or attach cells to an extracellular matrix; communicating junctions provide channels between cells for the movement of small molecules.

Preview: Epithelial tissue is attached to the underlying extracellular matrix by cell junctions (Module 20.4.).

NOTE: Provide examples of the importance of these cell junctions for the human body. For example, the role of tight junctions in the gastrointestinal tract, the role of anchoring junctions in keeping skin cells attached to each other and to the body (mention epidermolysis bullosa, a disease in which there is an inherited defect in anchoring junctions; ask what the result would be), and the role of communicating junctions in cardiac muscle contraction.

H. Eukaryotic organelles comprise four functional categories (*Module 4.20*; Table 4.20).

1. Manufacture: synthesis of macromolecules and transport within the cell.
2. Breakdown: elimination and recycling of cellular materials.
3. Energy processing: conversion of energy from one form to another.
4. Support, movement, and communication: relationships with extracellular environments.
5. Within each of the four categories there are structural similarities that underlie their functions.
6. All four categories work together as an integrated team, producing the emergent properties at the cellular level.

VI. All life forms share fundamental features (*Module 4.21*).

Review: The concept of the fundamental similarity of life is first discussed in Module 1.5.

A. The following features of cells are characteristic of life on Earth:
   1. Cells are highly structured units.
   2. Cell structure and function are related at the cellular and subcellular (and supracellular) levels.
   3. Cells are set off from their external environment by membranes.
   4. Each cell has DNA as the genetic material.
   5. Each cell carries out metabolism.

B. These features are likely to be characteristic of other life forms that may have evolved in our universe, although the materials and structures involved might be modified from the pattern seen on Earth.

CLASS ACTIVITIES

1. Bring in a large bowl of lime gelatin to use as an analogy to explain the physical structure of cytoplasm. Include in the gelatin mixture grapes, plastic wrap, marshmallows, and so on as organelles.

2. As time and resources permit, present views of living cells doing dynamic things. Protozoans, algae in pond water, and cheek cells are good examples of dynamic cells as opposed to the usual dead, stained, static cells students usually see.

3. After giving your cell tour lectures and reviewing the various cell organelles and their functions, show students a few previously unseen illustrations of cells of various types. Let students describe and name the dominant organelles they observe. Continue this activity as you proceed through the course, to help add some depth to the understanding of tissue and cell types in animals, plants, and other organisms.

RESOURCES AND REFERENCES

Allen, Robert Day. "The Microtubule as an Intracellular Engine." *Scientific American*, February 1987. A discussion of the role of microtubules in two-way transport of organelles and vesicles in the cytoplasm. Includes high-resolution LM views of the process.

DeDuve, C. *A Guided Tour of the Living Cell.* New York: Scientific American Books, 1986. A beautifully illustrated introduction to the cell by the discoverer of lysosomes.

Discovering the Cell. Washington, DC: National Geographic Society, 1990. A 28-minute videotape presenting an animated tour of cell components using computer-generated 3-D images.

Ezzel, C. "Sticky Situations." *Science News,* June 13, 1992. About the glue that holds animal cells together.

Goodsell, David S. "A Look Inside the Living Cell." *American Scientist,* September/October 1992. Unique views of cells using accurately scaled, space-filling diagrams of molecules in a simulated cross section of *E. coli* and a human red blood cell suspended in serum.

Hynes, Richard O. "Fibronectins." *Scientific American,* June 1986. The molecular bases for the functions of these proteins in the cytoplasmic skeleton.

Linder, Maurine E., and Alfred G. Gilman. "G Proteins." *Scientific American,* July 1992. How these plasma-membrane-bounded proteins coordinate cellular responses with external signals.

Sharon, N., and H. Lis. "Carbohydrates in Cell Recognition." *Scientific American,* January 1993. Sugars on cell surfaces: research and applications.

Stossel, T.P. "The Machinery of Cell Crawling." *Scientific American,* September 1994. How protein scaffolds play a role in cell movement.

Symmons, M., A. Prescott, and R. Warn. "The Shifting Scaffolds of the Cell." *New Scientist,* February 18, 1989. The dynamics of the cytoskeleton.

Videodiscovery. *Atoms to Anatomy.* Seattle, WA: Videodiscovery, 1993. Contains state-of-the-art visual images for teaching human anatomy and physiology (and introductory biology). High-resolution images—and animations—of molecules, organelles, cells, and tissues. Single- or double-sided CAV videodiscs and Macintosh HyperCard software.

Videodiscovery. *Cell Biology I Videodisc.* Seattle, WA: Videodiscovery, 1987. Produced from film by the Institute for Scientific Film in Germany, including considerable time-lapse and microcinematography of living cells. Structure and function of the nucleus; demonstrations of diffusion, osmosis, active transport, and endocytosis. Macintosh HyperCard stacks provide concepts, scientific names, microscopic techniques, comments, and user notes for more than 150 images and 80 films.

5

THE WORKING CELL

APPROACH

The concepts involved in molecular reactions and energetics are difficult for many nonscience majors, perhaps even more so than molecular structure. Energy relationships are concepts in physical chemistry, one level below the chemical hierarchy already introduced. Although everyday analogies can be used to clarify the laws of thermodynamics, they often contradict intuition, and the fact of biological organization sometimes seems to be a major counterargument.

Like cells, energy relationships are another great conceptual hinge in the study of biology, this time connecting the realm of chemistry to the study of organelles and organisms, and, most important, the realm of the organism to the biosphere. Critical to understanding the beauty that exists at these larger levels is a thorough understanding of three topics introduced in this chapter: energy conversions, enzyme function, and membrane structure and function.

Despite attempts to clarify these ideas for beginning students, you may still have some difficulty. Energy diagrams and the simplified diagrams of energy transformations within cellular reactions (such as Acetates 5.3A, B) will need to be explained to students who may not be familiar with X-Y graphs. Diagrams of chemical reactions, although they have been discussed before in terms of material changes of macromolecule building, will need to be reviewed. Basic terms and concepts can be confusing.

A critical idea to get across in this chapter is that the metabolic machinery and pathways illustrated are life's ways of getting the most usable energy into—and out of—its stored, chemical energy. A final point to stress is that living energy transformations are very efficient.

CHAPTER OBJECTIVES

Explain how the nature of energy transformations is guided by the two laws of thermodynamics. Explain how, while increasing their own organization, living organisms increase the disorganization of the universe, and thus do not violate the laws of thermodynamics.

Distinguish between endergonic and exergonic reactions.

Describe how ATP functions as a universal energy shuttle, carrying energy in usable chunks from large storage molecules to places in cells where work needs to be done.

Describe the structure of the enzyme-substrate interaction, and explain how enzymes catalyze biological reactions by lowering the energy of activation.

Describe the fluid mosaic model of membrane structure and the roles proteins in membranes play in the lives of cells.

Define the terms *diffusion, osmosis, facilitated diffusion, active transport, exocytosis, endocytosis, isotonic, hypotonic,* and *hypertonic.*

Preview and contrast the energy transformations and reactions involved in photosynthesis and cellular respiration, showing how these processes are related in a cycle.

LECTURE OUTLINE

I. Introduction.

A. Characteristics of organisms (the light of a firefly, the red pigments of a New England autumn, the trumpeting of an elk, the rank odor of mildew in a damp closet) are all the end-products of chemical reactions that occur in organisms and their cells (*Opening Essay*).

B. Organisms carry out chemical reactions for the purpose of energy transformation. In fact, all reactions involve some energy transformation.

C. This chapter covers several topics involved in how cells actually perform work: energy, enzymes, and membranes.

Preview: Some reactions are simply required for biosynthesis (for example, the digestion of food into smaller parts, the formation of pigments, the release of smelly waste products). Biosynthesis is discussed further in Module 6.17.

II. Energy is the capacity to perform work (*Module 5.1*).

A. Energy can only be described and measured by how it affects matter.

B. Energy comes in two forms.

1. Kinetic energy is the energy of motion. Heat is the kinetic energy associated with randomly moving molecules.

2. Potential energy is the stored capacity to perform work. The most important form of potential energy in living things is the potential energy stored in the arrangement of atoms in molecules. This is called chemical energy.

NOTE: Some everyday examples help clarify these early definitions, such as Figures 5.1A and B, or diagrams of water flowing downhill, dammed up, and flowing through a turbine or over a waterwheel.

III. Two laws govern energy conversion (*Module 5.2*).

A. Thermodynamics is the study of energy transformations that occur in matter.

Preview: As discussed in Module 36.11, there is a limit to the length of a food chain. These limits are the direct result of the laws of thermodynamics.

B. First law of thermodynamics (energy conservation).

1. The total amount of energy in the universe is constant; this energy can be transferred or transformed but neither created nor destroyed.

2. When a quantity of gasoline is ignited in the confines of a gasoline engine, the chemical energy ends up in three places: unspent chemical energy in unburnt gasoline, which exits in the exhaust; kinetic energy of the spinning engine; and heat. (This relates to the second law.)

C. Second law of thermodynamics (entropy increases).

1. Every energy change results in increased disorder, increased entropy (when looking at the state of the energy throughout the system studied).

2. When a quantity of gasoline is ignited, its potential (chemical energy) is released, causing a flash of light and burst of heat, which heats up the surrounding environment.

3. In much the same way, in biological systems, although some chemical energy may be channeled into useful work (light emission in the firefly,

moving vocal cords in the elk), there is always an increase in disorder; this almost always is an increase in the kinetic energy of molecules in cells (heat).

IV. Energy relationships in living things.

A. Life's chemical reactions either store or release energy (*Module 5.3;* Acetate 5.3A, B).

1. Endergonic reactions require an input of energy equal to the difference in the potential energy of the reactants and products.

2. *Preview:* Photosynthesis (Chapter 7) is an important process that is endergonic, requiring the energy of sunlight to cause energy-poor reactants to react to form energy-rich products.

3. Exergonic reactions result in an output of energy equal to the difference in the potential energy of the reactants and products.

4. Burning and cellular respiration are both exergonic processes by which the chemical energy of the reactants is released to form energy-poor products. In the case of burning, this happens all at once, with much "waste" of the chemical energy (gasoline, for example) to form heat and light.

5. *Preview:* Cellular respiration (Chapter 6) is an important biological process that releases the potential energy of sugar reactants, slowly, to form some energy-poor reactants and, most important, to convert the chemical energy of sugar into smaller, usable amounts of chemical energy in the form of ATP (*Module 5.4*).

6. Cellular metabolism is the sum total of all the endergonic and exergonic reactions in cells.

B. ATP shuttles chemical energy within the cell (*Module 5.4*).

1. Most endergonic cellular reactions require small amounts of energy, rather than the large amounts of energy available in food storage molecules.

2. Even a single glucose molecule contains too much energy. It's like a $50 or $100 bill: you want some $10s or $1s.

3. Adenosine triphosphate (ATP) is the energy-rich, spendable, "energy small change" of cellular reactions. It transfers usable amounts of energy from exergonic, food energy-releasing reactions to the endergonic reactions where cell work is done.

4. Various covalent bonds link the atoms in the parts of ATP, but the terminal bonds connecting the outer two phosphate parts are, at once, energy-rich and easily broken by hydrolysis.

5. The hydrolysis of ATP to release some of its chemical energy is an exergonic reaction.

6. When ATP gives up its energy, it forms ADP and an energy shuttle, the phosphate group (Acetate 5.4A, B).

7. The phosphate group is one of the reactants and the energy source for an endergonic reaction. This energizing process is known as phosphorylation. The products of the reaction hold chemical energy and are ready to do work.

The Working Cell

8. ATP regeneration is the reverse process. Endergonic reactions involved in cellular respiration phosphorylate (and energize) ADP in dehydration synthesis.

9. ATP is constantly being regenerated and used in a cycle involving endergonic dehydration synthesis and exergonic hydrolysis (Master 5.4C).

V. Enzymes.

A. What and where?

1. Enzymes are large protein molecules that function as biological catalysts. A catalyst is a chemical that speeds up the reaction without being consumed by it (Master 5.5A, Acetate 5.5B).

    NOTE: Enzyme names end in *-ase* and are often named after their substrates. For example, the enzyme that catalyzes the hydrolysis of sucrose is sucrase.

2. The energy of activation is the amount of energy, an "energy barrier," that must be put into an exergonic reaction before the reaction will proceed (analogy of the Mexican jumping beans, Master 5.5A; energy diagram, Acetate 5.5B).

3. Enzymes speed up the cell's chemical reactions by lowering energy barriers (*Module 5.5*).

B. A specific enzyme catalyzes each cellular reaction (*Module 5.6*).

1. The reactant in an enzyme-catalyzed reaction is the substrate.

2. One part of the enzyme binds to the substrate at the active site, holding the substrate in a specific position that facilitates the reaction (Acetate 5.6).

3. At the end of the reaction, the substrate changes into the product, and the enzyme is released, unchanged.

    NOTE: Use the "life of the party" analogy of enzyme action to explain enzyme function.

C. How enzyme activity is modified.

1. The cellular environment affects enzyme activity (*Module 5.7*). Such factors as temperature, pH, salt concentration, and the presence of cofactors often affect the way enzymes work.

    Preview: Magnesium is a cofactor that is essential for the proper functioning of chlorophyll (Module 32.6).

2. Enzyme inhibitors block enzyme action (*Module 5.8*). They do this by binding with the active site (competitive inhibitors) or some other site (noncompetitive inhibitors) on the enzyme, thus affecting the enzyme's ability to bind with the substrate (Master 5.8). Negative feedback is a type of inhibition whereby enzyme activity is blocked by one of the products of the reaction it catalyzes.

    NOTE: Negative feedback mechanisms are of major importance in the regulation of biological systems. A very clear example of this is seen in the regulation of female and male reproductive systems (Chapter 27).

3. Some pesticides and antibiotics inhibit enzymes (*Module 5.9*). For example, the pesticide malathion inhibits the enzyme acetylcholinesterase, involved in nerve transmission. The antibiotic penicillin interferes with an enzyme that helps build bacterial cell walls.

VI. Membrane structure.
   A. Membranes organize the chemical activities of cells (*Module 5.10*).
      1. They separate cells from their outside environments, including, in multicellular organisms, that environment in other cells that perform different functions.
      2. They control the passage of molecules from one side of the membrane to the other.
      3. In eukaryotes, they partition function into organelles.
      4. They provide reaction surfaces, and organize enzymes and their substrates.
      5. Membrane thickness cannot be seen in sections under the light microscope but can be resolved in TEMs (Figures 5.10A, B).
   B. Membrane phospholipids form a bilayer (*Module 5.11*; Acetate 5.11A, B).
      1. Phospholipids are like fats, with two nonpolar fatty acid "tails" and one polar phosphate "head" attached to the glycerol.
      2. In water, thousands of individual molecules form a stable bilayer, aiming their heads out and their tails in.
      3. The hydrophobic interior of this bilayer offers an effective barrier to the flow of most hydrophilic molecules.
   C. The membrane is a fluid mosaic of phospholipids and proteins (*Module 5.12*).
      1. It is a mosaic because the proteins form a "tiled pattern" in the "grout ground" of the phospholipid bilayer (Acetate 5.12).
      2. It is fluid (like salad oil) because the individual molecules are more or less free to move about laterally.
      3. The two sides of the membrane usually incorporate different sets of proteins: glycoproteins and glycolipids.
      4. Some proteins extend through both sides of the bilayer.
      5. Cholesterol is a common constituent of animal cell membranes and helps stabilize the fluidity at different temperatures.
VII. Membrane function.
   A. Proteins make the membrane a mosaic of function (*Module 5.13*).
      1. Identification tags: particularly glycoproteins (and nonprotein-containing glycolipids) (Acetate 5.12).
      2. Enzymes: catalyzing intracellular and extracellular reactions (Master 5.13A).
      3. Receptors: triggering cell activity when a messenger molecule attaches (e.g., signal transduction; Master 8.10B; Acetate 11.13).

      Preview: Signal transduction (Module 11.13)

      4. Cell junctions: either attachments to other cells or the internal cytoskeleton.
      5. Transporters: of hydrophilic molecules.
   B. Passive transport is diffusion across a membrane (*Module 5.14*).

1. Diffusion is the tendency for particles of any kind to spread out spontaneously from an area of high concentration to an area of low concentration.

2. Passive transport across membranes occurs (as diffusion does everywhere) when a molecule diffuses down a concentration gradient. At equilibrium, molecules continue to diffuse back and forth, but there is no net change in concentration anywhere (Master 5.14A).

3. Different molecules diffuse independently of one another (Master 5.14B).

4. Passive transport is an extremely important way for small molecules to get into and out of cells. For example, O_2 moves into red blood cells and CO_2 moves out of these cells by this process in the lungs.

C. Osmosis is the passive transport of water (*Module 5.15*; Acetates 5.15A, 5.15B).

1. If a membrane that is permeable to water but not to a solute separates an area of high solute concentration (hypertonic) from an area of low solute concentration (hypotonic), the water diffuses by osmosis to the hypertonic area until the concentrations are the same.

    NOTE: Osmosis can cause a physical force to be applied to the hypertonic solution. In the case shown in Acetate 5.15, this osmotic force raises the level of the solution on the right against the force of gravity, until the weight difference in levels equals the osmotic force.

2. The direction of osmosis is determined only by the difference in total solute concentrations.

3. Two solutions equal in solute concentrations so that osmosis does not occur between them are isotonic to each other.

D. Water balance between cells and their surroundings is crucial to organisms (*Module 5.16*; Master 5.16).

1. Cell membranes act as semipermeable membranes between the cell contents and its surroundings.

2. If a plant or animal cell is isotonic with its surroundings, no osmosis occurs, and the cells do not change. However, plant cells in such environments are flaccid or wilted, lacking the turgor that helps support some plant tissues.

3. An animal cell in a hypotonic solution will gain water and pop. A plant cell in a hypotonic solution will become turgid, as the cell wall counters the osmotic force of water moving in.

4. An animal cell in a hypertonic solution will lose water and shrivel. A plant cell in a hypertonic solution will lose water past the cell membrane but not the cell wall, resulting in the plasma membrane pulling away from the inside of the cell wall and the cell as a whole losing turgor.

5. *Preview:* The control of water balance, osmoregulation, is discussed in Module 25.5.

E. Specific proteins facilitate diffusion across membranes (*Module 5.17*; Master 5.17).

1. Facilitated diffusion occurs when a pored protein, spanning the membrane bilayer, allows a solute to diffuse down a concentration gradient.

2. No energy expenditure is required in this case.

3. The rate of facilitated diffusion depends on the number of such transport proteins, in addition to the strength of the concentration gradient.

F. Cells expend energy for active transport (*Module 5.18*).

1. Active transport involves the aid of a transport protein in moving a solute up a concentration gradient (Master 5.18, parts 1-3).

2. Energy expenditure in the form of ATP-mediated phosphorylation is required to help the protein change its structure and thus move the solute molecule.

3. Active transport proteins often couple the passage of two solutes in opposite directions across membranes. (Master 5.18, parts 4-6).

4. *Preview:* A very important example of a coupled active transport system is the Na^+ - K^+ pump, which functions in nerve impulse transmission (Modules 28.4 and 28.5).

G. Exocytosis and endocytosis transport large molecules (*Module 5.19*).

1. In exocytosis, membrane-bounded vesicles containing large molecules fuse with the plasma membrane and release their contents outside the cell (Master 5.19A, B).

2. In endocytosis, the plasma membrane surrounds materials outside the cell, closes around the materials, and forms membrane-bounded vesicles containing the materials.

3. Three important types of endocytosis are phagocytosis ("cell eating"), pinocytosis ("cell drinking"), and receptor-mediated endocytosis (Figure 5.19C).

H. Faulty membranes can overload the blood with cholesterol (*Module 5.20*; Master 5.20).

1. Cholesterol is carried in the blood by low-density lipoprotein (LDL) particles.

2. In people with normal cholesterol metabolism, excess LDL-bound cholesterol in the blood is eliminated by receptor-mediated endocytosis by liver cells.

3. In people with a genetic condition that results in hypercholesterolemia, fewer or no such receptor sites exist, and the people accumulate LDL-bound cholesterol, perhaps leading to heart disease.

4. *Preview:* The genetics of this disease is discussed in Module 9.10. As discussed in Module 21.20, hypercholesterolemia can also be a result of lifestyle.

VIII. Chloroplasts and mitochondria make energy available for cellular work (*Module 5.21; Acetate 5.21*).

A. The subjects of this chapter (energy, enzymes, and membranes) are important parts of the functioning of these two organelles and the processes they carry out (photosynthesis and cellular respiration).

B. Photosynthesis and cellular respiration are linked.

1. Solar energy is used to build energy-rich molecules in endergonic reactions in chloroplasts.

Preview: Photosynthesis is discussed in Chapter 7.

2. The energy-rich molecules release their energy to form ATP in mitochondria.

    Preview: Cellular respiration is discussed in Chapter 6.

3. The chemicals involved as the reactants in chloroplasts are the products in mitochondria, and vice versa.

CLASS ACTIVITIES

1. To demonstrate osmosis, take a limp piece of celery and place it in cold water (a hypotonic solution). The water will move into the celery by osmosis, and the resulting turgor pressure will stiffen the celery.

RESOURCES AND REFERENCES

Bretscher, M.S. "The Molecules of the Cell Membrane." *Scientific American*, October 1985.

Brown, M.S., and J.L. Goldstein. "How LDL Receptors Influence Cholesterol and Atherosclerosis." *Scientific American*, November 1984.

The Concept of Energy Flow. Princeton, NJ: Films for the Humanities and Sciences, 1993. A 10-minute video exploring basic concepts involved in energy relationships. Part of a larger series that examines energy at all levels of biology.

Dawson, Anthony. *Lysozyme.* Santa Barbara: Intellimation, 1992. A Macintosh computer-simulated exploration of the structure and function of enzymes.

Jacobson, K., E.R. Sheets, and R. Simson. "Revisiting the Fluid Mosaic Model of Membranes." *Science*, June 9, 1995. Review of current research on membrane structure and function.

Koshland, D.E. "Protein Shape and Biological Control." *Scientific American*, October 1973. A discussion of how enzymes are regulated.

Lipkin, R. "Controlling Life's Gateway: Opening and Closing Cell Membranes." *Science News*, September 24, 1994. Finding and using proteins to manipulate membrane function.

LoPresti, Vincent, and Fred Garafalo. "Visualizing Dynamic Molecular Geometry. Computer Animations for an Integrated Curriculum." *Journal of College Science Teaching*, May 1992. Discusses the pedagogical benefits of being able to demonstrate these dynamic processes.

MacNeil, P.L. "Cell Wounding and Healing." *American Scientist*, May-June 1991. How cell membranes repair breaks.

Schmaefsky, Brian. "Animated Enzyme Demonstrations." *Journal of College Science Teaching*, May 1991. A lecture demonstration involving student "bodies."

Videodiscovery. *Cell Biology I Videodisc.* Seattle, WA: Videodiscovery, 1987. Produced from film by the Institute for Scientific Film in Germany, including considerable time-lapse and microcinematography of living cells. Structure and function of the nucleus; demonstrations of diffusion, osmosis, active transport, and endocytosis. Macintosh HyperCard stacks provide concept, scientific names, microscopic techniques, comments, and user notes for more than 150 images and 80 films.

6

How Cells Harvest Chemical Energy

APPROACH

To the molecular biologist, the details of metabolic activities are the essence of beauty in biology. To the nonscience major, the details can be a considerable stumbling block. In addition to the challenge of new terms, metabolism is conceptually difficult. New concepts introduced here that are particularly troublesome for beginners are redox, electron carriers, energy cascades, a cycle of reactions, and rearranging molecular structures.

Be clear with students as to how much detail is required, and take particular care in developing the subject in an organized and logical manner. The pedagogical organization of the modules will help you in this respect. You can choose a subset of them to teach at your own level of detail.

As with Chapter 5, proceed cautiously at first, making sure students are comfortable with the terms and concepts introduced previously. The general organization of the chapter puts a human perspective on the process, details the ultimate mechanisms, reviews the subprocesses, and covers other related topics. This "spiral" approach makes sense. Make sure the overall message of cellular respiration—the orderly, stepwise transfer of stored energy to usable energy—is clear from your lectures and not hidden by the details.

Two aspects of intermediary metabolism that are not presented initially in the chapter should be discussed in lecture: some of the concepts behind the diagrams, and how these reactions are known. Students will not appreciate the "reality" behind the metabolic pathway diagrams unless you spend time discussing this aspect with them. Be sure to point out that in many diagrams, only the carbon chains are represented (usually as gray balls). Although methods used to study intermediary metabolism are not covered in the chapter, a brief diversion into this topic may help students appreciate the meaning behind the diagrams.

The money and economy analogies work well here—use them.

CHAPTER OBJECTIVES

Review covalent bonding as a means of storing energy, and review exergonic and endergonic reactions.

Contrast the phrase *cellular respiration* with the term *respiration*, which is often used to refer to the process of breathing.

Stress the relationship between form and function, particularly at the molecular and organelle levels. Show that some processes are best suited to occur in membranes, others in free solution.

Relate the caloric requirements at the human scale to the small amounts of energy needed to carry out individual reactions at the cellular scale.

Overview the workings of each phase (glycolysis, Krebs cycle, and electron transport chain), stressing the reactants, the products, the net production of ATP, and the cellular locations.

Discuss the cellular environments, scale, number, and pace of these reactions.

Explain how alcoholic fermentation and lactic acid fermentation can be used to generate ATP in the absence of oxygen.

Explain how the reactions of cellular respiration may function in a larger context of biodegradation and biosynthesis.

LECTURE OUTLINE

I. **Introduction.**

   A. *Review:* The definition of metabolism and what it entails (Chapter 5).

   B. Harvesting chemical energy from food molecules is one side of a cycle that, in eukaryotes, often involves mitochondria and chloroplasts (Acetate 5.21).

   C. This chapter covers the various metabolic pathways by which energy is released from food molecules. Slow burning of food molecules generates usable ATP.

   D. Yeast cells are simple eukaryotes that carry out all these pathways: With adequate supplies of O_2 (aerobic environments), they "burn" food molecules by cellular respiration; in environments without O_2 (anaerobic environments), they "ferment" food molecules (*Opening Essay*).

   NOTE: As a simple eukaryote with a variety of metabolic capabilities, bread-and-wine yeast (*Saccharomyces cerevisiae*) has played a major role as a research organism for studying these processes (Figure 6.0).

   E. Similar forms of the same two processes occur in human cells.

II. **Preview of cellular respiration in a human context.**

   A. Breathing supplies oxygen to our cells and removes carbon dioxide (*Module 6.1*).

   1. The oxygen needed to burn food by this process is outside the bodies of organisms.

   2. ATP is needed in cells in order to perform work.

   3. The O_2-requiring parts of cellular respiration require the cellular and mitochondrial environments to make ATP.

   4. In the process of breathing, the muscular, respiratory, and circulatory systems combine forces to bring reactants (food molecules and O_2) to cells and remove waste products (CO_2 and H_2O) from cells (Acetate 6.1).

   B. Cellular respiration banks energy in ATP molecules (*Module 6.2*; Master 6.2A).

   1. *Review:* The second law of thermodynamics. Remind students that the wasted energy is lost to each system as random kinetic energy, heat (Master 6.2B).

   2. Overall equation: $C_6H_{12}O_6 + 6O_2 \rightarrow 6CO_2 + 6H_2O$ + usable energy out.

   3. Compare the efficiency of the overall process in cells (about 40%) to the efficiency of energy use by an automobile (about 25%).

   C. The human body uses energy from ATP for all its activities (*Module 6.3*; Table 6.3).

   1. Energy is used for body maintenance and voluntary activities.

   2. A general estimate for an adult human of average weight for both types of energy expenditure is 2200 kcal per day.

III. **Some molecular basics of cellular respiration.**
   A. Cells tap energy from electrons transferred from organic fuels to oxygen (*Module 6.4*; Master 6.4).
      1. Discuss the rearrangements that have occurred in the locations of bonds in the reactants and products of cellular respiration.
      2. These movements of hydrogens represent movements of electrons.
      3. Cellular respiration involves a gradual series of steps, each coupling an exergonic with an endergonic reaction.
      4. Some of the energy that is released is stored in the bonds of phosphate bonds of ATP.

         Review: The coupling of the release of energy from ATP, an exergonic reaction, to provide energy to drive endergonic reactions is discussed in Module 5.4.
      5. At each step, electrons move from a chemical bond in a molecule where they have more energy to a bond where they have less energy.
      6. Oxygen atoms (from molecular oxygen, O_2) are the ultimate electron acceptor. When these oxygen atoms bind with the hydrogen atoms carrying the electrons, they form water molecules with relatively low-energy covalent bonds.

         NOTE: The majority of energy is released as heat. However, this should not be considered "wasted" energy since this heat is used to maintain body temperature and thus facilitate biochemical reactions.
   B. Hydrogen carriers such as NAD^+ shuttle electrons in redox reactions (*Module 6.5*; Master 6.5).
      1. The paired endergonic-exergonic reactions at each step in the transfer of energy are of a special kind known as redox (reduction-oxidation) reactions.
      2. Oxidation reactions involve electron loss and are the exergonic half.
      3. Reduction reactions involve electron gain and are the endergonic half.

         Note: A mnemonic for this is LEO - GER: Loss of Electrons, Oxidation; Gain of Electrons, Reduction.
      4. At each step in the breakdown of glucose, small redox reactions occur, involving an enzyme, dehydrogenase, and its coenzyme, NAD^+, which functions as an electron shuttle.
      5. During each step, the breakdown portion (glucose being stripped of its electrons) is oxidized while the NAD^+ is reduced, forming NADH.
   C. Redox reactions release energy when electrons "fall" from a hydrogen carrier to oxygen (*Module 6.6*).
      1. At the beginning of a different set of reactions, all the NADH generated as above gives up its energetic electrons, and NAD^+ is regenerated (Acetate 6.6A).
      2. These energetic electrons then pass from molecule to molecule in an "energy cascade," or electron transport chain. Each molecule is temporarily reduced by the oxidation of the previous molecule and, in turn, is oxidized when it reduces the next.

NOTE: This gradual release of energy can be analogized with a Slinky going down a flight of steps a step at a time.

    3. The ultimate electron acceptor in this part of the overall process is oxygen.

    4. During the cascade, small amounts of energy are released that can build ATP.

    5. Contrast the stepwise transfer of energy to oxygen against the direct reaction of hydrogen and oxygen (Figure 6.6B).

  D. Two mechanisms generate ATP (*Module 6.7*).

    1. Chemiosmosis is a process involving the electron transport chain and ATP synthases (protein clusters extending through the membrane). The electron transport chain temporarily produces potential energy in the form of an increase in H^+ concentration on one side of a membrane; the ATP synthases use the potential energy to generate ATP (from ADP and phosphate) by H^+ ions that flow through them down the concentration gradient (Master 6.7A).

    2. Substrate-level phosphorylation involves neither the electron transport chain nor membranes and takes place when a phosphorylated reactant gives up its covalently bonded phosphate to ADP with the aid of an enzyme (Master 6.7B).

IV. Overview: Respiration occurs in three main stages (*Module 6.8*).

  A. To summarize and preview the overall process, cellular respiration is composed of glycolysis (in the cytoplasm), the Krebs cycle (in the mitochondrial matrix), and the electron transport chain (on the inner mitochondrial membrane). These three parts are interconnected, as shown in Figure 6.8, which also shows the places where ATP is generated (Acetate 6.8).

  B. Glycolysis harvests chemical energy by oxidizing glucose to pyruvic acid (*Module 6.9*).

    1. This process occurs in the cytoplasm.

    2. Overall there are nine chemical steps, the net result of which is to split one six-carbon sugar molecule into two three-carbon pyruvic acid molecules (Acetate 6.9A).

    3. Each of the nine intermediate steps involves a separate enzyme (Acetates 6.9B start and 6.9B cont.).

    4. In addition to glucose, ADP, phosphate, and NAD^+ are required as reactants. ATP is also required because in order to get some intermediates formed, energy must be expended.

    5. Glycolysis can be broken into two phases: Steps 1–4 are preparatory and require ATP input (Acetate 6.9B start); Steps 5–9 are energy releasing, producing ATP and NADH (Acetate 6.9B cont.).

    6. Net energy production for glycolysis: 2 ATP (immediately usable for cellular work) and two NADH for each glucose entering the process.

  C. Pyruvic acid is chemically groomed for the Krebs cycle (*Module 6.10*; Master 6.10).

    1. This process occurs in the mitochondrial matrix (the fluid within the inner mitochondrial membrane).

    2. Oxidized, reducing NAD^+ to NADH.

3. Stripped of a carbon, releasing CO_2.

4. Complexed with coenzyme A, resulting in the molecule acetyl coenzyme A (acetyl CoA), the high-energy (but not as high as glucose) fuel for the Krebs cycle.

5. Net energy production for this step: 2 NADH for each glucose entering the process.

D. The Krebs cycle completes the oxidation of organic fuel, generating many NADH and $FADH_2$ molecules (*Module 6.11*).

1. This process occurs in the mitochondrial matrix.

2. Overall there are eight chemical steps, the net result of which is to disassemble one two-carbon acetyl CoA into two CO_2 molecules (Acetate 6.11A).

3. Each of the eight intermediate steps involves a separate enzyme (Acetate 6.11B).

4. In addition to acetyl CoA, ADP, phosphate, NAD^+, FAD (another energy shuttle), and oxaloacetic acid are required as reactants.

5. The eighth intermediate reaction regenerates oxaloacetic acid. This molecule is required at the beginning, and thus the intermediate steps cycle.

6. Coenzyme A is also released at the first step; it goes back to groom more pyruvic acid.

7. Net energy production for the Krebs cycle: 2 ATP (immediately usable), 6 NADH (not immediately usable), and 2 $FADH_2$ (not immediately usable) for each glucose entering the whole cellular respiration process.

E. Chemiosmosis powers most ATP production (*Module 6.12*; Acetate 6.12).

1. The electron transport chain is a series of protein complexes built into the cristae (inner mitochondrial membrane).

2. Each protein in the chain oscillates between reduced and oxidized states as the energetic electrons from NADH and $FADH_2$ pass through their region.

3. As redox occurs, H^+ ions are actively transported from inside the cristae to the intermembrane space.

4. The resulting H^+ gradient drives the production of ATP in the matrix, as the H^+ ions are transported through the ATP synthase.

5. Net energy production for the electron transport chain: 32 ATP (immediately usable) for each glucose entering the whole cellular respiration process.

6. These ATPs are only produced if O_2 is available as a terminal electron acceptor.

F. Certain poisons interrupt critical events in cellular respiration (*Module 6.13*; Master 6.13).

1. Rotenone (a plant product commonly used to kill fish and insect pests), cyanide, and carbon monoxide block various parts of the electron transport chain.

Preview: Biological magnification can be a consequence of the use of such poisons (Module 38.12).

2. The antifungal antibiotic oligomycin blocks passage of H^+ ions through the ATP synthase molecule.

3. "Uncouplers," such as dinitrophenol, cause the cristae to leak H^+ ions so that the H^+ gradient is not maintained and chemiosmosis cannot occur.

G. *Review:* Each molecule of glucose yields many molecules of ATP (*Module 6.14;* Acetate 6.14).

1. Glycolysis in cytoplasm yields some ATP in the absence of O_2 but mostly prepares for further steps in the mitochondria that require O_2.

2. The Krebs cycle in mitochondrial matrix yields some ATP directly but strips out CO_2, producing energy shuttles.

3. The electron transport chain produces lots of ATP, but only in the presence of O_2.

4. 3 ATP are produced for each NADH, and 2 ATP are produced for each $FADH_2$ introduced to the electron transport chain.

NOTE: This is a bit of a simplification. The general rule is that for each NADH sent to the ETC (electron transport chain), 3 ATP are produced. However, there is an exception, for each NADH produced by glycolysis and introduced to the ETC, either 2 or 3 ATP may be produced. Thus, the estimate of the total yield of ATP generated by the aerobic respiration of 1 glucose is put at 36 to 38. Also note that this is a good time to discuss the meaning behind the diagrammatic representations of metabolism and how the processes are studied. Reactions proceed from one "pool" of a compound to the next, depending on concentration gradients and the presence of the correct enzymes. The reactions are all happening in many places at the same time. Research into these pathways involves the introduction of radioactive isotope-labeled reactants followed by the recovery of the labeled products (Modules 2.5 and 7.3).

V. Fermentation is an anaerobic alternative to aerobic respiration (*Module 6.15*).

NOTE: Fermentation refers to energy-releasing molecular rearrangements in the absence of oxygen.

A. In the two cases reviewed in this module, the role of fermentation is to recharge NAD^+ so that glycolysis can continue to proceed in the absence of O_2. In addition, products are produced that are reduced and still energy-rich.

1. Alcoholic fermentation, characteristic of some yeasts and bacteria, results in one two-carbon ethanol. This product is toxic, and high concentrations will ultimately kill the cells that produce it (Acetate 6.15A).

NOTE: Different strains of yeast are killed by concentrations of up to 20%.

2. Lactic acid fermentation, characteristic of many organisms including animals and bacteria, results in one three-carbon lactic acid molecule. Although the accumulation of lactic acid causes muscle fatigue in animals, it is less toxic than alcohol and can be removed from the affected cells and detoxified by the liver (Acetate 6.15B).

NOTE: You might want to tell the story of the man who never drank alcohol yet got drunk whenever he ate. What happened is that outpouchings in his intestines (an anaerobic environment) contained yeast that produced ethanol by fermentation whenever he ate. A humorous way to finish this story is to ask the students to picture him getting arrested for driving under the influence of food.

B. Organisms that can live only in environments that lack oxygen are known as strict anaerobes. These organisms lack the necessary molecular and cellular equipment with which to carry out cellular respiration.

NOTE: Since aerobic photosynthesis evolved earlier than aerobic respiration, the oxygen that was produced was a toxin.

C. Organisms that can live in environments either lacking or containing oxygen are known as facultative anaerobes.

VI. Cells use many organic molecules as fuel for cellular respiration (*Module 6.16; Master 6.16*).

A. Free glucose is not the most common source of fuel in most animal diets, including the human diet.

B. Each of the basic food types can be used as a source of energy.

1. Carbohydrates such as polysaccharides and glycogen are usually hydrolyzed by digestive enzymes (or liver enzymes) to glucose, which enters glycolysis.

2. Proteins must first be digested to their constituent amino acids. The amino acids are then transformed into various compounds, which enter the middle of glycolysis or the Krebs cycle. Toxic parts of amino acids are stripped off and eliminated in urine.

3. Lipids contain almost twice as much energy per unit weight than carbohydrates. They must first be digested to glycerol, which enters in the middle of glycolysis, and fatty acids, which are converted to acetyl CoA and, thus, enter the Krebs cycle.

4. *Preview:* Human nutrition and the fate, following digestion, of many of the types of basic foods introduced here (and in the next module) are the subjects of Chapter 21.

VII. Food molecules provide raw materials for biosynthesis (*Module 6.17; Master 6.17*).

A. Cells and bodies obtain some raw materials directly from the digestion of the macromolecules in food.

B. The processes that produce new molecules are often the reverse of processes that break the same class of molecules down and feed their parts into the cellular respiration process, as discussed in Module 6.16.

C. ATP is required in biosynthetic pathways and produced by degradative pathways.

VIII. The fuel for respiration comes from photosynthesis (*Module 6.18*).

A. Cells of all living things can harvest molecular energy (by either cellular respiration or fermentation).

B. The ability to store molecular energy is not shared by all organisms.

CLASS ACTIVITIES

1. The production of CO_2 as a by-product of cellular respiration can be demonstrated by using a straw to blow bubbles into a pH indicator solution. CO_2 will make the solution acidic.

RESOURCES AND REFERENCES

Angier, N. "A Stupid Cell with All the Answers." *Discover*, November 1986. The important roles that yeast is playing in basic biological research.

Cellular Respiration. Princeton, NJ: Films for the Humanities and Sciences, 1993. A series of six 10-minute videotapes exploring this topic. Development is historical and includes animated sequences illustrating some of the facets.

Derr, M. "The End of the Road." *Scientific American*, April 1995. A debilitating disease in which mitochondria fail to make enough ATP for body exercise.

Harold, F.M. *The Vital Force: A Study of Bioenergetics*. New York: W. H. Freeman, 1986. A challenging introduction to energy and life, and how ATP is made by chemiosmosis.

Turney, Tully. *Metabolic Pathways*. Santa Barbara, CA: Intellimation, 1992. A supplementary review and lab tool for biology and biochemistry in a Macintosh HyperCard environment. Details metabolic pathways covered in this chapter, plus many other pathways in intermediary metabolism. Each reaction is annotated, and by stepping quickly from one molecule to the next and back, the user can clearly see molecular changes.

7

PHOTOSYNTHESIS: USING LIGHT TO MAKE FOOD

APPROACH

As stated at the end of this chapter, it can easily be argued that photosynthesis is the most important biological process. A firm understanding of it will strengthen a student's awareness of how metabolism proceeds and provide the basis for understanding the overall structure, organismal function, and ecological roles of plants and other producers. It will also provide the student with a framework for the brief introduction to bacterial chemosynthesis in Module 17.10.

This chapter starts with a historical perspective. Like Chapter 6, it then follows a "building" pattern, discussing the overall reaction, the several ultimate processes involved, and, finally, the subphases and related details. You should take a similar approach in your lectures. It may help to stop periodically to recapitulate what you have just covered, checking with students to be sure they are still with you.

There are many parallels between the processes of photosynthesis and cellular respiration, particularly in the molecular locations of the subprocesses, and in the flow of electrons. This similarity can be used to strengthen student understanding of the basic themes of structure/function and energy flow. And it may make it easier to remember the details if one process (or subprocess) can be seen to be the reverse of the other. But the parallels may also make it more difficult for students to sort out the two. So be careful how you treat the parallels, and consider omitting them the first time through.

CHAPTER OBJECTIVES

Give the overall equation for photosynthesis and contrast it to that for cellular respiration. Indicate where the process of photosynthesis occurs.

Describe the cellular locations and organisms in which the process occurs.

Describe the basic processes involved: splitting water to get electrons, redox reactions of electron transport chains, shuttling of energy and electrons by means of ATP and NADPH, and the coupling of the light reactions and the Calvin cycle.

Explain how pigments capture light energy and energize electrons.

Distinguish the structural and functional differences between photosystem I and photosystem II, and cyclic and noncyclic photophosphorylation.

Outline the steps in the cyclic fixation of carbon (Calvin cycle), and discuss the location of these reactions in the Calvin cycle.

Contrast the C_4 and CAM photosynthetic systems with the simpler system in C_3 plants, and explain how they are adaptations to hot, dry climates.

Show how photosynthesis relates to global warming.

LECTURE OUTLINE

I. Introduction.

    A. Review the overall equation for photosynthesis, and note that it is just the reverse of cellular respiration (Master 7.0).

B. A series of increasingly probing studies have provided knowledge about how photosynthesis works *(Opening Essay)*.

NOTE: The introductory module of this chapter discusses experiments of scientists studying metabolic reactions. If you have not done so, take this opportunity to review some of the ways scientists study metabolic processes.

1. Early experimentation followed thousands of years of casual observations of plant activity. One person (van Helmont) developed some early ideas about where plants get the materials they need to grow. Early experiments worked from the outside in, first testing the nature of the reactants and products of photosynthesis.

2. van Helmont's experiment was elegantly simple and disproved the then-current idea that soil provided everything a plant needs. He concluded that water provided the substance for plant growth.

3. In the early 1770s, Priestly showed that air (O_2) was restored by growing plants.

4. Ingenhousz soon demonstrated that air was restored only when plants were exposed to light.

5. Not until well into the 1900s were the details of the process worked out.

C. Autotrophs are the producers of the biosphere *(Module 7.1)*.

1. *Autotroph* means "self-feeding," and the term is applied to any organism that makes its own food without eating, decomposing, or absorbing other organisms or organic molecules.

    Preview: Autotrophs produce the biosphere's food supply (Modules 36.8 - 36.11 and 36.15).

2. Photosynthetic autotrophs (that "feed" themselves with light energy) include plants, photosynthetic protists (algae), and photosynthetic bacteria (Figure 7.1).

    NOTE: We will return to the term *producer* in our discussion of community ecology and the flow of energy among different kinds of organisms (Chapter 36).

D. Photosynthesis occurs in chloroplasts *(Module 7.2;* Master 7.2; review Figure 2.1).

1. This is true for all photosynthetic organisms except monerans, and it is true for all green parts of plants.

    NOTE: It is also true for the not-quite-so-green parts of other photosynthetic eukaryotes.

2. In most plants, the leaves and, specifically, mesophyll cells are the dominant photosynthetic locations.

3. Other structures in leaves provide entries and exits for the reactants and products of the process: CO_2 in and O_2 out through stomata; H_2O in through veins from the roots.

4. The green pigment that absorbs light energy is chlorophyll (Master 2.1A), located in the membranes (thylakoids in stacks, grana) within the chloroplasts.

5. Within the stroma carbon dioxide is built up into sugars.

NOTE: Ask students to note the parallels between photosynthesis and cellular respiration, particularly in the types of underlying processes and the locations in which these processes occur, but be careful not to confuse the two sequences.

II. General underlying processes.

A. Plants produce O_2 gas by splitting water (*Module 7.3*).

1. Experiments in the 1950s tested the early hypothesis of Ingenhousz that the oxygen given off in photosynthesis came from the reactant CO_2. Two experiments used ^{18}O-labeled reactants as tracers (see Module 2.5; Master 7.3B).

   NOTE: The splitting of water in photosynthesis is *the* major source of O_2 in the atmosphere.

2. A plant given $C^{18}O_2$ did not give off $^{18}O_2$.

3. A plant given $H_2^{18}O$ did give off $^{18}O_2$.

4. Additional experiments have confirmed where other atoms in the products come from (Master 7.3C).

   NOTE: In Figures 7.3 B and C, the overall equation for photosynthesis is written a bit differently, showing that water is both a reactant and a product. Because it takes 2 water molecules to get enough oxygen atoms to make 1 oxygen molecule, and since 6 molecules of oxygen are generated for each molecule of glucose formed, 12 water molecules are needed and some new water is formed at the end, with oxygen from the CO_2 and hydrogens from the original water molecules.

B. Photosynthesis is a redox process, as is cellular respiration (*Module 7.4*; Master 7.4A, B).

1. When H_2O molecules are split, yielding O_2, the water molecules are oxidized, giving up their electrons (and H^+ ions).

2. At the same time, CO_2 molecules are reduced to glucose as electrons and H^+ ions are added to them.

3. Compare this to the reverse overall reaction in cellular respiration (Module 6.5), where glucose is oxidized and oxygen is reduced.

4. In photosynthesis, the electrons travel "uphill" from the water to the glucose, adding the light energy captured by chlorophyll.

5. In cellular respiration, the electrons travel "downhill" from the glucose to the water, giving up their energy to ATP.

C. Overview: Photosynthesis occurs in two stages linked by ATP and NADPH (*Module 7.5*; Acetate 7.5).

1. Light reactions: steps that convert light energy to chemical energy and produce O_2 gas as a waste product. These reactions occur in the thylakoid membranes and produce energy shuttles in the form of ATP and energized electron shuttles in the form of molecules of NADPH. Light is required for these steps.

2. Calvin cycle: a cyclic series of steps that assemble glucose from CO_2 molecules. These reactions occur in the stroma (the fluid outside the thylakoids but inside the inner chloroplast membrane) and use the energy and electrons from ATP and NADPH in "carbon fixation." Light is not

directly required, but because production of the shuttles requires light, the Calvin cycle steps usually run during daytime.

III. The light reactions.

   A. Visible radiation drives the light reactions (*Module 7.6*).

      1. Light is a type of energy called electromagnetic radiation, which travels in rhythmic waves (Master 7.6A).

      2. Only a small fraction of electromagnetic radiation can be perceived by organisms. Humans perceive light of different wavelengths as different colors.

      3. During the light reactions, a leaf absorbs some light wavelengths (blue-violet and red-orange) and not others (what we see as green) (Master 7.6B).

      4. A variety of pigments are involved in absorbing light of different wavelengths (in plants, chlorophyll *a*, chlorophyll *b*, and carotenoids).

      5. In plants, only the chlorophyll *a* participates directly in the light reactions. The other pigments function to broaden the range of energy absorbed and convey this additional trapped energy to the chlorophyll *a*.

   B. Photosystems capture solar power (*Module 7.7*).

      1. Light also behaves as discrete packets of energy called photons.

      2. When a pigment absorbs a photon, the energy of one of the pigment's electrons is raised to an excited, unstable state.

         NOTE: The form of chlorophyll *a* (Master 2.1) is perfectly suited as a light trapper, containing many double bonds that expose many electron clouds to the passing radiation.

      3. In some cases, if the pigment is isolated from its surrounding molecular environment, the excited electron will lose its energy, return to the normal level, and emit heat or light (fluorescence). For instance, chlorophyll *a* fluoresces red (Master 7.7A, B).

      4. In contrast, in intact chloroplasts, the excited electrons are passed (the chlorophyll at the reaction center is oxidized) to a neighboring molecule, the primary electron acceptor (reduction) (Master 7.7A, B).

      5. Within the thylakoid membranes, many pigment molecules (200–300) are grouped with associated proteins into an antenna assembly, but only a pair of chlorophyll *a* molecules act as the reaction center (Master 7.7C).

      6. Two photosystems (antenna assembly + primary electron acceptor) have been identified, which differ in the wavelengths of light absorbed: photosystem I (P700) and photosystem II (P680).

   C. In the light reactions, electron transport chains generate ATP, NADPH, and O_2 (*Module 7.8*; Acetate 7.8).

      1. The kinetic energy of light is absorbed.

      2. The absorbed energy excites electrons.

      3. The excited electrons are passed along an electron transport chain in a series of redox reactions.

      4. The energy released by these redox reactions is used to generate ATP, NADPH, and O_2.

5. The production of NADPH requires 2 electrons. Photosystem I gets these electrons from photosystem II. Photosystem II gets its electrons from the splitting of water, a process that also produces $2H^+$ and $\frac{1}{2}O_2$.

D. Chemiosmosis powers ATP synthesis in the light reactions (*Module 7.9*; Acetate 7.9).

1. H^+ ions from the splitting of water and pumped from the stroma across the thylakoid membrane by the energy released from the electron transport chain end up inside the thylakoids in high concentration.
2. ATP synthase provides a port through which the H^+ ions can diffuse back into the stroma, releasing energy and phosphorylating ADP to ATP in the process.
3. This process is very similar to the ATP-generation mechanism in the mitochondrion (Module 6.12), but here it is known as photophosphorylation.

IV. Carbon fixation.

A. ATP and NADPH power sugar synthesis in the Calvin cycle (*Module 7.10*; Acetate 7.10A, B).

1. The net result of the steps of the Calvin cycle is the creation of phosphorylated, three-carbon molecules from carbon dioxide and the energy and electrons provided by ATP and NADPH from the light reactions.
2. Each CO_2 molecule is added to a five-carbon intermediate (RuBP, for ribulose biphosphate) catalyzed by the enzyme RuBP carboxylase (rubisco).
3. A number of rearrangements of molecules occur in many steps, some involving the use of energy from ATP, some oxidizing the NADPH (the reactants in these being reduced at the same time).
4. The last step of the cycle is the regeneration of the RuBP. The reactions involve considerable rearrangements of structure; all are proceeding at once, and since the steps ultimately regenerate one of the starting reactants, they can be regarded as occurring in a cycle.
5. It takes three molecules of RuBP entering into the cycle for every phosphorylated three-carbon molecule released out of the chloroplast.
6. The Calvin cycle takes place in the chloroplast stroma.
7. The product is used to synthesize glucose in the cytoplasm.
8. Plants that use only the Calvin cycle to fix carbon are known as C_3 plants.

B. *Review:* Photosynthesis uses light energy to make food molecules (*Module 7.11*; Acetate 7.11).

1. Sugar molecules a plant produces are the plant's own food supply, expended during cellular respiration.
2. Plants use sugars as building blocks for other organic compounds, including cellulose.
3. Plants, and other photosynthesizers, are the ultimate source of food for all other organisms.

V. C_4 and CAM plants have special adaptations that save water (*Module 7.12;* Master 7.12A, B, C).

1. When normal C_3 plants try to conserve water by closing their leaf pores, oxygen is fixed to RuBP by rubisco rather than CO_2, since new CO_2 is not able to enter the plant. This is called photophosphorylation, and it yields no sugar molecules and produces no ATP.

2. C_4 plants have special adaptations that conserve water and prevent photorespiration. These adaptations involve producing four-carbon compounds with a special enzyme in separate cells during hot, dry weather when the stomata are closed and the CO_2 concentration is much lower than the O_2 concentration. In other cells where the Calvin cycle is still operating, the four-carbon compounds are broken down to release CO_2 to complete the cycle. C_4 metabolism is found in corn, sorghum, and sugarcane.

3. CAM (crassulacean acid metabolism) plants form CO_2 into four-carbon compounds with another special enzyme at night, when temperatures are lower, humidity higher, and CO_2 more available. During the day, the four-carbon compounds are released to the Calvin cycle. CAM is found in several different types of succulent plants, such as cacti, pineapples, and jade plants.

VI. Global warming and forests.

A. Photosynthesis moderates the greenhouse effect; deforestation can intensify it (*Module 7.13*).

1. In the atmosphere, CO_2 retains heat from the sun that would otherwise radiate back into space. This is the basis for the greenhouse effect (Master 7.13B).

2. Burning fossil fuels (oil, coal, gas) and wood releases excess CO_2, which may be causing global warming.

Preview: Logging (Module 18.10) and the greenhouse effect (Module 38.13).

3. Replacing old forests with younger forests that grow more rapidly (and use up CO_2 more rapidly) may help slow down the rate of global warming. However, a good argument can also be made for keeping the old growth, since much of it rapidly ends up as CO_2 when burned, decomposed, or made into paper. There are also other, more compelling reasons to save these forests.

Preview: The greenhouse effect and global warming are featured again in Module 38.13 (Acetate 38.13B, C).

B. Talking About Science: Ecologist Margaret Davis studies ancient forests (*Module 7.14*).

1. Dr. Davis of the University of Minnesota studies ice age forests. By collecting fossilized plant material, such as pollen grains, Davis can determine the distribution of different plant species through time. She has determined that the distribution patterns of species vary through time as climate changes.

2. This work can be used to predict the effects of global warming on plant species. Davis has found that plant species are sensitive to rapid changes in temperature. Therefore, a worst-case scenario of rapid global warming (several degrees in less than a century) would result in the widespread

extinction of plant species and thus a loss of diversity of all forest species. Davis views such human-induced extinction as arrogant and immoral.

Preview: Global warming (Module 38.13).

3. Davis feels that managed forests, where commercially valuable trees are grown, will not be impacted as significantly as unmanaged forests.

4. Unmanaged forests that are a source of a significant amount of lumber and are nature preserves will be at great risk. Davis is concerned that human intervention to preserve these forests, especially at northern latitudes, will not be very successful.

CLASS ACTIVITIES

1. With a prism, demonstrate the spectrum of wavelengths in visible light.

2. Set up a demonstration using paper or thin-layer chromatography to separate the pigments in a leaf, to take about an hour. The visual spread of pigments supports the multipigment makeup of photosystems, a fact that is not immediately apparent.

3. Trace the flow of reactants and products of photosynthesis using a large, three-dimensional model of plant leaf cross sections and chloroplasts.

4. If the facilities are available, demonstrate some of the controlling factors in photosynthesis, using *Elodea* (or other aquatic plants available from aquarium supply stores). Set this up several hours before lecture. Trap the *Elodea* and its emitted oxygen bubbles in inverted, water-filled test tubes. One experimental setup could contain boiled water (to remove the CO_2). Use an unfiltered bright light, several different cellophane filters, and aluminum foil around different test tubes to show the efficiency of photosynthesis at different wavelengths. A glowing splint thrust into the gas will demonstrate its chemical makeup. In introducing these experiments, be sure to discuss experimental procedure, including the use of controls.

RESOURCES AND REFERENCES

Bazzazz, F.A., and E.D. Fajer. "Plant Life in a CO_2-Rich World." *Scientific American*, January 1992. How will increasing atmospheric CO_2 and global warming affect the relative success of C_3 and C_4 plants?

Can Polar Bears Tread Water? The Changing Climate. Deerfield, IL: MTI/Coronet, 1991. Investigates the greenhouse effect and the factors that influence it. Also available from Laser Learning Technologies.

Culotta, E. "Will Plants Profit from High CO_2?" *Science*, May 5 1995. Analysis of possible plant responses to increasing atmospheric CO_2.

Govindjee and W.J. Coleman. "How Plants Make Oxygen." *Scientific American*, February 1990.

Harmon, Mark E., William K. Ferrell, and Jerry F. Franklin. "Effects on Carbon Storage of Conversion of Old-Growth Forests to Young Forests." *Science*, February 9, 1990. Simulations of carbon storage suggest that conversion will not decrease atmospheric CO_2.

Hendry, George. "Making, Breaking, and Remaking Chlorophyll." *Natural History*, May 1990. Describes world stocks of chlorophyll and the rates and mechanisms of cyclical turnover.

Lawson, Anton E., Steven W. Rissing, and Stanley H. Faeth. "An Inquiry Approach to Nonmajors Biology," *Journal of College Science Teaching*, May 1990. Dealing with photosynthesis (one involving a historical perspective) in a nonscience majors class that puts ecology first.

Monastersky, Richard. "The Deforestation Debate." *Science News*, July 10, 1993. The latest FAO estimates of remaining forest areas and rates of deforestation in the world's tropical forests suggest previous estimates may have been somewhat high.

Photosynthesis. Cambridge, MA: Logal Software, Inc. A computer simulation of the process of photosynthesis. Includes simulations of the rate of photosynthesis, comparison of cellular respiration and photosynthesis, plant color, light dependent reactions, Calvin-Benson cycle, environmental factors, comparison of sun and shade plants, and fossil fuels.

Photosynthesis. Princeton, NJ: Films for the Humanities and Sciences, 1993. A series of six 10-minute videotapes detailing the phases of photosynthesis. Begins with a historical review of early discoveries and animated demonstrations of the processes.

Photosynthesis: Life Energy for Survival. Washington, DC: National Geographic Society, 1983. Scientists introduce the mechanics of photosynthesis and how it supports the food pyramid and fuels our industry.

Richter, Erwin W. "Demonstrating Absorption Spectra in the Classroom." *Journal of College Science Teaching*, February 1989. This simple technique requires an overhead projector, a solution of chlorophyll, and a diffraction grating.

Shmona, Kiryat. *Biology Explorer. Photosynthesis*. Scotts Valley, CA: Sunburst Communications, 1991. Part of a series of interactive, experimental simulations, this Macintosh program allows the user to make hypotheses, plan experiments, observe results, and draw conclusions about the various facets of the photosynthetic process.

Youvan, Douglas C., and Barry L. Marrs. "Molecular Mechanisms of Photosynthesis." *Scientific American*, June 1987. Details of the multimolecular structure of one particular bacterial photosynthetic reaction center and the timing of electron transport to the primary electron acceptor following photon absorption (all taking place in four trillionths of a second).

8

THE CELLULAR BASIS OF REPRODUCTION AND INHERITANCE

APPROACH

This chapter is the first of a new unit on genetics, one area of biology to which nonscience majors relate well. It is important to spend a few moments at the beginning of this set of lectures introducing the general topics of this unit: cellular basics, transmission genetics, molecular genetics, control of gene expression, gene engineering, and human genetics.

Growth, reproduction, heredity, and development are all fascinating topics that lend themselves to discussion with nonbiologists. Everyone is interested in how body repairs are made, how we received the traits we did from our parents, what our offspring are going to look like, how a single cell of a multicellular animal (or plant) is able to direct the development of the whole multicellular organism, and the wonders and ethics of genetic engineering. Understanding one of the most important medical concerns of modern society—cancer—involves a thorough understanding of the cell cycle.

The molecular aspects of these topics are not the ones students will immediately relate to. Therefore, start from the topics they know or have perceived (such as growth, reproduction, and trait inheritance). This is the overall approach of the unit. Advances in molecular genetics have and will affect all our lives. This is an extremely important topic in a nonmajors course.

It is important to firmly base students in an understanding of genetics at the cellular level because understanding at the transmission—and molecular—levels depends so much on this knowledge. Most important in this respect is a clear understanding of the consequences of meiosis, probably the main stumbling block of this chapter. The concept of homologous chromosomes is particularly difficult to get across at the beginning. Using the technique outlined in Activity 2 below should help. Make sure students have understood this critical point before proceeding with the details of close pairing and crossing over.

CHAPTER OBJECTIVES

Introduce the concept of a life cycle as a repeating series of processes and phases from one generation to the next.

Describe binary fission in bacteria, and distinguish this process from cell division in eukaryotes.

Describe the structures that play roles in the mitotic phase of the cell cycle: the centrioles, spindle microtubules, and chromosomes.

Outline the phases of the cell cycle, including mitosis, but focus on the overall results and the dynamic, repeating, and continuous nature of the process.

Describe the factors controlling cell growth and how cancer results following a breakdown of this control.

Outline the roles of the mitotic cell cycle in growth, cell replacement, and sexual reproduction.

Outline the general progression and overall results of meiosis, contrasting them with mitosis.

Explain the ways meiosis provides possibilities for genetic recombination.

LECTURE OUTLINE

I. **Introduction.**

   A. Introduce the general topics of reproduction, genetics, and inheritance, perhaps tracing the pedagogical development of the chapters in this unit.

   B. A life cycle is the sequence of life forms (and the processes forming them) from one generation to the next. Depending on one's needs, a life cycle can be more or less detailed. For example, use the sea star life cycle (*Opening Essay*).

   1. Adult sea star (sexual reproduction) → Fertilized egg (development) → Morula (development) → Young sea star (development) → Reproductively mature, adult sea star → etc. Possible asexual reproduction by fragmentation is mentioned (Figures 8.0A, B).

   2. Sexual reproduction involves passing traits from two parents to the next generation.

   3. Asexual reproduction involves passing traits from only one parent to the next generation.

   4. Cell division is the basis of all of the processes (developmental or reproductive) that link the phases in a life cycle.

   NOTE: There are two conflicting events in the whole life cycle progression: How, during reproduction, are faithful copies of organisms assured? How, during development, are subtle changes to the cells of a multicellular organism introduced?

   C. Like begets like, more or less (*Module 8.1*).

   1. This is true only for organisms reproducing asexually.

   2. Single-celled organisms, like protozoans, can reproduce asexually by dividing in two. Each daughter cell receives an identical copy of the parent's genes (Figure 8.1A).

   3. For multicellular organisms (and many single-celled organisms), the offspring are not genetically identical to the parents, but each is a unique combination of the traits of both parents (Figure 8.1B).

   4. Breeders of domestic plants and animals manipulate sexual reproduction by selecting offspring that exhibit certain desired traits. In doing so, the breeders reduce the variability of the breed's population of individuals.

   NOTE: You might want to discuss the ethics of selective breeding as well as the impact of reduced variability on a population's survivorship. For example, some species have reduced genetic variability due to being pushed to the verge of extinction by human behaviors.

   5. *Preview:* Observations of the work of breeders were part of the data Charles Darwin used in developing his theory of natural selection (Module 14.4).

   D. Cells arise only from preexisting cells (*Module 8.2*).

   1. This principle was formulated in 1858 by German physician Rudolf Virchow.

   2. Cell reproduction is called cell division.

   3. Cell division has two major roles. It enables a fertilized egg to develop through various embryonic stages, and for an embryo to develop into an

adult organism. It ensures the continuity from generation to generation; it is the basis of both asexual reproduction and sperm and egg formation in sexual reproduction.

II. Bacteria reproduce by binary fission (*Module 8.3*).

A. Bacterial chromosomes.

1. Genes of bacteria are carried on one circular DNA molecule that is up to 500 times the length of the cell.

2. Packaging is minimal: The DNA is complexed with a few proteins and attached to the plasma membrane at one point.

3. Most of the DNA lies non-membrane bounded, in the center of the cell.

B. Binary fission (Master 8.3A).

Preview: Fission in sea anemones is discussed in Module 27.1.

1. Prior to cell division, an exact copy of the chromosome is made. The attachment point divides so that the two new chromosomes are attached on adjacent parts of the plasma membrane.

2. As the cell elongates and new plasma membrane is added, the attachment points of the two chromosomes move apart.

3. Finally, the plasma membrane and new cell wall "pinch" through the cell, separating the two chromosomes into two new, genetically identical cells.

III. Background for understanding eukaryotic cell division.

A. Eukaryotes have multiple chromosomes that are large and complex (*Module 8.4*).

1. Whereas a typical bacterium might have 3000 genes, human cells, for example, have 50,000–100,000.

2. These genes are organized into several separate, linear chromosomes that are found inside the nucleus.

3. The DNA in eukaryotic chromosomes is complexed with protein in a much more complicated manner. This organizes and allows expression of much greater numbers of genes (Acetate 11.7A).

4. The chromosomes are only visible under the light microscope just prior to—and during—cell division (Figure 8.4).

5. In multicellular plants and animals, the body cells (somatic cells) contain twice the number of chromosomes as the sex cells. Humans have 46 chromosomes in their somatic cells and 23 chromosomes in their sex cells. Different species may have different numbers of chromosomes.

B. Chromosomes duplicate and then split in two as a cell divides (*Module 8.5*).

1. The DNA molecule in each chromosome is copied prior to the chromosomes' becoming visible.

2. As the chromosomes become visible, each is seen to be composed of two identical sister chromatids, attached at the centromere (Figure 8.5A).

3. It is the sister chromatids that are parceled out to the daughter cells (the chromatids are then referred to as chromosomes). Each new cell gets a complete set of identical chromosomes (Master 8.5B).

C. The cell cycle multiplies cells (*Module 8.6*).

1. Most cells in growing, and fully grown, organisms divide on a regular basis (once an hour, once a day), although some have stopped dividing.
2. Such dividing cells undergo a cycle, a sequence of steps that is repeated from the time of one division to the time of the next (Acetate 8.6).
3. Interphase represents 90% or more of the total cycle time and is divided into G_1, S, and G_2 subphases.
4. During G_1, the cell increases its supply of proteins and organelles and grows in size.
5. During S, DNA synthesis (replication) occurs.
6. During G_2, the cell continues to prepare for the actual division, increasing the supply of other proteins, particularly those used in the process.
7. Cell division itself is called the mitotic phase and involves two subprocesses, mitosis (nuclear division) and cytokinesis (cytoplasmic division).

    Preview: Ask what the result would be if mitosis occurred without cytokinesis. Skeletal muscle fibers are multinucleate (Module 30.8).

8. The overall result is two daughter cells, each with identical chromosomes.
9. Mitosis is very accurate. In yeasts, one error occurs every 100,000 divisions.
10. *Preview:* The molecular mechanism by which DNA is copied prior to mitosis is discussed in Chapter 10.

IV. Mitosis.

A. Cell division is a continuum of dynamic changes (*Module 8.7*; Acetates 8.7 start and 8.7 cont.).

1. If possible, show a video or film clip of the process.
2. Stress the dynamic, repeating, and continuous nature of mitosis, pointing out that biologists divide the overall process into what appear to be natural phases, to make it easier to follow.

B. Following are the established dividing points for the phases of the cell cycle (mitosis includes prophase through telophase):

1. Interphase: duplication of the genetic material; ends when chromosomes begin to become visible.
2. Prophase: mitotic spindle is forming, emerging from mucrotubule-organizing centers (MTOCs). Prophase ends when the chromatin has completely coiled into chromosomes; nucleoli are gone; nuclear membrane and nucleoli have dissolved.
3. Metaphase: spindle fully formed; chromosomes are aligned single file with centromeres on the metaphase plate (the plane that cuts the spindle's equator).
4. Anaphase: chromosome separation, from centromere dividing to arrival at poles.
5. Telophase: the reverse of prophase.
6. Cytokinesis: the division of the cytoplasm.

NOTE: The concept that a single chromosome can consist of a single chromatid or two chromatids and that when two chromatids separate they are then independent chromosomes can be confusing. The way to determine the number of chromosomes a cell contains is to count the centromeres.

- C. Cytokinesis differs for plant and animal cells (*Module 8.8*; Master 8.8A, B).
  1. In animals, a ring of microfilaments contracts around the periphery of the cell, forming a cleavage furrow that eventually cleaves the cytoplasm.
  2. In plants, vesicles containing cell wall material collect among the spindle microtubules, in the center of the cell, then gradually fuse, from the inside out, forming a cell plate which gradually develops into a new wall between the two new cells. The membranes surrounding the vesicles fuse to form the new parts of the plasma membrane.
- D. Anchorage, cell density, and chemical growth factors affect cell division (*Module 8.9*; Master 8.9A,B).
  1. To grow and develop, or replenish and repair tissues, multicellular plants and animals must control when and where cell divisions take place.
  2. Most animal and plant cells will not divide unless they are in contact with a solid surface; this is known as anchorage dependence.
  3. Laboratory studies show that cells usually stop dividing when a single layer is formed and the cells touch each other. This density-dependent inhibition of cell growth is controlled by the depletion of growth factor proteins in masses of crowded cells.
- E. Growth factors signal the cell-cycle control system (*Module 8.10*; Master 8.10A,B).
  1. The cell-cycle control system regulates the events of the cell cycle. Three major checkpoints exist: (a) at G_1 of interphase, (b) at G_2 of interphase, and (c) at the M phase.

     Preview: This regulation is a type of signal transduction (Module 11.13; also see Module 5.13).
  2. If, at these checkpoints, a growth factor is released, the cell cycle will continue. If a growth factor is not released, the cell cycle will stop.

     NOTE: Nerve and muscle cells are nondividing cells stuck at the G_1 checkpoint.
- F. Growing out of control, cancer cells produce malignant tumors (*Module 8.11*).

  NOTE: Cancer is a general term for many diseases in multicellular animals and plants involving uncontrolled cell division with the resultant tumor metastasizing (Master 8.11).

  Preview: Lifestyle and cancer are discussed in Modules 13.18 and 21.20.
  1. Cancer cells grown in culture are not affected by the growth factors that regulate density-dependent inhibition of cell division.
  2. A malignant tumor consists of cancerous cells. These tumors metastasize. This is in contrast to benign tumors, which do not metastasize.

     NOTE: When someone dies of cancer, they rarely die as a result of the primary tumor; it is usually the metastases that kill them.
  3. Cancers are named according to the tissue or organ of origin.

4. Some cancer cells actually continually synthesize factors that keep them dividing. Thus, unlike normal mammalian cells (in culture), there is no limit to the number of times cancer cells can divide.

5. Radiation and chemotherapy are two treatments for cancer. Radiation disrupts the process of cell division, and since cancer cells divide more often than most normal cells, they are more likely to be affected by radiation. Chemotherapy involves drugs that, like radiation, disrupt cell division. Some of these drugs—for example, taxol—target the mitotic spindle.

G. Review of the function of mitosis: Growth, cell replacement, and asexual reproduction (*Module 8.12*).

V. Meiosis.

A. Chromosomes are matched in homologous pairs (*Module 8.13*; Master 8.13).

1. Homologous chromosomes share shape, genetic loci, and carry genes controlling the same inherited characteristics.

2. Each of the homologues is inherited from a separate parent.

   NOTE: The sets are combined in the first cell following fertilization and passed down together from cell to cell during growth and development by mitosis.

3. In humans, 22 pairs, found in males and females, are autosomes. Two other chromosomes are sex chromosomes.

4. In females, there are two X chromosomes; in males, an X and a Y.

5. *Preview:* Sex chromosomes, sex determination, and sex chromosome anomalies are discussed further in Modules 9.18, 9.19, 13.5, and 13.9.

B. Gametes have a single set of chromosomes (*Module 8.14*; Acetate 8.14).

1. Adult animals have somatic cells with two sets of homologues (diploid, 2*n*).

2. Sex cells (gametes = eggs and sperm) have one set of homologues (haploid, *n*). These cells are produced by meiosis.

3. Sexual life cycles involve the alternation between a diploid phase and a haploid phase.

4. The fusion of haploid gametes in the process of fertilization results in the formation of a diploid zygote.

C. Meiosis reduces the chromosome number from diploid to haploid (*Module 8.15*).

1. Meiosis occurs only in diploid cells.

2. Like mitosis, meiosis is preceded by a single duplication of the chromosomes.

3. The overall result is four daughter cells, each with half the number of chromosomes.

4. Again, the process is dynamic but may stop at certain phases for long periods of time.

5. The process includes two consecutive divisions (meiosis I and meiosis II).

   NOTE: With this first pass through the process, avoid concentrating on the details of meiosis I (see Modules 6.16–6.19).

6. The halving of the chromosome number occurs in meiosis I (Acetate 8.15 start).

7. Sister chromatids separate in meiosis II (Acetate 8.15 cont.).

Preview: Gamete formation by meiosis is discussed in Module 27.4.

D. *Review:* A comparison of mitosis and meiosis (*Module 8.16;* Acetate 8.16).

1. The cell diagrammed has four chromosomes, two homologous pairs.

2. All the events unique to meiosis occur in meiosis I. In prophase I, homologous chromosomes pair to form a tetrad, and crossing over occurs between homologous chromatids.

NOTE: This results in the formation of unique genetic combinations (Module 8.19).

3. Meiosis II is virtually identical to mitosis (except the cells are haploid).

4. Mitosis results in two daughter cells, each with the same chromosomes as the parent cell. Mitosis can happen in diploid or haploid cells.

5. Meiosis results in four daughter cells (or, at least, nuclei), each with half the number of chromosomes as the parent cell. Meiosis happens only in diploid cells.

E. Independent orientation of chromosomes in meiosis and random fertilization lead to varied offspring (*Module 8.17;* Acetate 8.17).

1. During prophase I of meiosis, each homologue pairs up with its "other."

2. When they separate at anaphase I, maternally and paternally inherited homologues move to one pole or the other independently of other pairs.

3. Given n chromosomes, there are 2^n ways that different combinations of the half-pairs can move to one pole.

4. In humans, there are 2^{23} combinations of combining an individual's maternally inherited and paternally inherited homologues.

5. Combining gametes into zygotes suggests there are $2^{23} \times 2^{23}$ combinations in the zygote (but see the next two modules).

Preview: The consequences of the large amount of genetic variation generated by sexual reproduction is contrasted with the lower levels of genetic variation associated with asexual reproduction in Module 27.1.

F. Homologous chromosomes carry different versions of genes: A closer look (*Module 8.18*).

1. Example: coat color and eye color in mice.

2. *C* (agouti = brown) and *c* (white) for different versions of the coat-color gene and *E* (black) and *e* (pink) for different eye-color genes (Master 8.18A).

3. In this example, with the information up to this point, there would be two possible outcomes for the genes on the two chromosomes in a gamete (2^1).

G. Crossing over further increases genetic variability (*Module 8.19*).

1. Crossing over is the exchange of corresponding segments between two homologues. The site of crossing over is called a chiasma (Master 8.19A).

2. This happens between chromatids within tetrads as homologues pair up closely during synapsis (prophase I).

3. Crossing over produces new combinations of genes (genetic recombination) (Figure 8.19B).

4. Because crossing over can occur several times in variable locations among thousands of genes in each tetrad, the possibilities are much greater than calculated above. Essentially, two individual parents could never produce identical offspring from two separate fertilizations.

H. *Preview:* The mechanisms discussed here that result in new genetic combinations, meiosis and fertilization, do not occur in bacteria. However, there are several processes in which bacteria engage that result in the production of new genetic combinations (Modules 12.2 and 12.3).

CLASS ACTIVITIES

1. Show a video or film clip that illustrates the dynamic development of the early, few-cell to larval stages of a tadpole or echinoderm as a basis from which to start this lecture. Several sources are given below and in the resource list in Appendix B. Time-lapse movies of mitosis are also valuable for stressing the dynamic nature of this process. Be sure to describe any change in the time frame used in the films. Ask students to watch for points where there are natural hesitations in the flow of activity, or events that might be used to divide the process into phases. Then go over the process with acetates from the text, pointing out the fact that biologists have divided up the process into what appear to be natural phases, to make it easier to follow.

2. In developing the background for meiosis, actively engage your class in determining chromosome number for several generations of the human life cycle, using mitosis instead of meiosis (Acetate 8.14). Introduce the concept of a "reduction division" to halve the chromosome set in each gamete (without going through the details of orderly segregation of homologues). This will set up the discussion of homologues, beginning in Module 8.13.

RESOURCES AND REFERENCES

BBC Television. *What You Never Knew About Sex*. Deerfield, IL: Coronet Film and Video, 1993. Explores different methods of sex determination in species other than humans.

Blackburn, Elizabeth H. "Telomeres and Their Synthesis." *Science*, August 3, 1990. A discussion of the specialized DNA-protein structures found at the ends of every eukaryotic chromosome, which stabilize the chromosomes.

"Chromosomes." *Science*, December 8, 1995. Special reports.

"Frontiers in Biology. The Cell Cycle." *Science*, November 3, 1989. Six review articles in a special issue on the individual phases in the cell cycle and how they are integrated and controlled.

Glover, David M., Cayetano Gonzalez, and Jordan W. Raff. "The Centrosome." *Scientific American*, June 1993. Describes the recent work toward understanding the microtubule-organizing region around the centrioles, including its role in directing cell division, motility, and shape.

Hartwell, L.H., and M.B. Kastan. "Cell Cycle Control and Cancer." *Science*, December 15, 1994. How genetic changes can disrupt the cell cycle and lead to cancer.

Joyce, Christopher. "Taxol: Search for a Cancer Drug." *BioScience*, March 1993. A recent report on the development of this botanical medicine. Includes a historical survey of the ethnobotanical uses of yew and how the material is gathered and purified.

Liotta, Lance A. "Cancer Cell Invasion and Metastasis." *Scientific American*, February 1992. Improved understanding of how these cancer cells invade tissues is leading to new treatments.

Living Body Series. Growth and Change. Princeton, NJ: Films for the Humanities and Sciences, 1987. A 28-minute videotape uses a circus analogy to demonstrate cell growth and includes time-lapse photography of bone and tooth growth.

Mange, E.L., and A.P. Mange. *Basic Human Genetics*. Sunderland, MA: Sinauer Associates, 1994. A basic text on human genetics.

McIntosh, J. Richard, and Kent L. McDonald. "The Mitotic Spindle." *Scientific American*, October 1989. Describes the dynamic activities of the spindle as it helps parcel out the chromosomes during mitosis.

Moyzis, Robert K. "The Human Telomere." *Scientific American*, August 1991. Although this specialized DNA cap at each end of the chromosome carries no genes, chromosomes need their telomeres to remain intact.

Murray, Andrew W., and Marc W. Kirschner. "What Controls the Cell Cycle." *Scientific American*, March 1991. Details the action of one protein (cyclin) in regulating the recursive events leading to the birth of new cells.

Nova. *The Miracle of Life.* Boston, MA: Carolina Biological Supply, 1991. A classic film, video, or videodisc, that records the first second of human conception and follows the complex developments that culminate in the birth of a newborn baby. Videodisc and Macintosh HyperCard stack interface available from Education Express.

Organic Evolution Series. The Meiotic Mix. Princeton, NJ: Films for the Humanities and Sciences, 1993. A 10-minute videotape illustrating the difference between mitosis and meiosis and discussing the significance of both forms of cell division to organic evolution.

Scott, Douglas. *Mitosis and Meiosis*. Santa Barbara, CA: Intellimation, 1992. An annotated animation of mitosis and meiosis in animal cells, allowing the user to flip back and forth between comparable phases of the two types of division. For the Macintosh only.

Suzuki, D., A. Griffiths, J. Miller, and R. Lewontin. *An Introduction to Genetic Analysis*, 4th ed. New York: Freeman, 1989. A good genetics textbook, including Mendelian, molecular, and population genetics.

Videodiscovery. *Cell Biology I Videodisc*. Seattle, WA: Videodiscovery, 1987. Wonderful time-lapse films of echinoderm development and mitosis, plus other segments for use in this unit.

Videodiscovery. *Life Cycles Videodisc*. Seattle: Videodiscovery, 1985. A visual record of animal and plant reproduction with nearly 2000 color images, graphics, illustrations, and footage from the Oxford Scientific Film Collection. HyperCard stacks for Macintosh also available from Videodiscovery.

Watson, J.D., N.H. Hopkins, J.W. Roberts, J.A. Steitz, and A.M. Weiner. *Molecular Biology of the Gene*, 4th ed. Menlo Park, CA: Benjamin/Cummings, 1987. The classic textbook on genetics at the molecular level.

INTERNET RESOURCES FOR UNIT II
Cooperative Human Linkage Center

http://www.chlc.org

Human gene maps.

Ethical and Social Issues (HUM-MOLGEN)

http://www.informatik.uni-rostock.de/HUM-MOLGEN/ethic/ethic.html

Bioethics.

Genetics and Public Issues (GPI) Program at NCGR

http://ncgr.org/gpi/Index.html

Bioethics.

The Genome Database

http://gdbwww.gdb.org

Human gene maps.

Human/Mouse Homology Relationships

http://www3.ncbi.nlm.nih.gov/Homology

Comparison of human and mouse gene maps. Also of use for Unit III (Concepts of Evolution).

OMIM Home Page—Online Mendelian Genetics in Man

http://www3.ncbi.nlm.nih.gov/Omim

Gene maps.

Virtual FlyLab

http://vflylab.calstatela.edu/edesktop/VirtApps/VflyLab/IntoVflyLab.html

On-line mating of fruit flies.

9

PATTERNS OF INHERITANCE

APPROACH

Students are very interested in inheritance. Once students learn the specialized vocabulary and conventions, most of the concepts introduced in this chapter are easy to grasp. The explanation of Mendelian principles by the behavior of chromosomes should strengthen a student's ability to understand the basic principles of genetics.

Most of the specialized terminology appears in Module 9.3. As you cover this material, check periodically to ensure your students are with you.

Make sure students understand that Mendel had no preconceptions about chromosome behavior during sexual reproduction; he based his principles solely on the numbers of offspring he observed. The scientific process is used to structure the presentation of Mendel's observations and principles. You might wish to develop this theme in your lectures.

Although gene frequencies and probability calculations can be used with great success to generate predictions about multiple gene behavior (as in Module 9.7), if you introduce this technique, be prepared to spend some time in the beginning going over the rules and their applications. Probability calculations are not usually intuitive to nonscience majors.

Specialized conventions that indicate multiple-allele genotypes, sex chromosomes, and sex-linked genes are explained but may be confusing to students. To clarify, use the symbolic notation introduced earlier in the chapter.

CHAPTER OBJECTIVES

Describe Mendel's experimental framework and approach, stressing that his ideas were based solely on observations of phenotypic characters.

Explain the distinction between the terms in each of these pairs: *gene* and *allele*, *dominant* and *recessive*, *homozygous* and *heterozygous*, *phenotype* and *genotype*, and *monohybrid cross* and *dihybrid cross*.

Explain the experimental background of Mendel's principle of segregation; define this principle, and explain its chromosomal basis.

Explain the experimental background of Mendel's principle of independent assortment; define this principle, and explain its chromosomal basis.

Review the four patterns of inheritance that do not follow Mendel's principles, and give an example of each.

Explain how chromosomes (including sex chromosomes) determine gender differently in a variety of organisms.

Describe the pattern of inheritance of sex-linked characteristics.

LECTURE OUTLINE

I. Introduction.

　　A. Close observations of breeding organisms and their offspring show patterns in the inheritance of characteristics (*Opening Essay*; chapter opening photo; Master 9.0A, B).

B. We will see that these patterns of inheritance can be explained by the behavior of chromosomes during meiosis and fertilization.

C. The science of genetics has ancient roots (*Module 9.1*).

   1. The ancient Greeks believed in pangenesis, the idea that particles governing the inheritance of each characteristic collect in eggs and sperm and are passed on to the next generation.

   2. But many, including Aristotle, realized there were problems with this idea: The potential to produce characteristics is inherited, not pieces of the characteristics themselves. Reproductive cells are not changed by the development or activity of other cells.

   3. Based on artificial breeding, nineteenth-century observers believed in the "blending" hypothesis, in which characteristics from both parents blend in the offspring.

   4. *Preview:* Not only did plant and animal breeders provide data for hypotheses concerning inheritance, but this information greatly influenced the ideas of Charles Darwin and Alfred Wallace, at about the same time (Chapter 14).

D. Experimental genetics began in an abbey garden (*Module 9.2*).

   1. Mendel was university-trained in precise experimental technique. He studied peas because they offered advantages over other organisms. Peas grow easily, have relatively short life spans (one year), and have numerous and distinct characteristics (Master 9.2D), and the mating of individuals can be controlled so that the parentage of offspring can be known for certain.

      NOTE: This is a place to talk about the fact that good biological experimentation often results from the choice of suitable study organisms that enable the experimenter to focus on particular questions.

   2. Mendel's paper, published in an obscure journal in 1866, argued that there are discrete, heritable factors (what we call genes) that retain their individuality when transmitted from generation to generation.

E. Additional background and terms to use in discussing Mendel's experiments.

   1. Mendel could intentionally self-fertilize a flower by covering it with a bag, or cross-fertilize two different plants by dusting the carpels of one with the pollen of another (Master 9.2C).

      NOTE: The life history of flowering plants, for our purposes here, is similar to that of most animals, with male and female gamete-producing organs found in flowers (Master 9.2B).

   2. By continuous self-fertilization for many generations, Mendel developed breeds of plants that bred true (continued to show a characteristic when self-fertilized) for each of the characteristics he followed. He found seven characteristics, each of which came in two distinct forms.

   3. Mendel developed two principles based on two types of experiments. In one type (monohybrid crosses), he hybridized true-breeding plants for each of the two forms of a characteristic. In a second type (dihybrid and trihybrid crosses), he hybridized plants that combined two or more of the seven characteristics.

4. In these experiments, the true-breeding parents are the P (parental) generation, their hybrid offspring is the F_1 (first filial) generation, and the offspring of the mating of two F_1s is the F_2 (second filial) generation.

II. Mendel's principle of segregation describes inheritance of a single characteristic (*Module 9.3*).

A. The principle stated in modern language: Pairs of genes segregate (separate) during gamete formation; the fusion of gametes at fertilization pairs genes once again.

B. What Mendel did and observed (Acetate 9.3A, B).

1. Monohybrid cross with the flower-color characteristic.

2. Results: Out of 929 F_2 offspring, 705 were purple, and 224 were white.

   NOTE: The proportions are not exactly ¾ and ¼ because mating involves probabilities. See below.

3. Mendel observed that each of the seven characteristics exhibited the same inheritance pattern.

C. Mendel developed four hypotheses (in modern language):

1. There are alternative forms of genes, the units that determine heritable characteristics. These alternative forms are called alleles.

2. For each inherited characteristic, an organism has two genes, one from each parent. They may be the same allele or different alleles.

3. A sperm or egg carries only one allele for each characteristic because the allele pairs segregate from each other during gamete production.

4. When the two alleles are different, the one that is fully expressed is said to be dominant and the one that is not noticeably expressed is said to be recessive.

D. Some simplified ways of expressing this information (Acetate 9.3A, B).

1. Conventions for alleles: *P*, the dominant (purple) allele, and *p*, the recessive (white) allele. P generation: *PP* × *pp*; their gametes: *P* and *p*; F_1 generation: *Pp*.

2. Homozygous dominant, homozygous recessive, and heterozygous refer to the genotypes (the nature of the genes as inferred from observations and knowledge of how the system works). The phenotypes are what we see.

3. The Punnett square is used to keep track of the gametes (two sides of the square) and offspring (cells within the square).

   NOTE: This pattern, whereby each gamete contains a single copy of each gene, is stated by Mendel's principle of segregation and is based on the events of meiosis (anaphase I and anaphase II; Module 8.15).

E. Homologous chromosomes bear the two alleles for each characteristic (*Module 9.4*).

1. *Review:* Homologous pairs (Module 8.13).

2. Although Mendel knew nothing about chromosomes, our knowledge of chromosome arrangements (in homologous pairs) strongly supports the principle of segregation.

3. Alleles of a gene reside at the same locus on homologous chromosomes.

NOTE: One of the chromosomes illustrated was inherited from the female parent, the other from the male parent (Master 9.4).

III. The principle of independent assortment is revealed by tracking two characteristics at once (*Module 9.5*).

    A. The principle stated in modern language: Each pair of alleles segregates independently during gamete formation.

    B. Experimental procedure.

        1. Breed two strains true, each exhibiting one of the two forms of two characteristics (in the example used, round yellow seeded plants [*RRYY*] and wrinkled green seeded plants [*rryy*]).

        2. Hybridize these two strains as P generation resulting in hybrid offspring (F_1: *RrYy*).

        3. Then allow the F_1 to self-fertilize (*RrYy* × *RrYy*).

        NOTE: Each of these individuals produces the same four gametes: *RY*, *Ry*, *rY*, and *ry*. Taking one gamete from each individual means that there are $4^2 = 16$ possible gametic combinations.

        4. Two hypotheses: The characteristics are inherited either dependently or independently of each other (Acetate 9.5A, left).

    C. Results.

        1. The F_1 generation exhibits only the dominant phenotype (this is expected).

        2. The F_2 generation exhibits a phenotypic ratio of 9:3:3:1 (round yellow : round green : wrinkled yellow : wrinkled green).

        NOTE: $9 + 3 + 3 + 1 = 16$, the same as the number of possible gametic combinations. That the phenotypic ratio adds up to the number of possible gametic combinations serves as a check of the results of a cross.

        3. Use a Punnett square to analyze these results, with the sides of the square representing the male and female gametes possible if alleles of two characteristics segregate independently. Notice that the genotypes that produce the same genotype are not all the same (Acetate 9.5A, right).

        4. Feather color in budgies follows this pattern of assortment if pure strains of green-feathered and white-feathered birds are used as the P generation. *B* allele, yellow pigment; *C* allele, melanin. *B_C_*, green; *B_cc*, yellow; *bbC_*, blue; *bbcc*, white. Explain the use of blank lines to indicate unknown alleles that do not change the phenotype. This is a somewhat complicated example: Although the *B* and *C* genes segregate independently, the dominant phenotypes are expressed together as a blend (Master 9.5B).

IV. More ways to explain Mendelian ("transmission") genetics.

    A. Geneticists use the testcross to determine unknown genotypes (*Module 9.6*; Master 9.6).

        1. A testcross involves crossing an unknown genotype (expressing the dominant phenotype) with the recessive phenotype (by necessity, homozygous).

        2. Each of two possible genotypes (homozygous or heterozygous) gives a different phenotypic ratio in the F_1 generation. Homozygous dominant gives all dominant. Heterozygous gives half recessive, half dominant.

NOTE: This technique uses phenotypic results to determine genotypes.

B. Mendel's principles reflect the rules of probability (*Module 9.7;* Master 9.7).

1. Events that follow probability rules are independent events; that is, one such event does not influence the outcome of a later such event. If you flip a coin four times and get four heads, the probability for tails on the next flip is still ½.

2. The probability of two events occurring together is the product of the probabilities of the two events occurring apart (the rule of multiplication).

3. Thus, when studying how the alleles of two (or more) genes that segregate independently behave, use the probabilities of how they behave individually.

   NOTE: The probability of a recessive phenotype occurring in a monohybrid cross is 1 out of 4. The probability of two recessives occurring together in a dihybrid cross is ¼ × ¼, or 1 out of 16 (recall 9:3:3:1). In a trihybrid cross, as mentioned, the probability of a triple recessive is 1 out of 64.

4. If there is more than one way an outcome can occur, these probabilities must be added, as in the case of determining the chances for heterozygous mixtures (the rule of addition).

C. Mendel's principles apply to the inheritance of many human traits (*Module 9.8*).

1. Many human characteristics are thought to be determined by simple dominant-recessive inheritance, and sometimes the ratio of dominant-to-recessive phenotype exhibits a Mendelian ratio (Figure 9.8).

2. The terms *dominant* and *recessive* refer only to whether or not a characteristic is expressed in the heterozygote, not to whether it is the most common.

   NOTE: It is really only the distribution of phenotypes in the offspring of one couple of known phenotype or genotype that will follow Mendelian principles.

3. With practice, the principles and techniques outlined above can be used to determine many interesting things about the genotypes of individuals. This information, in turn, can be used to predict future characteristics in offspring.

   Preview: In large populations, the commonality of dominant and recessive characteristics may depend on whether one or the other allele confers advantages or disadvantages on those who have it. This subject will be discussed in Chapter 14 on population genetics.

V. Situations that do not behave in a Mendelian manner.

A. The relationship of genotype to phenotype is rarely simple (*Module 9.9*).

1. The inheritance of many characteristics among all eukaryotes follows the principles that Mendel discovered.

   NOTE: Discussing these principles first has allowed us to focus on the conventions and basic functioning of the system that underlies inheritance patterns.

2. However, most characteristics are inherited in ways that follow more complex patterns.

3. Before looking at the chromosomal explanation of Mendel's principle of independent assortment, we will look at four such complex patterns: incomplete dominance, multiple alleles at a gene locus, pleiotropy, and polygenic inheritance.

4. These patterns are extensions of Mendel's principles, not exceptions to them.

B. Incomplete dominance results in intermediate phenotypes (*Module 9.10*).

1. Incomplete dominance describes the situation where one allele is not completely dominant in the heterozygote; the heterozygote usually exhibits characteristics intermediate between both homozygous conditions.

2. Snapdragon color is a good example of how this works. Note that the possibilities of each genotype are the same as in a case of complete dominance, but the phenotypic ratios are different (Acetate 9.10A).

3. Another example: the inheritance of alleles that relate to hypercholesterolemia. Normal individuals, *HH*, have normal amounts of LDL receptor proteins; *hh* individuals (rare in the population, about one in 1 million) have no receptors and five times the amount of blood cholesterol; *Hh* individuals (one in 500) have half the number of receptors and twice the amount of blood cholesterol (Acetate 9.10B).

Preview: Lifestyle can also lead to hypercholesterolemia (Module 21.20).

4. *Preview:* In this last example, the relative numbers of each phenotype in the population depend on the manner in which genes are inherited in populations, the subject of Chapter 14.

C. Many genes have more than two alleles (*Module 9.11*; Acetate 9.11).

1. The ABO blood groups in humans follow this pattern, in which individuals can have two alleles from a set of three possible alleles.

2. These blood-type alleles code for two carbohydrates (or the absence of any carbohydrate) on the surface of red blood cells (a total of three alleles). There are six possible genotypes and four possible phenotypes.

3. When blood is transfused, recipients develop antibodies (discussed further in Chapter 24) for the types of carbohydrate on the donor red blood cells that the recipients lack.

4. Type O (universal donor) has neither carbohydrate and can receive no other type. Type AB (universal recipient) has both carbohydrates and can receive any type. Type A has carbohydrate A and can receive A or O. Type B has carbohydrate B and can receive B or O.

5. Blood types can be used to disprove or suggest parentage in paternity suits.

D. A single gene may affect many phenotypic characteristics (*Module 9.12*).

1. This common situation is known as pleiotropy.

2. A relatively uncommon example is the inheritance of an allele that codes for abnormal hemoglobin and, in the homozygous condition, causes sickle-cell anemia.

Preview: The allelic variant that is responsible for sickle-cell anemia is discussed in Module 10.16.

3. The sickle shape of the red blood cells confers a whole suite of symptoms on homozygous individuals, attributable to three underlying difficulties resulting from the abnormal cell shape (Master 9.12).

4. The normal and abnormal alleles are codominant, so heterozygous individuals (carriers) can exhibit some symptoms, although normally they are healthy.

5. The incidence of the allele is relatively high in individuals of African descent (one in 10 African Americans is heterozygous), because sickle-cell carriers are somewhat protected from malaria, a protozoan-caused disease prevalent in tropical regions.

Preview: This is an example of the action of natural selection (Chapter 14).

NOTE: Another good example of pleiotropy is Marfan's syndrome, which results in long, loose-jointed limbs, cardiovascular defects, and several other problems.

E. A single characteristic may be influenced by many genes (*Module 9.13*; Acetate 9.13).

1. This situation is known as polygenic inheritance.

2. Skin pigmentation is just such a phenotypic character whose underlying genetics has not been completely determined. Acetate 9.13 is hypothetical, showing the phenotypic outcome of mixtures of three genes, each with two alleles coding for "additive units," which produce the overall characteristic.

VI. Chromosome behavior accounts for Mendel's principles (*Module 9.14*).

A. While the existence and behavior of chromosomes was not appreciated by Mendel himself, the significance of his work was understood later, in the late 1800s and early 1900s.

B. We have already seen that the fact that there are homologous pairs of chromosomes accounts for the principle of segregation.

C. The fact that there are several sets of homologous pairs of chromosomes accounts for the principle of independent assortment (Acetate 9.14).

NOTE: Mendel's seven garden pea characteristics all sorted independently of each other because the genes governing each characteristic are all on separate chromosomes.

D. Genes on the same chromosome tend to be inherited together (*Module 9.15*; Acetate 9.15).

1. Linked genes are located close together on the same chromosome.

2. The inheritance of such genes does not follow the pattern described by the principle of independent assortment because the two genes are normally inherited together on the adjoining portions of the same chromosome.

3. The phenotypic ratios of such dihybrid crosses approach that of a monohybrid cross (3:1), rather than the typical pattern of the dihybrid cross (9:3:3:1).

E. Crossing over produces new combinations of alleles (*Module 9.16*).

1. In the case of many linked genes (for example, pollen grain length and flower color in sweet peas), there are some offspring of a dihybrid cross that do involve independent assortment of the genes (Acetate 9.15).

2. These situations of recombination are accounted for by crossing over, which occurs between the genes on homologous pairs of the same chromosome during some meiotic divisions, but not all (Acetate 9.16A, B).

Review: Crossing over and genetic recombination are discussed in Module 8.19.

Preview: Introns may increase the number of points at which crossing over occurs (Module 11.9).

3. Early examples of recombination were demonstrated in fruit flies by embryologist T. H. Morgan and colleagues in the early 1900s.

F. Geneticists use crossover data to map genes (*Module 9.17*).

1. The study of fruit-fly genetics resulted in considerable additional understanding of genetic principles.

NOTE: This is another example of the use of an experimental organism that lends itself to study. Fruit flies have many phenotypic characters, are easily raised and bred in captivity, and have a short life cycle. In addition, they have only four chromosomes (simplifying the situation), and these chromosomes can be easily visualized in nondividing cells in the salivary glands.

2. A. H. Sturtevant, one of Morgan's colleagues, developed a technique of using crossover data to map the locations of genes on chromosomes on which they were linked.

3. Sturtevant assumed that the rate of recombination is proportional to the distance apart two genes are on a chromosome (Master 9.17B).

VII. The genetics of sex chromosomes.

A. Chromosomes determine sex in many species (*Module 9.18*; Master 9.18).

1. Sex chromosomes in humans are nonidentical members of a homologous pair.

2. In humans, *XX* individuals are female, and *XY* are male.

3. In other species, other patterns of sex chromosomes exist.

4. In some species, sex is determined by chromosome number rather than chromosome type. In some invertebrates, diploid individuals are female and haploid are male.

B. Sex-linked genes exhibit a unique pattern of inheritance (*Module 9.19*).

1. Sex chromosomes contain genes specifying sex and other genes for characteristics unrelated to sex.

2. Because of linkage and location, the inheritance of these characteristics follows peculiar patterns.

3. Examples using eye color in fruit flies (*X*-linked recessive for white eyes; Figure 9.19A). Depending on the genotypes of the parents, three patterns emerge (Master 9.19B, C, D).

4. In humans, most sex-linked characteristics result from genes on the *X* chromosome.

Preview: Thus, mostly males are affected (Module 13.9).

5. Examples of such characteristics are red-green color blindness, a type of muscular dystrophy, male-pattern baldness, and hemophilia.

6. Because the male has only one X chromosome, his recessive X-linked characteristic will always be exhibited.

    NOTE: The DNA in the Y chromosome has recently been completely sequenced under the auspices of the Human Genome Project (see Chapter 13).

CLASS ACTIVITIES

1. Remind students that most of the details covered in this chapter can be deduced by observation of patterns of inheritance across generations. Make an accounting of the ratios of some simple, Mendelian characteristics among your class (earlobes, mid-digital hair, tongue curling, etc.). You may want to see if it is possible to determine the genotypes of students for these characteristics by comparing how they express the characteristic with how each of their parents expresses that characteristic. Examine the distribution of phenotypes of those illustrated in Figure 9.8 in your class by asking for a show of hands. You may want to explain that the class represents a "freely interbreeding" population in which the 3:1 ratio of dominant-to-recessive phenotypes is maintained only if there is no relative advantage of one characteristic over the other. This demonstration helps preview population genetics (Chapter 14).

RESOURCES AND REFERENCES

Charlesworth, B. "The Evolution of Sex Chromosomes." *Science*, March 1, 1991. How did the X-Y mode of sex determination evolve?

Erickson, John. *Introduction to Genetics*. Campton, NH: Trinity Software, 1992. A modular computer tutorial for the IBM PC that reviews many aspects of Mendelian/transmission genetics. Although designed to support a beginning genetics course, the examples are well developed and would support this chapter and Chapter 13.

Genetics. Cambridge, MA: Logal Software, Inc. Computer simulations of monohybrid and dihybrid crosses, sex linkage, incomplete dominance, codominance, multiple alleles, linkage and crossing over, population genetics, gene mapping, founder effects, environmental effects, and human genetic counseling.

Gillis, A.M. "Turning Off the X Chromosome." *BioScience*, March 1994. Inactivation of one X chromosome in the cells of females.

Hartl, D.L. *Essential Genetics*. Jones and Bartlett Publishers: Boston, 1996. A good genetics text covering Mendelian and molecular genetics to evolution.

Kinnear, Judith. *Catgen*. Iowa City, IA: Conduit, 1990. One of the best basic software simulations of Mendelian/transmission genetics. Available for a variety of computer platforms.

Miller, K.R. "Wither the Y?" *Discover*, February 1995. Commentary on the evolution of the human sex chromosomes.

Nathans, Jeremy. "The Genes for Color Vision." *Scientific American*, February 1989. A discussion of the evolution of normal color vision and the genetic basis of color blindness, including details of the color-detecting proteins coded for by the X-linked gene and others involved in some forms of color blindness.

"Organisms That Change Sex." *BioScience*, July/August 1987. A special issue with six articles concerning various odd and flexible mechanisms of sex determination in different groups of plants and animals.

Peters, J.A., ed. *Classic Papers in Genetics*. Englewood Cliffs, NJ: Prentice-Hall, 1959. English translations of Mendel's work and other classics.

Shmona, Kiryat. *Biology Explorer. Genetics*. Scotts Valley, CA: Sunburst Communications, 1991. Part of a series of interactive, experimental simulations, this Macintosh program enables the user to make hypotheses, plan experiments, observe results, and draw conclusions about the various facets of genetics.

Thelen, Thomas. *Hyperfly*. Santa Barbara, CA: Intellimation, 1991. A simulated genetics lab for genetics data gathering. The program shows results of mating experiments between a wide variety of *Drosophila* mutants the student chooses and the wild-type strain. F_1 can be crossed also. Analysis of the numbers must be done on paper. Macintosh version only.

10

MOLECULAR BIOLOGY OF THE GENE

APPROACH

Molecular biology is at the heart of modern biology and behind much of the current work in biomedicine and applied aspects of recombinant DNA technology. Students must assimilate the background and concepts in this chapter in order to get the most out of the rest of the text, and to fully understand many of the important concerns of our time.

In addition, the historical development of molecular biology is a wonderful example of how science operates. Tracing this development in the early part of the chapter has become a standard approach, and it works well.

The molecular mechanisms described are excellent examples of many previous concepts developed in Chapter 3 (macromolecular structure), Chapter 4 (cellular locations of transcription and translation), Chapter 5 (enzymes and energetics), Chapter 8 (chromosome structure and function in the cell cycle), and Chapter 9 (Mendelian genetics). Be sure to review and emphasize these concepts as you come across them.

We are beginning to understand a good deal about the molecular orientations of the chemical players involved in transcription and translation. Illustrations in the text, along with resources given below, will help students gain an appreciation for the beauty of the molecular mechanisms involved in genetics. Seeing nucleic acids and enzymatic proteins in the configurations they take during the processes described in this chapter will help students understand molecular function in general and begin to grasp protein synthesis, which can be very abstract.

The point that proteins are the molecular basis of phenotype is made in Module 10.5. Concentrate on this idea a bit, reviewing the roles proteins play in organisms (Module 3.11).

CHAPTER OBJECTIVES

Introduce the viral life cycle in the context of early experiments showing DNA to be the material of genes, and as part of a simple experimental system used to study molecular genetics.

Describe the data that led Watson and Crick to suggest their model of DNA structure.

Show how the double helix model of DNA conformed to prior observations of the chemistry of the molecule and, at the same time, suggested mechanisms for the roles of DNA in gene expression and chromosome replication.

Define and explain the terms *replication, transcription,* and *translation.*

Trace the steps of protein synthesis, emphasizing the locations of the processes, the involvement of complex molecular machinery, and the requirement for inputs of energy.

Explain why the DNA code must involve a reading frame of three bases.

Describe various viral life cycles, including a complete description of the life cycle of HIV.

LECTURE OUTLINE

I. Introduction.

A. The chromosome theory of inheritance set the historical and structural stage for the development of a molecular understanding of the gene.

B. Many of the basics of molecular biology began to be understood by studying viruses that infect bacteria (*Opening Essay*; Master 10.0).

1. *Review:* Are viruses living things? Recall some of the characteristics of life that viruses do not exhibit, particularly cellular structure and metabolism.

2. Viruses are composed of a protein coat and internal DNA (or RNA), and they depend on the metabolism of their host to make more viral particles (Master 10.1C).

3. All living things are infected by viruses. Bacterial viruses are known as bacteriophages (or more simply, phages).

4. Experimental systems using phages were a logical choice for early experiments on the molecular biology of the gene. Phages are simple, with simple genes infecting relatively simple and easily manipulated bacteria.

C. Experiments showed that DNA is the genetic material (*Module 10.1*).

1. In 1928, English bacteriologist Frederick Griffith showed that some substance (he did not know what) conveyed traits (pathogenicity) from heat-killed bacteria to living bacteria without the trait.

2. Evidence gathered during the 1930s and 1940s showed it was DNA rather than protein (both complex macromolecules found in chromosomes) that was the genetic material.

3. In 1952, American biologists Alfred Hershey and Martha Chase, using T2 phage, showed that the radioactive isotope of sulfur (found only in proteins) was not transferred into new viral particles, whereas the radioactive isotope of phosphorus (found only in DNA) was (Master 10.1A, B).

II. The structure and duplication of DNA.

A. DNA and RNA are polymers of nucleotides (*Module 10.2*).

1. *Review:* The polymeric nature of DNA and RNA polynucleotides (Module 3.20, Acetate 3.20A, B).

2. Focus on the chemical structure of the three components of each monomer nucleotide: a phosphate group, acidic; deoxyribose, a five-carbon sugar; and nitrogenous bases (Acetate 10.2A).

3. Mention the presence of ribose, rather than deoxyribose, in RNA (Figure 10.2D).

4. Briefly discuss the structural similarities and differences between the four nitrogenous bases (thymine, cytosine, adenine, and guanine) that occur in DNA and the one, uracil, that occurs instead of thymine in RNA, noting their commonly used abbreviations (Acetate 10.2B, C).

B. DNA is a double-stranded helix (*Module 10.3*).

1. Some of the data that went into the Watson-Crick model: the chemical structure of DNA, including that of the component structures; Rosalind Franklin's X-ray crystallographs (from which one can deduce helical form

and width and repeating length of the helix); Erwin Chargaff's chemical analysis showing that the amounts of A and T, and G and C, were always equal, and previous knowledge that the ratios of A + T to G + C varied from species to species.

2. The model that fit all the observations was a double helix (a twisted rope ladder) with sugar backbones on the outside and hydrogen-bonded nitrogenous bases on the inside (Master 10.3C, Acetate 10.3D).

3. A always bonds with T, and G always bonds with C, but there are no restrictions on the linear sequence of nucleotides along the length of the helix.

NOTE: The story of American James Watson's and his English colleague Francis Crick's discovery of the structure of DNA includes many aspects of great scientific discoveries: making the necessary observations, careful thought as to what the observations mean, insightful formulation of a hypothesis (model) based on the analysis of the observations, and being in the right place at the right time. There is also some controversy about the manner in which some of the story unfolded.

4. The Watson-Crick model was proposed in a short paper in 1953 and almost immediately led to proposed mechanisms about DNA function.

C. DNA replication depends on specific base pairing (*Module 10.4*).

1. The nature of the reproductive process, and of the cell cycle involved in it, requires that complete and faithful copies of DNA be produced (replicated).

2. Watson and Crick stated that their model suggested a copying mechanism.

3. The mechanism proposed and confirmed by the end of the 1950s involved each half of the double helix functioning as a template upon which a new, missing half is built (Master 10.4A).

4. The actual mechanism involves a complex arrangement of molecular players, the help of enzymes, particularly DNA polymerases, and some geometric contortions including untwisting of the parent helix and retwisting the daughter helices (Acetate 10.4B).

5. Despite its speed (50-500 pairs per second), replication is very accurate, with approximately one mistaken nucleotide pair in a billion.

D. DNA replication: A closer look (*Module 10.5; Acetate 10.5*)

1. Replication occurs simultaneously at many sites (replication bubbles) on a double helix. This allows DNA replication to occur in a shorter period of time than replication from a single origin would allow.

2. DNA polymerases can only attach nucleotides to the 3' end of a growing daughter strand. Thus replication always proceeds in the 5' to 3' direction.

NOTE: Within the replication bubbles, one daughter strand is synthesized continuously while the other daughter stand must be synthesized in short pieces which are then joined together by DNA ligase.

3. DNA polymerases also proofread the new daughter strands.

III. The expression of genes on DNA.

A. The DNA genotype is expressed as proteins, which provide the molecular basis for phenotypic traits (*Module 10.6*).

1. *Review:* The roles that proteins play in organisms (Module 3.11).
2. The molecular basis of phenotypic traits are the proteins an organism can make (with a variety of functions).
3. The molecular basis of genotype is now recognized to be DNA.
4. The one gene–one enzyme hypothesis was formulated in the 1940s by American geneticists George Beadle and Edward Tatum, who were studying nutritional mutants of the mold *Neurospora*. Beadle and Tatum found that genetic mutants lacked single enzymes needed to complete metabolic pathways (Figure 10.6B).
5. This idea was soon extended to include all proteins (adding a variety of structural types) and later restricted to individual polypeptides (because some proteins are composed of several distinct polypeptide chains).
6. The flow of information in gene expression is from DNA structure to RNA (transcription) to polypeptide structure (translation) (Acetate 10.6A).

B. Genetic information is written as codons and translated into amino acid sequences (*Module 10.7*).
1. This flow is now known to occur in two stages: transcription of the genetic code in the nucleus to a messenger RNA molecule, and translation of the mRNA message in the cytoplasm (Acetate 10.7).
2. The nucleotide monomers represent letters in an alphabet that can form words in a language. Each word codes for one amino acid in a polypeptide.
3. Ask students how many monomer letters would be required to create enough words in the amino acid vocabulary.
4. Recall the discussion of probability from Module 9.7. There are four letters (A, T, G, and C) and 20 amino acids. One-letter words would create 4 distinct words. Two-letter words would create a vocabulary of 16 words (4×4). Three-letter words would create a vocabulary of 64 words ($4 \times 4 \times 4$).
5. Triplets of bases are the smallest words of uniform length that can specify all the amino acids. These triplets are known as codons.

C. The genetic code is the Rosetta Stone of life (*Module 10.8*).
1. The first codon was deciphered by American biochemist Marshall Nirenberg in 1961.
2. He added polyuracil (an artificially made RNA polynucleotide) to a mixture containing ribosomes and other cell fractions required for translation. The polypeptide polyphenylalanine was produced, indicating UUU was the codon for phenylalanine.
3. The code was completely known by the end of the 1960s. It shows redundancy but no ambiguity (Acetate 10.8A).
4. Make a polypeptide using an arbitrary sequence of bases (Acetate 10.8B).
5. The code is virtually the same for all organisms. Thus, bacterial cells can translate the genetic messages of human cells, and vice versa.
6. *Preview:* Recombinant DNA techniques enable biologists to transfer genes of one organism to another and have them expressed (Chapter 12).

IV. Transcription produces genetic messages in the form of RNA (*Module 10.9*).

   A. Messenger RNA copies of the DNA genes are made, nucleotide by nucleotide (codon by codon).

       1. RNA polymerase and a complex physical and chemical arrangement in the nucleus are involved (Master 10.9A).

       2. Transcription is initiated from one strand of the DNA as indicated by a promoter region, the DNA unwinds, and RNA polymerization and elongation occur. Finally, the mRNA sequence is terminated when the process reaches a special terminator region of the DNA (Master 10.9B).

       NOTE: Transcription means copying a message into a new medium.

       Preview: The regulation of this process is discussed in Chapter 11.

   B. Two other types of RNA, rRNA and tRNA, play a role in translation and are transcribed by this process.

V. Genetic messages are translated in the cytoplasm (*Module 10.10*).

   A. In prokaryotes, transcription and translation both occur in the cytoplasm.

   B. In eukaryotes, transcription and translation are separated by membranes of the nucleus.

       1. The fact that a completed mRNA molecule leaves the nucleus and the message is translated in the cytoplasm can be demonstrated using radioactive tracer RNA nucleotides (Figure 10.10).

       2. The players in the translation process include ribosomes, tRNA molecules, enzymes and protein factors, and sources of cellular energy.

       NOTE: Translation means rewording a message into a new language. The new language in this case is the linear sequence of amino acids in polypeptides.

   C. Transfer RNA molecules serve as interpreters during translation (*Module 10.11*).

       1. Amino acids that are to be joined in correct sequence cannot recognize the codons on the mRNA.

       2. Transfer RNA molecules, one or more for each type of amino acid, match the right amino acid to the right codon (Acetate 10.11A, B).

       3. Each tRNA contains a region (the anticodon) that recognizes and binds to the correct codon for its amino acid on the mRNA.

       4. The right tRNA for each amino acid and its codon are temporarily joined by the aid of a specific enzyme (at least one for each tRNA–amino acid complex) and the expenditure of one ATP molecule (Figure 10.11C).

   D. Ribosomes build polypeptides (*Module 10.12*; Acetate 10.12A, B, C).

       1. Ribosomes are composed of rRNA and protein, arranged in two subunits.

       2. The shape of ribosomes provides a platform on which protein synthesis can take place. There are locations for the mRNA, two tRNA–amino acid complexes, and the growing polypeptide.

   E. An initiation codon marks the start of an mRNA message (*Module 10.13*).

       1. Translation can be divided into the same three phases as transcription: initiation, elongation, and termination.

2. An mRNA molecule is longer than the genetic message it contains. It contains a starting nucleotide sequence that helps in the initiation phase and an ending sequence that helps in the termination phase (Acetate 10.13A, B, part 1 and Acetate 10.15, part 5).

3. During initiation, the initial sequence helps bind the mRNA to the small ribosomal subunit, a specific start codon binds with an initiator tRNA anticodon, and the large ribosome binds to the small subunit as the initiator tRNA fits into the P site on the large subunit (Acetate 10.13B, part 2).

F. Elongation adds amino acids to the polypeptide chain until a stop codon terminates translation (*Module 10.14*; Acetate 10.14).

1. Elongation involves three steps. The anticodon of an incoming tRNA–amino acid complex binds with the codon at the ribosome's A site. A polypeptide bond is formed between the growing polypeptide (attached to the tRNA at the P site) and the new amino acid. The P-site tRNA leaves the complex, and the A-site tRNA–polypeptide chain complex moves to the P site.

2. The formation of the polypeptide bond is catalyzed by an enzyme within the ribosome structure.

3. Elongation continues until a special stop codon (UAA, UAG, or UGA) causes termination of the process. The finished polypeptide is freed, and the ribosome splits into its two subunits.

4. A polypeptide develops its tertiary structure both during and after translation.

5. Following their synthesis, several polypeptides may come together to form a protein with quaternary structure.

Review: Levels of protein structure are discussed in Modules 3.15 - 3.18.

VI. Summary: The flow of genetic information in the cell is DNA —> RNA —> protein (*Module 10.15*; Acetate 10.15).

A. The synthesis of a strand of mRNA complementary to a DNA template is transcription.

B. The conversion of the information encoded within a strand of mRNA into a polypeptide is translation.

VII. Mutations can change the meaning of genes (*Module 10.16*).

A. Many differences in inherited traits in humans have been traced to their molecular causes.

B. A change in the nucleotide sequence of DNA is known as a mutation.

C. Certain substitutions of one nucleotide base for another will lead to mutations, resulting in the replacement of one amino acid for another in a polypeptide sequence.

1. Such a substitution is known to account for the type of hemoglobin produced by the sickle-cell allele (Master 10.16A).

Review: The phenotypic effects of the sickle-cell allele are discussed in Module 9.12.

2. Base substitutions usually cause a gene to produce an abnormal product, or they result in no change if the new codon still codes for the same amino acid.

3. Rarely, base substitutions lead to improved or changed genes that may enhance the success of the individual in which it occurs. These types of mutations provide the genetic variability that may lead to the evolution of new species (Chapter 14).

D. Other types of mutations occur when a nucleotide base is deleted from or inserted into the DNA sequence.

1. Because this type of mutation results in a phase shift of the three-base reading frame, all codons past the affected one are likely to code for different amino acids (Acetate 10.16B).

2. The profound differences that result will almost always result in a nonfunctional polypeptide.

E. Mutagenesis can occur spontaneously or because of physical (radiation) or chemical mutagens. Such mutagenesis may result in cancer (Module 13.18).

VIII. Viruses.

A. Viral DNA may become part of the host chromosome (*Module 10.17*).

1. Viruses depend on their host cells for the replication, transcription, and translation of their nucleic acid.

2. Some bacterial phages are known to reproduce in two ways.

3. In the lytic cycle, a phage immediately directs the host cell to replicate the viral nucleic acid, transcribe and translate its protein-coding genes, assemble new viruses, and cause host cell lysis, releasing the reproduced phages.

4. In the lysogenic cycle, a phage's DNA is inserted into the host cell DNA by recombination and becomes a prophage. This DNA sequence is replicated with the host cell's DNA over many generations. Finally, some environmental cue directs the prophage to switch to the lytic cycle. Such prophages may cause the host cell to act differently than if the prophage were not there (Acetate 10.17).

B. Many viruses cause disease in animals (*Module 10.18*).

1. Viruses have a great variety of infectious cycles in eukaryotes. Those that infect plants or animals can cause disease.

   NOTE: Organisms from all kingdoms have viruses that infect their cells.

2. In one type (enveloped RNA virus, such as the virus that causes mumps), the viral genes are in the form of RNA, which functions as a template to make complementary RNA. This RNA, in turn, functions either as mRNA to direct virus protein synthesis directly or as a template from which more viral RNA is made. Newly assembled viral particles leave the cell by enveloping themselves in host plasma membrane (Master 10.18B).

3. Other viruses of eukaryotes, such as the herpesviruses that cause chickenpox, shingles, mononucleosis, cold sores, and genital herpes, reproduce inside the host cell's nucleus and can insert as a provirus in the host DNA, much like a prophage in the lysogenic cycle.

   NOTE: Viruses that cause cold sores and genital herpes are different strains.

4. *Preview:* Animals defend against viruses through their immune systems. Vaccines, which induce the immune system's delayed responses to viral coat molecules, offer a possible defense against future viral infection (Chapter 24).

C. Plant viruses are serious agricultural pests (*Module 10.19*).

1. Most plant viruses are RNA viruses.
2. Insects, farmers, and gardeners may all spread plant viruses.
3. Infected plants may pass viruses to their offspring.
4. The are no cures for most viral diseases of plants. Research has focused on prevention and the selective breeding of resistant varieties.

D. Emerging viruses threaten human health (*Module 10.20*).

1. HIV, the virus that causes AIDS, is an example of an emerging virus, as are *Ebola* and hantavirus.

NOTE: Emerging viruses (as well as the emerging antibiotic-resistant bacteria) are predictable by evolutionary theory. Further, emerging viruses are one of the consequences of habitat destruction by humans (Chapter 38).

E. The AIDS virus makes DNA on an RNA template (*Module 10.21*).

1. The virus that causes AIDS is human immunodeficiency virus, or HIV (Acetate 10.21A).
2. HIV particles are enveloped, like those that cause mumps. Although they carry genes in the form of RNA, these genes are expressed by being first transcribed back to DNA (reverse transcription), at which time they enter the host cell's chromosomes as a provirus and remain unexpressed for several years. HIV finally becomes active by using the host cell's machinery to direct its reassembly, much like a DNA virus (Acetate 10.21B).
3. *Preview:* HIV infects cells involved in the human immune system and is discussed in greater detail in Chapter 24.

F. Virus research and molecular genetics are intertwined (*Module 10.22*).

1. On the one hand, virus research played an important early role in experiments on the molecular structure and activity of genes, and continues to do so.
2. On the other hand, viruses cause some of the worst diseases we are dealing with today.

NOTE: Most antibiotics have no effect on viruses.

CLASS ACTIVITIES

1. The mechanics of DNA processing and protein synthesis can be made into an engaging play involving selected members of a large class. Act as the play's director, and use placards to label the cast members.

RESOURCES AND REFERENCES

Apple Multimedia Lab. *Life Story*. Scott's Valley, CA: Sunburst Communications, 1993. A CD-ROM (optional videodisc), multimedia version of the story of the discovery of the structure of DNA. It includes clips from the BBC dramatization of James Watson's book *The Double Helix* about DNA structure. Supporting material includes interviews with the real players, molecular animations, and historical photographs. Macintosh version only.

BioMedia Associates. *The Living Cell/Nucleus*. Deerfield, IL: Coronet Film and Video, 1993. A videodisc that examines the structure and function of plant, animal, and protist cells; introduces the structure and function of the nucleus; and gives the details of DNA molecules and protein synthesis.

Bronowski, Jacob. *Generation Upon Generation*. New York: Time-Life Video, 1974. Episode 12 of the "Ascent of Man" PBS series. An engaging discussion of genetics from the experiments of Mendel to those of molecular geneticists.

Darnell, J.E., Jr. "RNA." *Scientific American*, October 1985. The role of RNA in protein synthesis and its relationship to DNA.

"Disease Ecology." *BioScience*, February 1996. A special issue highlighting how disease organisms affect the global environment.

Eigen, Manfred. "Viral Quasispecies." *Scientific American*, July 1993. A new view of how viruses act by a biochemist and Nobel laureate. Because of imprecise replication of their genomes, many virus "species" are far more genetically diverse than species of true organisms. Examination of the nature of this diversity holds clues to understanding all viruses and defeating those that pose serious threats to humankind.

EME Corporation. *DNA—The Master Molecule, Levels I & II*. Danbury CT. Autotutorial software that teaches the structure of DNA, DNA replication, mutation, and gene expression (transcription and translation).

Hogle, James M., Marie Chow, and David J. Filman. "The Structure of Poliovirus." *Scientific American*, March 1987. The virus known for its devastating effects has become a model for investigating the molecular links between form and function. Analysis of its structure will enlarge the scope of viral research. Includes computer-generated molecular models of the virus.

Judson, H.F. *The Eighth Day of Creation: Makers of the Revolution in Biology*. New York: Simon & Schuster, 1979. A history of molecular biology.

Langone, J. "Emerging Viruses." *Discover*, December 1990.

Miller, Julie Ann. "Diseases for Our Future. Global Ecology and Emerging Viruses." *BioScience*, September 1989. HIV has called attention to the continuing emergence of apparently new viruses. This article discusses some potentially even more dangerous viruses that now exist in isolated pockets of the world.

"The New Face of AIDS." *Science*, June 28, 1996. Special reports.

Protein Synthesis. Princeton, NJ: Films for the Humanities and Sciences, 1993. A series of six 10-minute videos detailing all aspects of this chapter, from DNA structure to protein synthesis. Illustrated with simple computer animations that clarify all processes.

Tiollais, Pierre, and Marie-Annick Buendia. "Hepatitis B Virus." *Scientific American*, April 1991. This small, extraordinary virus causes liver diseases and a common form of cancer. New vaccines produced by genetic engineering hold the promise of eventually eradicating both.

Watson, James D. *The Double Helix*. New York: Atheneum, 1968. The brash, controversial best-seller by the codiscoverer of the double helix.

"What Science Knows About AIDS." *Scientific American*, October 1988. An entire issue devoted to AIDS. For additional resources on AIDS, see Chapter 24.

Wuethrich, B. "Army Scientists Isolate Deadly Virus." *Science News*, August 21, 1993. A brief report on the hantaviral cause of a dangerous disease in rural areas across the U.S.

11

THE CONTROL OF GENE EXPRESSION

APPROACH

The development of multicellular organisms from a single, totipotent cell is one of the miracles of life and one of the major unsolved mysteries of current biological study. Two aspects of life particularly important to modern society, cancer and aging, are closely involved with gene expression. Among students there will likely be considerable interest in the cancer connection to this topic. Be sensitive to the fact that many students will have had some experience of a close relative's dying from cancer.

The control of gene expression is essentially gene expression in multicellular organisms, but this fact may not be evident to nonscience majors. Dealing with gene control will ensure that students finally understand basic chromosomal structure and the molecular machinery described in the previous chapters.

In your presentation of this material, be sure to set the stage by discussing just how amazing gene control is. Media clips of developing frog or echinoderm embryos, or other developmental sequences (time-lapse photography of plant growth, flowering, etc.), will help focus attention on some of the beautiful organismal-level topics of the chapter.

CHAPTER OBJECTIVES

Introduce the concept of the control of gene expression involving molecular control of protein synthesis, and differentiate the level of complexity of this control in prokaryotes and eukaryotes.

Explain the operon model of gene control.

Trace the information flow from DNA to protein, and note which structures and functions along the route have been shown to control the flow of information.

Describe the genetic evidence for master control genes that determine the development of multicellular animals, and explain how these homeotic genes likely function and their evolutionary significance.

LECTURE OUTLINE

I. Introduction.

   A. What happens when a multicellular organism does not develop into a normal individual, as in *Drosophila melanogaster* (Master 11.0)? How does a developing zygote and descendant cells turn off and on their genes so that collections of cells with different structures and functions form in just the right place at just the right time? What happens when a cell line develops out of control? These are some of the fascinating topics involved in multicellular development and the more general topic of the control of gene expression (*Opening Essay*).

   B. Two approaches are used to study development.

      1. *Preview:* Recombinant DNA techniques are used to study the parts of gene regulation (Chapter 12).

2. Trace the genetic causes of abnormal development as clues to how genes control normal development. This is how *D. melanogaster* mutants are studied.

C. In both eukaryotes and prokaryotes, cell specialization depends on the selective expression of genes (*Module 11.1*).

1. Turning on and off genes is the main way that gene expression is regulated.

    Review: This involves regulation of the transcription of RNA (Module 10.9).

2. Single-celled organisms (including bacteria) go through metabolic changes during their lives and must turn on and off gene activity to cause changes in their protein contents.

    NOTE: This is true for all organisms.

3. Thus, our earliest understanding of gene control came not from the study of eukaryotes, but from study of the bacterium *Escherichia coli*.

II. The operon model of gene control found in prokaryotes.

A. Proteins interacting with DNA turn prokaryotic genes on or off in response to environmental changes (*Module 11.2*).

1. This model of gene control was first proposed as a hypothesis in 1961 by French biologists François Jacob and Jacques Monod, for the control of lactose utilization enzymes in *E. coli*.

    NOTE: Much experimental evidence has since confirmed the existence of this and other operons in many bacteria.

2. Important features of the model: An operon consists of several DNA sequences coding for different enzymes, all involved in the same cellular process. Expression of the operon is controlled as a unit. Other DNA sequences in and near the operon control the operon's expression. The presence or absence of the enzyme's substrate turns on or off the controls.

3. Operon expression normally starts with RNA polymerase binding at the promoter region (the first nongene region of the operon) and moving along and transcribing each gene in the operon.

4. When the *lac* operon is "turned off," a regulator gene is transcribed and translated into a repressor protein. The repressor protein binds with the operator region of the operon, repressing the transcription of the genes further along the operon (Acetate 11.2A).

5. When the *lac* operon is "turned on," the regulator gene continues to be transcribed and translated into repressor, but the presence of substrate (lactose) interferes with the binding of the repressor to the operator. This permits the expression of the remainder of the operon. Expression continues until the substrate is used up. Then the repressor is free to repress the operator, and the operon turns off as above (Acetate 11.2B).

B. The operons of prokaryotes come in several different varieties (*Module 11.3*; Master 11.3).

1. The *lac* operon is repressed when lactose is absent and transcribed when lactose is present.

2. Another operon, the *trp* operon, is transcribed when tryptophan is absent and repressed when tryptophan is present. The enzymes expressed by *trp* help synthesize tryptophan.

3. A third type of operon uses activators rather than repressors. Activators are proteins produced by the regulatory genes that make it easier for RNA polymerase to bind to the promoter region of an operon.

III. The pattern of gene expression in eukaryotes.

A. Differentiation produces a variety of specialized cells in eukaryotes (*Module 11.4*).

1. Producing eukaryotic organelles and regulating their functions require a much more complex network of gene control, even in single-celled protists.

2. In multicellular eukaryotes, there is the added complexity of regulating what kinds of cells are produced when and where.

3. Muscle, nerve, sperm, and blood cells (and other cell types) of a single animal all are derived by repeated cell divisions from the zygote.

4. The structure of each different cell type is visibly different, reflecting its function (Figure 11.4A-D).

B. Specialized cells may retain all of their genetic potential (*Module 11.5*).

1. Experimental evidence supports the retention of all of a multicellular organism's DNA in each of its differentiated cells, in most cases.

2. In the 1950s, American embryologists Robert Briggs and Thomas King transplanted nuclei from differentiated cells lining a frog tadpole's intestine to unfertilized, enucleated frog eggs. Many such treated eggs developed into normal tadpoles (Master 11.5A).

3. In some naturally occurring situations, differentiated cells' DNA may "dedifferentiate" to give rise to other cell types. Many animals can regenerate lost parts from differentiated cells that remain nearby. Many plants can regenerate completely from differentiated cells. Plant tissue culture on sterile culture media is now used widely to produce hundreds or thousands of genetic clones of domestic plants (Master 11.5B).

4. In many cases, particularly in animals, differentiated cells do not normally dedifferentiate. For instance, most animals lack the ability to regenerate whole bodies from single isolated cells.

5. Most evidence suggests that cellular differentiation does not involve changes to the DNA.

C. Each type of differentiated cell has a particular pattern of expressed genes (*Module 11.6*; Master 11.6).

1. As a developing embryo undergoes successive cell divisions, different genes are active in different cells at different times.

2. Some genes (for example, those involved in the glycolysis pathway) are active in all metabolizing cells.

3. Other genes are turned on in only one cell line (for example, the crystallin gene in lens cells).

IV. The mechanisms of gene expression in eukaryotes.

A. DNA packing in eukaryotic chromosomes affects gene expression (*Module 11.7*).

1. The total DNA in a human cell's 46 chromosomes would stretch 3 meters.

   NOTE: This amount of DNA is packed in cell nuclei as small as 5 µm in diameter, a reduction factor of almost 1 million in scale!

2. All the DNA fits because of elaborate packing: wrapping around histones and other proteins into nucleosomes, coiling, supercoiling, and additional folding into chromosomes (Acetate 11.7A).

   NOTE: During interphase, chromosomes of most cells are more loosely packed than the metaphase chromosome shown in Acetate 11.7A.

3. DNA packing must control the expression of genes, but there is little experimental evidence of how this happens.

4. An interesting known example of the role of DNA packing in the control of expression is the formation of Barr bodies from X chromosomes in the cells of female animals. Certain cell lines have one or the other X chromosome (inherited from the individual's mother or father) inactivated; thus, there can be a random mosaic of expression of these two X chromosomes, as is seen in tortiseshell cats (Figure 11.7B).

B. Complex assemblies of proteins control eukaryotic transcription (*Module 11.8*).

   1. Giant chromosomes in the salivary glands of fruit-fly larvae contain hundreds of replicated copies of each DNA strand (Figure 11.8A).

   2. At different times, different regions of these chromosomes "puff" or loop out, exposing parts of the DNA.

   3. These puffs correspond to regions of RNA synthesis, implying that just the puffs are being transcribed at the time.

   4. Prokaryotic and eukaryotic gene regulation are fundamentally similar. However, whereas prokaryotes combine several regulated genes into one operon, eukaryotes apparently tend to regulate individual genes. Thus, in eukaryotes there are many more regulatory proteins involved and a greater degree of complex interactions than in prokaryotes.

   5. Activation appears to be of greater importance in the regulation of eukaryotic gene expression than is repression. Transcription factors (of which activators are an example) interact with enhancer sites in regulating the binding of RNA polymerase to a gene's promoter (Acetate 11.8B).

   6. Repressor protein interaction with silencer sites on DNA inhibits the start of transcription.

C. Eukaryotic RNA is processed to remove noncoding segments and to add a cap and a tail (*Module 11.9*).

   1. Structural compartmentalization of eukaryotes offers opportunities for the post-transcriptional control of gene expression.

   2. The noncoding stretches of eukaryotic genes are called introns, and the parts that are expressed are called exons.

   3. Both introns and exons are transcribed. Before leaving the nucleus, the introns are removed from the mRNA transcript, and the remaining exons are spliced together (Master 11.9A).

   4. **Introns have been shown to function in gene regulation in several ways (and also probably function in ways not yet known). Some introns appear to include sequences that function at the transcription level in gene regulation and are not needed to translate into protein structure. In other**

cases, the remaining exons can be spliced in different ways, to provide a cell with several possible products from one gene region (Master 11.9B). Finally, it has been suggested that introns make genes longer, thereby increasing the possibility of crossovers between exons, and providing another mechanism to increase genetic diversity.

    5. Eukaryotic mRNA is capped, tailed, and sometimes edited. These actions occur while the spliced (intron-less) mRNA is still in the nucleus. Additions of cap and tail seem to help protect the mRNA from attack by cellular enzymes and may enhance translation.

D. Translation and later stages of gene expression are also subject to regulation (*Module 11.10*).

    1. The lifetime of mRNA molecules varies, controlling the amount of protein translated from a single transcription and post-transcriptional processing event. In nonmammalian vertebrates, red blood cells lose their nuclei, but not their ribosomes and mRNAs, which can continue to translate into hemoglobin for a month or more.

    2. Some inhibitory control of the process of translation is known, such as the inhibitory action of a protein found in red blood cells when heme subunits are not available (Master 11.10A).

    3. Post-translational control mechanisms in eukaryotes often involve cutting polypeptides into smaller, active final products (Master 11.10B).

    4. Another post-translational control affects how fast protein products are degraded.

E. *Review:* Multiple mechanisms regulate gene expression in eukaryotes (*Module 11.11*; Acetate 11.11).

V. Cascades of gene expression and cell-to-cell signaling direct development of an animal (*Module 11.12*).

A. Homeotic genes are master controls that function during embryonic development in animals to determine the developmental fates of different groups of cells destined to become different tissues.

Preview: Embryonic development is discussed in detail in Chapter 27.

B. Their improper functioning can lead to bizarre changes in morphology (Figure 11.12A).

C. An example of these cascades can be seen in the determination of which end of a fruit-fly egg cell will become the head and which end will become the tail. These events occur within the ovaries of the mother fly and involve the following series of events (Master 11.12B):

    1. The egg cell produces (gene activation) a protein that signals the adjacent follicle cells.

    2. These follicle cells are stimulated (gene activation) to produce proteins that provide feedback to the egg cell.

    3. As a result, microtubules within the egg cell are oriented along what will become the new fly's head-to-tail axis, with different types of mRNA located at the two different ends.

    4. After fertilization, repeated mitosis results in the development of the embryo from the zygote. Translation of the head mRNA results in the production of a protein concentration gradient from head to tail. This

protein concentration gradient corresponds to a gradient of gene expression.

5. This gradient of gene expression results in the development of the fly's body segments.

6. The cascade continues as this gradient of gene expression results in further differentiation and specialization of the body segments.

7. The genes that regulate these major features of the body plan (body segments and the body parts that develop at each segment) are called homeotic genes.

VI. Signal transduction pathways convert messages received at the cell surface into responses within the cell (*Module 11.13*; Acetate 11.13).

A. As shown in Module 11.12, the gene expression of one cell can affect the gene expression of other cells. This is the result of signal transduction pathways.

B. *Review:* Signal transduction plays a major role in the regulation of the cell-cycle (Module 8.10). *Preview:* The importance of signal transduction pathways is also shown in the discussions of cancer (Module 13.16) and control systems (Chapters 26 and 28).

C. The main components of a signal transduction pathway are:

1. The signaling cell secretes signal molecules.

2. The signal molecules bind to receptors on the target cell's plasma membrane.

3. This results in a cascade of events that leads to the activation of a specific transcription factor.

4. The transcription factor activates a specific gene, resulting in the expression of the protein for which the gene codes.

VII. Key developmental genes are very ancient (*Module 11.14*; Acetate 11.14).

A. Virtually every homeotic gene found in fruit flies contains a common 180-nucleotide sequence.

B. Very similar sequences have been found in virtually all eukaryotic organisms studied.

1. These organisms range from unicellular organisms such as yeast to plants, earthworms, frogs, chickens, mice, and humans.

   NOTE: This is evidence for the common origin of life and is relevant to the material in Module 1.5 (Unity in Diversity) and Unit III (Concepts of Evolution).

2. These nucleotide sequences, called homeoboxes, translate into a small polypeptide sequence that binds to specific DNA sequences and thereby regulates their expression.

CLASS ACTIVITIES

1. Link gene regulation to the students' development from a single undifferentiated cell (the zygote) through all the stages of their lives: relate this to embryonic development (Chapter 27).

2. Point out how lactose intolerance is overplayed in the media. However, do link lactose intolerance to development. Babies nurse, therefore they are less likely to be lactose

intolerant than are adults—especially those whose ancestors are from regions of the world where adults traditionally drink little milk. Ask what happened to the "lactose-digesting gene."

RESOURCES AND REFERENCES

BBC Television. *Patterns of Development*. Seattle, WA: Laser Learning Technologies, 1991. A videodisc that provides considerable background material for the study of multicellular development. Molecular and genetic techniques and their results are illustrated in frogs, fruit flies, and other organisms.

BBC Television. *Shaping Up*. Deerfield, IL: Coronet Film and Video, 1993. A 24-minute videodisc that covers the developmental processes of morphogenesis in constructing and maintaining adult organisms, including cell migration, regeneration, and cell division; polarity of regeneration; and body-patterning roles played by epithelial cells, neurons, and morphogens.

Beardsley, Tim. "Trends in Biology. Smart Genes." *Scientific American*, August 1991. A well-illustrated review of current knowledge about gene activation during development. Includes examples from experiments on *Drosophila* and other animals.

Bishop, J. Michael. "The Molecular Genetics of Cancer." *Science*, January 16, 1987. Diverse examples of genetic damage found in various cancer cells may hint at the biochemical explanations for cancer cell growth.

DeRobertis, E., G. Oliver, and C. Wright. "Homeobox Genes and the Vertebrate Body Plan." *Scientific American*, July 1990.

Dorit, Robert L., Lloyd Schoenbach, and Walter Gilbert. "How Big Is the Universe of Exons?" *Science*, December 7, 1990. Statistical estimates of the number of repeating exons.

Grunstein, M. "Histones as Regulators of Genes." *Scientific American*, October 1992. Histones were once dismissed as little more than packing material for nuclear DNA. In fact, these proteins can both repress and facilitate the activation of many genes.

Holliday, Robin. "A Different Kind of Inheritance." *Scientific American*, June 1989. The methylation of DNA may be a major epigenetic mechanism by which gene activity patterns—as opposed to genes per se—are passed from one generation of cells to another during development.

Jazwinski, S.M. "Longevity, Genes, and Aging." *Science*, July 5, 1996. The new revolution in aging research.

Lowenstein, J. "Genetic Surprises." *Discover*, November 1990. What biologists are learning about the organization of eukaryotic genomes.

McKnight, Steven Lanier. "Molecular Zippers in Gene Regulation." *Scientific American*, April 1991. Recurring copies of the amino acid leucine in proteins can serve as teeth that "zip" two protein molecules together. Such zipping plays a role in turning genes on and off.

Pennisi, Elizabeth. "Tracking RNA." *Science News*, March 20, 1993. A review of current knowledge about transcriptional and post-transcriptional RNA processing as a means of gene control.

Ptashne, M. "How Gene Activators Work." *Scientific American*, January 1989. How regulatory proteins bind to DNA in eukaryotes.

Radetsky, P. "Genetic Heretic." *Discover*, November 1990. The story of the discovery that RNA can act as an enzyme in RNA splicing.

Ross, Jeffrey. "The Turnover of Messenger RNA." *Scientific American*, April 1989. The synthesis rate of many proteins is influenced more by how rapidly their mRNA templates decay, or turn over, than by how rapidly the RNAs are synthesized. This article discusses the regulation of this decay.

Tjain, R. "Molecular Machines That Control Genes." *Scientific American*, February, 1995. How genes are regulated in eukaryotes.

12
DNA Technology

APPROACH

This chapter presents one of the most important practical aspects of biology that nonscience majors will learn from your course. Many medical procedures and ethical questions they will face as citizens will involve these complex procedures. Learning about the technical aspects of genetic engineering will also help students fully understand how genes work on a molecular level.

Recombinant DNA technology is used in many ways as a practical extension of a variety of biological processes. Therefore, summarize the topics that will be covered in detail later in the text. At this point, it is best to be brief in your discussions of antibodies, vaccines, and nitrogen fixation (among other broad topics introduced here). Allow students to have a general understanding of these subjects rather than detailed knowledge.

There are daily advances in medicine and agriculture using recombinant DNA technology. Students may already have some misinformation from the media coverage of the more dramatic results. Ask students during your introduction what they know about this work and the controversies surrounding it. This will help you determine how much they have assimilated from the previous chapters on genetics, and how accurate the recent media coverage of the work and the controversies has been. You may hear some fast-breaking stories you had not heard before.

CHAPTER OBJECTIVES

Describe the three known mechanisms by which bacteria recombine with DNA from other bacteria, emphasizing that sexual reproduction does not occur as a single process in bacteria.

Explain the role of bacterial plasmids in carrying some nonessential genes, and how plasmid DNA is passed between mating bacteria during conjugation.

Explain how restriction enzymes cut DNA, leaving sticky ends.

Describe how reverse transcriptase is used to produce specific DNA sequences from mRNA.

Explain how molecular probes are used to choose the right recombinant from a collection of recombinants.

Explore some of the technical aspects of work in molecular genetics, such as the use of electrophoresis and the polymerase chain reaction (PCR).

Describe how electrophoresis, PCR, and restriction fragment length polymorphisms (RFLP) analysis are used in criminal and paternity cases.

Outline the current uses of recombinant DNA technology, particularly in medicine (drugs and vaccines), agriculture, and biological research.

Explain the risks and ethical questions involved in the use of recombinant DNA technology.

State the guidelines used by Americans to minimize the risks of recombinant DNA technology.

LECTURE OUTLINE

I. Introduction.

A. Recombinant DNA technology (genetic engineering) involves combining genes from different sources into new cells that express the genes.

B. Recombinant DNA technology has had—and will have—many important applications.

  1. More efficient methods of basic and applied research into molecular genetics.

  2. Mass-production by bacteria of biochemicals needed by other species.

  3. Creation of new strains of plants and animals.

C. Recombinant DNA techniques are based on bacterial mechanisms (*Opening Essay*).

  1. In 1946, American geneticists Joshua Lederberg and Edward Tatum discovered that *Escherichia coli* has a sexual mechanism.

  2. They combined *E. coli* strains, each of which required a different amino acid to grow. Cells of a new strain appeared in the cultures that did not require the addition of either amino acid.

II. The basics of bacterial genetics.

A. Barbara McClintock discovered jumping genes (*Module 12.1*).

  1. In 1940, American geneticist Barbara McClintock discovered that some traits of corn change in a way implying that genes move from one chromosome location to another, even between different chromosomes.

  2. When these transposons move, they either disrupt the expression of other genes (Figure 12.1B) or change the transmission pattern of sets of genes.

  3. One type of transposon codes for the enzyme that catalyzes the movement of the transposon (Master 12.1C).

  4. Transposons are found in most, if not all, organisms.

  5. Transposons play a role in cancer.

  6. Transposons have uses in genetic engineering.

  NOTE: Not only do transposons seem to have great significance in natural evolution, but their properties would appear to make them suitable agents for artificially modifying the structure and expression of genes.

B. Bacteria can transfer DNA in three ways (*Module 12.2*; Master 12.2A–C).

  1. *Review:* In sexually reproducing organisms, new genetic combinations are the result of meiosis and fertilization (Chapter 8). The mechanisms discussed in this module are the ways by which bacteria produce new genetic combinations.

  2. *Review:* Studies by Frederick Griffith showing nonpneumonia-causing strains of *Pneumococcus* becoming disease-causing in a culture medium that previously contained the disease-causing strain (Module 10.1).

  3. Transformation is the taking up of DNA from the nonliving environment around a bacterium. Transformation caused the results Griffith observed.

4. Transduction is the transfer of bacterial genes from one bacterium to another by a phage.

5. Conjugation is the process by which two bacteria mate. Conjugation is initiated by "male" cells (gene donors) that recognize "female" cells (gene recipients) by means of the male sex pili. After the initial male-female recognition, a cytoplasmic bridge forms between two cells. Replicated DNA from the male passes through this bridge to the female.

6. In all three mechanisms, the new DNA is integrated into the existing DNA in the recipient by a crossover-like event that replaces part of the existing DNA (Master 12.2D).

7. These mechanisms are not reproductive. Sexual reproduction does not occur in bacteria, unlike the situation in plants and animals.

C. Bacterial plasmids can serve as carriers for gene transfer (*Module 12.3*; Masters 12.3A, B).

1. The F (fertility) factor is a portion of *E. coli* DNA that carries genes for making sex pili and other requirements for conjugation.

2. The F factor may be integrated into the main bacterial DNA, or it may exist as a separate, circular DNA fragment, a plasmid, that is free in the cytoplasm. Plasmids replicate separately from the main DNA.

3. If the F factor is integrated into the donor's main DNA, replication begins in the middle of the F-factor region. The replicated length of DNA is transferred from the donor to the recipient but usually breaks before the remaining F factor is transferred. Thus, the recipient does not receive the F-factor genes, and it and its descendants remain female.

4. If the F factor exists as a separate plasmid, it replicates into a linear DNA molecule that is entirely transferred to the recipient. The recipient and all its descendants become male.

5. Plasmids that carry genes other than those needed for conjugation are called vectors. For example, R plasmids are a class of plasmids that carry genes for antibiotic resistance. The widespread use of antibiotics in medicine and agriculture has tended to kill bacteria that lack R plasmids and favor those bacteria that have R plasmids.

NOTE: The ease of transmission of plasmid DNA has been implicated in the rapid transfer of DNA among bacteria, even between different species. Transfers such as these are partly responsible for the spread of multidrug-resistant bacteria.

6. Plasmids have important places among the techniques of genetic engineers.

III. Techniques used in genetic engineering.

A. Molecular mechanisms from a variety of prokaryotes and eukaryotes have been combined into sets of techniques for manipulating the genes of all organisms.

B. Plasmids are used to customize bacteria (*Module 12.4*; Acetate 12.4).

1. Plasmids are isolated from one strain of bacterium.

2. DNA that encodes useful proteins or traits is removed from another organism.

3. The plasmid DNA and gene of interest are joined and returned to the bacterial cells.

4. The bacteria are grown in culture to produce many copies of the isolated gene or its product.

5. Such engineered bacteria play a role in the manufacture of drugs such as human insulin and human growth hormone.

C. Enzymes are used to "cut and paste" DNA (*Module 12.5*).

1. Restriction enzymes were first discovered in bacteria in the late 1960s.

2. In nature, bacteria use restriction enzymes to cut up intruder DNA from phages and from other organisms into nonfunctional pieces. The bacteria first chemically modify their own DNA so that it will not be cut.

3. Restriction enzymes recognize, and cut, specific nucleotide sequences (usually palindromes).

4. Several hundred different restriction enzymes have been discovered that recognize about 100 different sequences.

5. Restriction enzymes are used as tools as follows: The same enzyme is allowed to cut two different DNA molecules. The fragments are mixed so that complementary, single-stranded sticky ends recombine by base pairing. Finally, the gaps in the sugar-phosphate backbones are "repaired" with enzymes known as DNA ligases and the result is recombinant DNA (Acetate 12.5).

Review: DNA ligase is normally used in DNA replication (Module 10.5).

D. Genes can be cloned in recombinant plasmids (*Module 12.6*; Master 12.6).

1. The system for cloning genes combines these last two techniques.

2. Plasmid DNA and the DNA of the cell containing the gene of interest are each cut with the same restriction enzyme. The new gene is inserted into the plasmid. The new plasmid is returned to a bacterium by transformation.

3. The example in Master 12.6 uses a hypothetical situation where human gene V is cloned.

E. Cloned genes can be stored in genomic libraries (*Module 12.7*; Master 12.7).

1. Recombining genes of interest from humans or other sources with recipient bacteria involves several steps.

2. The first step is to isolate the gene of interest.

3. Using a shotgun approach to do this, scientists cut up target DNA into thousands of fragments, each of which carries one or a few genes of unknown identity (one or more fragments will carry the gene of interest).

4. These fragments are temporarily stored in a "library" of plasmids in separate bacterial cells, or in separate phages.

F. Reverse transcriptase helps make DNA for cloning (*Module 12.8*).

1. One problem of using the shotgun approach described above is that eukaryotic DNA contains noncoding introns. Fragments with whole genes may be too long to work with effectively.

NOTE: Another problem is that many fragments in shotgun libraries are meaningless. It takes extra work to sort through all these for the right one.

2. Special enzymes called reverse transcriptases are found in retroviruses. These enzymes make DNA from viral genome RNA (Module 10.21).

NOTE: An example of such a retrovirus is HIV.

    3. A different, more precise way to get genes of interest out of eukaryotes is to work backward from the mRNA being transcribed by a cell expressing the gene of interest.

    4. Already processed to remove introns, mRNA is isolated from a cell expressing the gene desired. Reverse transcriptase is mixed with the mRNA to produce DNA coding for the gene of interest (Master 12.8).

    5. The DNA fragments are "cleaned up," intron-less, sequences coding for whatever proteins the cell had been making. These DNA fragments are again temporarily stored in plasmid or phage libraries.

G. Molecular probes identify clones carrying specific genes (*Module 12.9*).

    1. If some of the bacterial clones in the genomic library actually produce the product expressed by the gene of interest, the right clone can be isolated by testing the medium they are growing in for the product.

    2. If this cannot be done, scientists use radioactively labeled single-stranded DNA probes, which pair with selected regions of the gene of interest. The cells or phages in the genomic library that hold onto the radioactive label are the locations of the gene in question (Master 12.9).

    3. The probes can be assembled artificially if some sequence in the target protein (and hence a corresponding sequence of nucleotides) is known.

H. Automation makes rapid DNA synthesis and DNA sequencing possible (*Module 12.10*).

    1. DNA fragments can be synthesized by highly automated machines.

    2. Nucleotides sequences in a stretch of DNA can also be determined using automated machines. If the DNA is long, it is first cut into smaller fragments by restriction enzymes, and the fragments are sequenced separately.

    NOTE: The individual, shorter sequences often overlap regions of the original whole gene but can be combined into whole-gene sequences using the analytical powers of computers.

    3. Sequences are stored in computer data banks and are available on a world-wide basis to scientists interested in the genes of particular organisms.

I. Restriction fragment analysis is a powerful method that detects differences in DNA sequences (*Module 12.11*; Master 12.11D).

    1. Nucleotide sequences of all but identical twins are different.

    2. Extracted DNA from a person's cells can be cut up into a set of fragments by reacting the DNA with a series of different restriction enzymes (Master 12.11A; recall Module 12.5).

    3. Sets of these fragments (restriction fragment length polymorphisms, or RFLPs) differ in length and number between different, nonidentical twin individuals.

    4. These DNA fragments are positively charged, and different DNA lengths will migrate different distances toward a negative charge in an electrophoretic gel. The darkness of the developed spot is an indication of the amount of each fragment size (Master 12.11B,C).

5. RFLP analysis was used to enable workers studying Huntington's disease to find a marker closely associated to the HD gene on chromosome 4.

   Preview: Huntington's disease, and other genetic diseases, are discussed in greater detail in Chapter 13.

6. Once the marker is known for a particular disease, RFLP analysis can be used to test for it.

J. The PCR method is used to amplify DNA samples (*Module 12.12*; Master 12.12).

1. The polymerase chain reaction (PCR) is a technique for copying a single DNA sequence many times.

2. A mixture of the DNA, DNA polymerase, and nucleotide monomers will continue to replicate, forming a geometrically increasing number of copies.

3. This technique has revolutionized DNA work because sequences can now be obtained from extremely small samples. Prehistoric DNA from a number of sites has been cloned into partial genomes in this way (prehistoric men, mammoths, and 30-million-year-old plants, but not dinosaurs).

   NOTE: Discuss some of the techniques developed here (particularly RFLP analysis and PCR).

IV. DNA technology and the law (*Module 12.13*).

1. An individual's band pattern from RFLP analysis is called a DNA fingerprint. Like traditional fingerprints, DNA fingerprints are unique.

2. DNA fingerprinting uses RFLP analysis to identify small amounts of DNA from blood, tissue, or semen.

3. DNA lasts in the environment much longer than proteins. DNA can be extracted from dried blood stains that are 3–4 years old.

4. In criminal investigations RFLPs are compared with those of other sources.

5. In paternity cases, RFLPs from DNA fragments of a suspected parent are compared with those of the child and the other parent (Figure 12.13B).

6. The technology of DNA fingerprinting has been shown to be reliable. However, there are those who fear that it may unduly influence a jury and therefore feel there must be no possibility of error before its use in court is allowed. Of major concern is the possibility of misinterpretation of the results (human error).

7. Ethical issues concerning the use of DNA fingerprinting remain unsettled (Modules 12.19 and 13.20). Many of these questions center on the issues of invasion of privacy and discrimination.

V. Practical applications of recombinant DNA technology.

A. Bacteria, yeasts, and mammalian cells are used to mass-produce gene products (*Module 12.14*; Table 12.14).

1. Bacteria are the host of choice for making large amounts of eukaryotic gene products because bacteria are simple and can be grown rapidly and cheaply.

2. The single-celled fungus *Saccharomyces cerevisiae* (bread and wine yeast) is one of the simplest eukaryotes that can be grown rapidly and cheaply.

Yeast also has plasmids that can be used as vectors. In some cases, yeast does a better job than bacteria in expressing eukaryotic genes.

3. Some products are best made by mammalian cells. The genes of interest are often cloned in bacterial plasmids first and then introduced to the final host.

4. Among these mammalian cell products are monoclonal antibodies, which are glycoproteins used widely in research on cellular structures and functions. The sugar chain part of monoclonal antibodies can be added to the protein part only under the control of the eukaryotic gene expression system.

5. *Preview:* Chapter 24 on the immune system discusses monoclonal antibodies further (Module 24.12).

B. DNA technology is revolutionizing the pharmaceutical industry and human medicine (*Module 12.15*).

1. Human insulin (Humulin®) was one of the first commercially produced recombinant DNA products, and it is much more effective at treating diabetes than the previously used insulins from pigs and cattle.

2. Human growth hormone is a large molecule coded for by 600 nucleotides, but it has been successfully produced with recombinant *E. coli*.

3. Vaccines are harmless or derivative variants of proteins produced on the surfaces of pathogens. Exposing a person to the vaccine (by injection, usually) primes the person's immune system to recognize and destroy the pathogen in the case of future infection. Vaccines are particularly important in the defense against many viral diseases.

NOTE: Normally vaccines are made using natural mutant forms of pathogens or proteins extracted from the pathogens.

4. Recombinant DNA techniques can be used to make vaccines in many ways: mass-production of vaccine proteins; assembling artificial mutant pathogens; and adding proteins from several pathogens to the coat of the natural mutant smallpox virus, previously used to successfully eradicate smallpox.

5. "Genetic drugs" are a promising avenue of research for the development of treatments for viral diseases, cancers, and inherited disorders.

C. Agriculture will make increasing use of genetic engineering (*Module 12.16*).

1. Nitrogen fixation is the conversion of atmospheric nitrogen to biologically usable nitrogen in the form of amino acids. In nature, nitrogen-fixing genes are found in a few bacteria living in soil, roots, lichens, and aquatic habitats.

Preview: Plants' dependence on bacteria for nitrogen is discussed in Modules 32.13 and 32.14.

2. One promising area where DNA technology may play a future role is in the production of plants that have the nitrogen-fixing genes of prokaryotes. Only partial progress has been made in moving this system to eukaryotic plants.

NOTE: Nitrogen fixation requires not only specific enzymes but the specific molecular and physical environment of the cells where it naturally occurs.

3. Other features of plants that have been improved by genetic engineering are delayed ripening and resistance to disease.

4. The bacterium *Agrobacterium tumefaciens*, which is pathogenic to a number of plant hosts, is used to transfer genes between plants, in recombined form with the bacterium's Ti plasmid. The resulting transgenic plant cells are cloned in culture and grown into adult plants (Master 12.16).

    NOTE: This process can be done much more rapidly than by depending on a full season for natural breeding to produce hybrids, and genes can be transferred between unrelated plants.

5. Progress in the production of transgenic, grain-producing plants has been slow because *A. tumefaciens* does not naturally grow in them.

6. *Preview:* Nitrogen fixation is an important part of the nitrogen cycle (Module 36.16).

7. The FlavrSavr® tomato is currently being marketed.

D. Transgenic animals are especially useful in research (*Module 12.17*).

1. Transgenic animals are created by introducing the genes of interest (from a third "parent") into previously fertilized eggs in the laboratory. A fraction of the modified eggs develop and are placed in foster mothers to gestate.

2. Transgenic animals have been developed experimentally, in an attempt to produce animals that grow faster. So far, these animals do grow faster, but they exhibit other traits that cause them to be less healthy.

3. Transgenic mice, modified to be susceptible to HIV, should help research in fighting AIDS.

NOTE: Ask students if they feel there is any ethical difference between the use of recombinant DNA technology and the use of classical techniques of artificial breeders to create new life forms.

E. Genetic engineering involves some risks (*Module 12.18*).

1. American scientists have developed safety guidelines administered by the U.S. government to minimize the risks involved in genetic engineering.

    NOTE: American guidelines tend to be stricter than those of other countries.

2. The guidelines designate laboratory safety procedures for various types of experiments, including procedures normally used to protect scientists who study natural pathogens.

3. The guidelines also specify the kinds of microorganisms that can be used, often requiring that the recipient organisms be genetically altered so that they cannot live outside the laboratory.

    NOTE: Certain kinds of experiments are forbidden or strictly regulated, such as working with human cancer genes or genes of extremely virulent pathogens.

4. Nevertheless, controversies have developed over the release of genetically altered organisms. An early controversy arose over the release of bacteria altered to function as ice-nucleation centers at a lower temperature than their natural relatives. These bacteria were field-tested only after a lengthy court battle, but are now used commercially (Frostban®).

VI. DNA technology raises important ethical questions (*Module 12.19*).

  A. The long-term effects (including safety issues) of using recombinant products or transgenic species are not yet known.

  B. There are ethical questions regarding our roles in making new organisms.

  C. There are ethical questions regarding the use of these techniques to provide genetic cures of human diseases. For example, giving growth hormone to short but hormonally normal children.

CLASS ACTIVITIES

1. Developments are occurring constantly in biotechnology and are often reported in the daily press. Have students find and critique articles from different sources for class discussion. Journalistic bias and misinformation are often easily caught by students who understand from this chapter the way the techniques are actually used. Some very real societal concerns are highlighted in these articles. In large classes, this activity would best be done in discussion sections.

2. Class discussion can also focus on the ethical issue of the use of these techniques on humans. Should parents be allowed to genetically engineer a child to be a star athlete? a genius? Would this be acceptable as long as the gametes are left alone? Should gametes be engineered so that a disease such as Huntington's is eliminated? What about the future? A gene that is harmful in the present environment may be of benefit in a different environment.

RESOURCES AND REFERENCES

Beardsley, T. "Genes in the Not So Public Domain." *Scientific American*, April, 1995. Who owns the base sequences in DNA databases?

Cavalieri, Liebe F. "Scaling-Up Field Testing of Modified Microorganisms." *BioScience*, September 1991. Describes the problems with applying small-scale field data to large-scale tests of genetically engineered microorganisms.

Fedoroff, Nina V. "Transposable Genetic Elements in Maize." *Scientific American*, June 1984. Further information on transposons, including some nice diagrams explaining how they cause traits at the tissue level.

Gasser, Charles. *Genetic Crop Improvement*. Bethesda, MD: National Library of Medicine, 1988. A videotape that presents the role of genetic engineering in improving plant breeds.

Gasser, Charles S., and Robert T. Fraley. "Transgenic Crops." *Scientific American*, June 1992. A discussion of the biology, ecology, and economics of creating new plant strains with recombinant DNA techniques.

Gilbert, W. "Toward a Paradigm Shift in Biology." *Nature*, January 10, 1991. How the ability to dissect genomes is changing the kinds of questions biologists ask.

The Life Revolution. Princeton, NJ: Films for the Humanities and Sciences, 1991. A series of twelve 26-minute videotapes on a variety of human genetics topics. Particularly good on the technical aspects of recombinant DNA technology and the practical results of such work. Also applicable to Chapter 13.

Marshall, E. "Gene Therapy's Growing Pains." *Science*, August 25, 1995.

Moyers, Bill. *Science and Gender with Evelyn Fox Keller*. Washington, DC: PBSA Video, 1989. Discusses the work of women in science, including the details of Barbara McClintock's contributions.

National Geographic Society. *Laboratory of Life*. Washington, DC: National Geographic Society, 1985. A 21-minute videotape narrated by Linus Pauling explains how DNA works in the cells. Shows how genetically engineered insulin is manufactured and how genetically engineered tPA may revolutionize the treatment of heart attacks.

"The New Harvest: Genetically Engineered Species." *Science*, June 16, 1989. A special issue with seven articles on this topic.

Schmidt, Karen F. "Evolution in a Test Tube." *Science News*, August 7, 1993. Describes the work involved in designing totally new proteins using the molecular machinery of genetic engineering.

Watson, J.D., J. Tooze, and G.T. Kurtz. *Recombinant DNA: A Short Course*, 2nd ed. New York: Scientific American Books, 1989. The principles, methods, and applications of recombinant DNA technology.

Weintraub, Pamela. "The Coming of the High-Tech Harvest." *Audubon*, July/August, 1992. An environmentalist's view of the brave new world of genetic farming. Discusses the promises and controversies of transgenic plant development and use.

13

THE HUMAN GENOME

APPROACH

This is one of the two or three topics that nonscience majors are automatically interested in. However, you will have to counteract some misinformation they may have.

Growing directly out of material introduced in Chapters 9–12, this chapter integrates the material with a human perspective, and it makes a nice review of the whole unit. As much as possible, use these lectures to double-check students on their mastery of such topics as meiosis, protein synthesis, and recombinant DNA technology. Mendelian genetics provides powerful analytical tools for understanding the inheritance of genetic traits among human families and should be used when discussing the appropriate material from this chapter.

It will help to use diagrams to go through the nondisjunction events that lead to the sex chromosome abnormalities listed in Table 13.5.

The controversial issue of abortion comes up in Modules 13.10 and 13.11. It is important to point out here, as with all such issues, that, as scientists, we cannot dictate how others should behave relative to their ethical and religious convictions.

Spend some time discussing Nancy Wexler's and others' research on Huntington's disease. This story is fascinating regarding the interplay between Mendelian and molecular techniques. Ethical dilemmas face humans who must decide whether to find out if they have a potentially lethal gene.

The importance of education and counseling to accompany genetic screening is discussed in the interview with David Satcher (Module 13.9).

CHAPTER OBJECTIVES

Name the human genetic abnormalities attributable to incorrect numbers of chromosomes, explaining their cause by errors in meiosis.

Introduce karyotyping and pedigree charts as commonly used tools of genetics counselors.

Name and describe some of the more common genetic disorders attributable to recessive, dominant, and sex-linked alleles, and review the patterns of inheritance of these traits.

Detail the course of research on Huntington's disease, noting the timing, the functioning of scientific inquiry, and the interplay between Mendelian and molecular techniques of genetic research.

Describe how cancer relates to gene expression.

Explain the relationship between lifestyle and the cancer risk.

Describe the Human Genome Project and the interlocking steps being undertaken to complete it.

Discuss the ethical questions associated with knowledge of the human genome.

LECTURE OUTLINE

I. Introduction.

   A. Mendelian and molecular genetics are powerful tools that provide humans with a means for understanding and predicting the consequences of matings.

   B. The added power of recombinant DNA technology promises the deciphering of the human genome and uncovering the molecular causes of many genetic afflictions (*Opening Essay*).

II. Physical description of the genome.

   A. Three billion nucleotide pairs are packed into the human genome (*Module 13.1*).

      1. The human genome is all the genes present in a haploid human cell.

      2. The total length of DNA in a diploid cell (with 6 billion nucleotide pairs) is 3 m, packed into 46 chromosomes.

      3. Although the amount of DNA in a human cell is 1000 times that in *E. coli*, and *E. coli* has about 2000 genes, the human genome probably only has between 50,000 and 100,000 genes (that is, 25–50 times the number of genes).

      4. This is because about 80% of the human genome is noncoding: introns, regulatory sequences, and structural sequences.

   B. A karyotype is a photographic inventory of an individual's chromosomes (*Module 13.2*; Acetate 13.2).

      1. Blood samples are cultured for several days under conditions that promote cell division of white blood cells.

         NOTE: Red blood cells lack nuclei and do not divide.

      2. The culture is treated with a chemical that stops cell division at metaphase.

      3. White blood cells are separated, stained, and squashed (to spread out the chromosomes) following the procedure.

      4. The individual chromosomes in a photograph are cut out and rearranged by number.

III. Genetic disorders due to abnormalities in chromosome number.

   A. An extra copy of chromosome 21 causes Down syndrome (*Module 13.3*; Figure 13.3A).

      1. In most cases, human offspring that develop from zygotes with an incorrect number of chromosomes abort spontaneously.

      2. Trisomy 21 is the most common chromosome-number abnormality, occurring in about one out of 700 births.

      3. Down syndrome includes a number and range of physical, mental, and disease-susceptibility features.

      4. The incidence of Down syndrome increases with the age of the mother (Master 13.3C).

         NOTE: The age of the father is also correlated with an increased incidence of Down syndrome.

   B. Accidents during meiosis can alter chromosome number (*Module 13.4*).

1. *Review:* Meiosis (Module 8.15).
2. Nondisjunction is the failure of chromosome pairs to separate during either meiosis I or meiosis II (Acetate 13.4A, B).
3. Fertilization of an egg resulting from nondisjunction with a normal sperm results in a zygote with abnormal chromosome number (Acetate 13.4C).
4. The explanation for the increased incidence of trisomy 21 among older women is not entirely clear but probably involves the length of time a woman's developing eggs are in meiosis. Meiosis begins in all eggs before the woman is born, and finishes as each egg matures in the monthly cycle following puberty. Eggs of older women have been "within" meiosis longer.

C. Nondisjunction can alter the number of sex chromosomes (*Module 13.5*).

1. Unusual numbers of sex chromosomes upset the genetic balance less than do unusual numbers of autosomes, perhaps because the Y chromosome carries fewer genes and extra X chromosomes are inactivated as Barr bodies in females.

   Review: X-chromosome inactivation is discussed in Module 11.7.

2. Abnormalities in sex chromosome number result in individuals with a variety of different characteristics, some more seriously affecting fertility or intelligence than others (Table 13.5).
3. The greater the number of X chromosomes (beyond 2) the more likely (and the greater the severity of) mental retardation.
4. These sex chromosome abnormalities illustrate the crucial role of the Y chromosome in determining a person's sex. A single Y is enough to produce "maleness," even in combination with a number of Xs, whereas the lack of a Y results in "femaleness."

IV. Genetic disorders due to abnormalities in chromosome structure.

A. Alterations of chromosome structure may cause serious disorders (*Module 13.6*; Master 13.6A, B).

1. Deletions, duplications, and inversions occur within one chromosome.
2. Inversions are less likely to produce harmful effects than deletions or duplications because all the chromosome's genes are still present.
3. Duplications, if they result in the duplication of an oncogene in somatic cells, may increase the incidence of cancer (Module 13.15).
4. Translocation involves the transfer of a chromosome fragment between nonhomologous chromosomes.
5. Translocations may not be harmful, but they may be associated with increases in cancer, if the translocated segment activates an oncogene in a somatic cell.

B. Pedigrees track genetic traits through a family history (*Module 13.7*; Acetate 13.7).

1. Point out the commonly used, symbolic conventions on the pedigree chart, showing the appearance of congenital deafness in a Martha's Vineyard family.
2. By applying Mendel's principles, one can deduce the information on the chart from the pattern of phenotypes.

3. Assuming that Jonathan Lambert inherited his deafness from his parents, the only explanation is that his deafness is caused by a recessive allele because neither of his parents was deaf. Because some of his children were deaf, his wife, Elizabeth Eddy, must have been a carrier. From this it follows that all their hearing children were carriers.

4. This final deduction shows the power of applying Mendelian principles to pedigrees and how to make predictions (see the next module).

   NOTE: Since the pattern in the pedigree is not tied to gender, the gene for congenital deafness is not sex-linked.

C. Many human disorders are inherited as Mendelian traits (*Module 13.8*; Master 13.8A).

   1. Over 1000 known genetic traits are attributable to a single gene locus and show simple Mendelian patterns of inheritance (Table 13.8).

   2. Most disorders are caused by recessive alleles and vary in the severity of the expressed trait.

   3. The vast majority of people afflicted with recessive disorders are born to normal, heterozygous parents.

   4. Cystic fibrosis is the most common lethal genetic disease in the U.S.

   5. Most genetic diseases of this sort are not evenly distributed across all racial and cultural groups because of the prior and existing reproductive isolation of various populations.

   6. Laws forbidding inbreeding may have arisen from observations that such marriages more often resulted in miscarriages, stillbirths, and birth defects. On the other hand, there is a debate over this issue because seriously detrimental alleles would likely be eliminated from populations when expressed in the homozygous embryo, and there are societies where inbreeding occurs without detrimental results.

   7. Some disorders are caused by dominant alleles. These disorders vary in how deadly they are. Some are nonlethal handicaps, some are lethal in the homozygous condition, and some are intermediate in severity.

   8. Achondroplasia, a type of dwarfism, is lethal in the homozygous condition; individuals who express the trait are heterozygous.

   9. Other conditions attributable to dominant alleles are lethal only in older adults, so the allele can be passed to children before it is realized that the parent has the condition.

   Review: These diseases either are, or in the future may be, treatable by use of the techniques of genetic engineering (Modules 12.14, 12.15, and 12.19).

D. Talking About Science: David Satcher: The CDC chief discusses sickle-cell disease and genetic screening (*Module 13.9*).

   1. As a child growing up on a farm in Alabama, David Satcher aspired to be a physician. He succeeded, earning a combined M.D.-Ph.D. degree.

   2. Dr. Satcher's early research concerned the effects of radioactivity on chromosome structure. He demonstrated that, to at least some degree, the extent of chromosomal damage exhibited a linear relationship to radiation dosage.

   3. Satcher was involved in the development of the King-Drew Sickle-Cell Research Center and served as its director. He is also a past President of

Meharry Medical College. Since 1993 he has been director of the Centers for Disease Control and Prevention (CDC).

4. A major focus for Satcher has been sickle-cell anemia: its treatment, diagnosis, and screening, and community and counseling-related issues. He emphasizes that genetic screening (or screening for diseases such as AIDS, which do not have a genetic basis) in itself is not sufficient. Education of the public and genetic counseling must accompany any such screening program.

E. Sex-linked disorders affect mostly males (*Module 13.10*).

1. Most known sex-linked traits are caused by genes (alleles) on the X chromosome.

2. When these traits are recessive (most are), males express them because they have only one X. Females who have the allele are normally carriers and will exhibit the condition only if they are homozygous.

3. Males cannot pass sex-linked traits to sons (who get a Y from their father).

4. Red-green color blindness is a complex of sex-linked disorders, each of which is caused by an allele on the X chromosome. The result is considerable variation in the changes in color perception.

5. Hemophilia is a sex-linked trait with a particularly well-studied history because of its incidence among the intermarrying royal families of Europe (Master 13.10B).

NOTE: Hemophilia contributed to the Russian revolution of 1917. Rasputin gained influence over the Czar Nicholas II and Czarina Alexandra by his apparent ability to control hemophilic episodes experienced by their child, Alexis.

6. Duchenne muscular dystrophy (DMD) is a severe disease affecting muscle tissue and has been traced to a particular nucleotide sequence (Master 13.9C). About 99% of this gene consists of introns; the remaining 1% codes for a protein that has been named dystrophin. Dystrophin is missing in individuals with DMD.

NOTE: Dystrophin is found in the plasma membrane (sarcolemma) of muscle fibers (Modules 20.6, 30.8, and 30.9). It appears that the result of not having dystrophin is an increase in Ca^{2+} levels in the sarcoplasm (cytoplasm of a muscle fiber). The excess Ca^{2+} appears ultimately to lead to degeneration of the muscle fiber.

V. How humans deal with genetic disorders.

A. Fetal testing can spot many disorders early in pregnancy (*Module 13.11*).

1. Amniocentesis involves removing the amniotic fluid that bathes the fetus, at 14–16 weeks. This fluid contains living fetal cells (from skin and the mouth cavity) and can be karyotyped. Some chemical tests can be performed on the fluid itself (Master 13.11A).

2. Chorionic villi sampling involves removing tissue from the fetal side of the placenta nurturing a fetus, at 8–10 weeks. These cells are rapidly dividing and can be immediately karyotyped. Some biochemical tests can be performed (Master 13.11B).

3. Ultrasound imaging of the fetus provides a noninvasive view inside the womb (Figures 13.11C, D).

4. Fetoscopy provides a more direct view of the fetus through a needle-thin viewing scope inserted into the uterus.

5. Amniocentesis, chorionic villi sampling, and fetoscopy carry a small risk and are reserved for situations with higher probabilities of disorders (for example, older parents or situations where genetic counseling has uncovered a higher risk).

6. If fetal testing suggests that there is a problem that cannot be helped by routine surgery or other therapy, the difficult choice must be made between terminating a pregnancy by abortion or carrying a defective baby to term.

B. Carrier recognition is a key strategy in human genetics (*Module 13.12*).

1. Most children with recessive genetic disorders (the most common type) are born to parents with normal phenotypes.

2. A number of biochemical tests exist for identifying carriers of some genetic disorders, such as PKU, Tay-Sach disease, and sickle-cell anemia.

3. A method similar to DNA probing (Module 12.9) will likely provide many more diagnostic procedures to help recognize carriers.

VI. Some recent developments in dealing with genetic disorders.

A. Talking About Science: Nancy Wexler: A top researcher discusses a mind-ravaging genetic disease (*Module 13.13*).

1. American human geneticist Nancy Wexler has chosen to work on Huntington's disease (HD) because of very personal reasons: Her mother had HD, and she has a 50% chance of contracting the disease. HD is characterized by a slow deterioration of the central nervous system (Module 13.8).

2. Since 1981, she and her associates have determined a pedigree for 10,000 Venezuelan people in several villages where the incidence of HD is high.

3. In 1983, by comparing the inheritance pattern of this disease with that of other traits known to be caused by genes on various chromosomes, they determined that the gene was on chromosome 4.

4. The next step was to search for genetic markers (either visual, already typified genes or molecular tags) that are so closely associated with the HD gene that they are almost always inherited together. Later in 1983, just such a marker was found. It was hoped that this would be followed by actual positioning and sequencing of the HD gene, thus providing the means for testing for the presence of the defective allele.

5. The gene has been found, but it took 10 years. The HD gene is a so-called expanding gene, containing a short nucleotide sequence (CAG) that is repeated from 6–35 times in the normal alleles, while the disease allele has 40–100+ CAG repeats.

6. Scientists are currently attempting to show what the normal allele does and what the molecular basis of the disease is, the ultimate goal being a treatment or cure for the disease.

7. Meanwhile, the existence of a test for HD has proved vexing to those who might have the disease: Is it better to be uncertain or to know for sure?

B. Gene therapy is already a reality, and it raises serious ethical questions (*Module 13.14*).

1. In certain cases where a disorder is due to a single gene, it is possible to replace defective genes with normal genes.

2. An example is a trial procedure that should cure individuals with an autosomal recessive allele that causes defective functioning of the immune system and is usually fatal. Bone marrow stem cells, which are essential for blood cell formation, are removed. By means of a retrovirus, the defective gene is replaced with the normal one. The recombinant cells are cloned in culture and reintroduced in the individual, after the natural bone marrow cells have all been killed (Master 13.14).

   Review: Module 10.21 discusses the process by which retroviruses insert genes into a host's genome.

3. The expense of these techniques raises ethical questions concerning who can have access to these therapies. Further ethical questions are raised concerning the use of gene therapy, not for treatment of disease but for enhancement of physical ability and appearance, as well as intelligence.

VII. Cancer

A. Cancer results from mutations in genes that control cell division (*Module 13.15*).

   1. In all its forms, cancer is a disease of gene expression.

      Preview: Diet influences cancer risk (Module 21.20).

   2. Viruses can cause cancer by inserting cancer-causing genes (oncogenes) into the host genome.

      Review: Modules 10.17–10.19 and 10.21 discuss how viruses insert genes into a host's genome.

   3. A normal gene with the potential to become an oncogene is called a proto-oncogene. Proto-oncogenes usually code for proteins that stimulate cell division or affect growth-factor synthesis or function. A mutation that results in a failure to regulate the production of these proteins will result in the conversion of a proto-oncogene into an oncogene.

   4. Most cancers occur in somatic cells, thus they are not inherited.

   5. Mutations in tumor-suppressor genes, genes whose products inhibit cell division, also contribute to uncontrolled cell division.

B. Oncogene proteins and faulty tumor-suppressor proteins interfere with normal signal transduction pathways (*Module 13.16*).

   Review: Regulation of the cell cycle (Module 8.10) and signal transduction pathways (Module 11.13).

   1. In the presence of growth factor, proto-oncogenes produce gene products that stimulate cells to divide.

   2. Oncogenes produce hyperactive versions of proteins that stimulate cell division, even in the absence of growth factor. Moreover, abnormal amounts or versions of growth factor, transcription factor, and so on, could all result in the abnormal excess production of proteins that stimulate cell division.

   3. Faulty tumor-suppressor genes produce faulty tumor-suppressor proteins that may fail to inhibit cell division.

   4. Figure 13.16 illustrates two different types of mutations (*p53* and *ras*) that have been implicated in many cancers.

C. Multiple genetic changes underlie the development of cancer (*Module 13.17*).

1. Current evidence suggests that more than one somatic mutation is required to produce a significant cancer. An example of this is the development of colon cancer.

2. Colon cancer first appears as an unusually high rate of cell division occurring in apparently normal cells. Next, a benign tumor (polyp) appears, followed by the development of this benign tumor into a malignant tumor.

3. Underlying these changes are changes at the DNA level (that are passed on to daughter cells: mitosis) to proto-oncogenes and tumor-suppressor genes. That several mutations are required explains why some cancers can take a long time to develop.

D. Changes in lifestyle can reduce the risk of cancer (*Module 13.18*).

Review: The cellular basis of cancer is discussed in Module 8.11.

1. Cancer-causing agents other than viruses are called carcinogens.

2. Mutagenic chemicals cause mutations. In general, mutagens are carcinogens. Two significant mutagens, X-rays and UV radiation, were mentioned in Module 10.16).

NOTE: Almost inevitably there will be a student in the class who goes to a tanning salon. Damage to the skin by exposure to UV radiation results in a tan. UV radiation is a mutagen that greatly increases the risk of skin cancer; and the younger the age at which the exposure occurs, the greater the risk. Depletion of the ozone layer increases the amount of UV radiation that reaches the Earth's surface.

3. The largest group of carcinogens are mutagens. Substances from tobacco are known to cause more cases and types of cancer than any other single agent (Table 13.18).

NOTE: Recent studies have shown that beta-carotene does not reduce the risk of lung cancer and death from lung cancer in nonsmokers and actually increases these risks in smokers.

4. Exposure to carcinogens is additive, so long-term exposure to these agents is more likely to cause cancer.

5. Tissues in which cells have a high rate of cell division are more likely to become cancerous.

6. Many factors that expose a person to cancer-causing agents involve voluntary behaviors. But other voluntary behaviors, such as choosing to include more fiber in one's diet, can lower the risk.

Preview: Diet can influence cancer risk (Module 21.20).

NOTE: There is much evidence that the tendency to get certain cancers is hereditary.

E. The Human Genome Project is a scientific adventure (*Module 13.19*).

1. This international project proposes to completely sequence the human genome in four interlocking lines of work.

2. The first will locate several thousand genetic markers spaced evenly along all chromosomes, to act as a set of references for other work.

3. The second will map sequenceable fragments on each chromosome.

4. The third will produce the exact order of the nucleotide pairs in each fragment and hence each chromosome.

5. Finally, by comparing sequences in the human genome with those of other species (*E. coli*, yeast, the plant *Arabidopsis*, *Drosophila*, and mice), it should be possible to begin to define the functions of some of the sequences.

6. Automation and sophisticated computer technology are required to rapidly sequence so many fragments and to handle the huge amounts of data (Figure 13.19).

7. The project has huge potential benefits: insight into embryonic development and evolution, and identification of genes that cause genetic disorders and genes that are partly implicated in more common diseases such as cancer, heart disease, diabetes, schizophrenia, alcoholism, and Alzheimer's disease.

F. The ethical questions raised by human genetics demand our attention (*Module 13.20*).

1. James Watson suggests that 3% of the NIH budget for the project be earmarked for research on ethical issues.

2. The Human Genome Project offers some potential for misuse of the information by governments, employers, and insurers, which we as citizens must be aware of and prevent.

3. The possibility of gene therapy suggests a return to the notion of eugenics (the systematic removal of deleterious genes from the human population).

4. Genetic discrimination by potential employers is a concern.

5. The best way we can help control the use of these techniques and the information they result in is to become aware of how they work.

CLASS ACTIVITIES

1. In introducing pedigree charts, try starting from a chart that shows only phenotype and gender. (Make a modified version of Acetate 13.7, or use a different example from a genetics text.) Have students work with you to determine the genotypes logically. Start by assuming that the allele for deafness is dominant (the wrong explanation), and work through some of the possibilities for genotypes in the second and third generations. Show how observed ratios more closely approximate those obtained by assuming that the allele is recessive. Also be sure to point out that the study referenced used many more pedigrees to avoid the problems inherent in small samples. When discussing genetic counseling, diagram all the possibilities using Punnett squares. Have the students make pedigree charts for their own families.

2. When discussing changes in lifestyle to reduce the risk of cancer, you may also want to take this opportunity to link this topic to the environmental concepts discussed in Chapter 38 by discussing the effect of ozone depletion on skin cancer rates. Also, be sure to poll your class to see how many students smoke or make use of tanning beds; point out that by doing so they are paying someone money to increase their cancer risk.

3. Have the class determine which members of the class are at the greatest risk of skin cancer and which are at the lowest risk. Lighter-skinned individuals are at greater risk than darker-skinned individuals. However, darker-skinned individuals are at greater risk of not getting sufficient vitamin D production in their skin. Point out the evolutionary (Unit III) tradeoff. Darker-skinned individuals have recent ancestors from regions where the sun is strong for most, if not all, of the year; therefore, the ancestral risk of skin cancer was greater than the ancestral risk of vitamin D deficiency. Whereas

lighter-skinned individuals have recent ancestors from more northern environments where the risk of vitamin D deficiency outweighed the risk of skin cancer. Point out that some of the highest rates of skin cancer are found among the Caucasians of Australia, whose ancestors were from the British Isles (greater risk of vitamin D deficiency, and thus fairer-skinned).

4. Further class discussion can focus on whether or not the hoped-for results of the Human Genome Project justify the expense.

5. Have the students gather information on a genetic characteristic or disorder using the Human Genome Internet sites listed at the end of Chapter 8 of this guide.

RESOURCES AND REFERENCES

Anderson, W. French. *Prospects for Human Gene Therapy*. Bethesda, MD: National Library of Medicine, 1988. A 52-minute videotape that explores the current directions in developing genetic cures for some of the disorders mentioned in this chapter.

Bishop, J. Michael. "The Molecular Genetics of Cancer." *Science*, January 16, 1987. Diverse examples of genetic damage found in various cancer cells may hint at the biochemical explanations for cancer growth.

"Cancer." *Science*, November 22, 1991. A special issue of seven articles on basic and applied cancer research.

"Cancer." *Scientific American*, September 1996. A special issue on cancer.

Cavanee, W.K., and R.L. White. "The Genetic Basis of Cancer." *Scientific American*, March 1995.

Cusack, Richard. *Lights Breaking: Ethical Questions About Genetic Engineering*. Oley, PA: Bullfrog Films, 1985. A 58-minute videotape that reviews the important ethical concerns.

Erickson, Deborah. "Hacking the Genome." *Scientific American*, April 1992. Includes background information, details of the methods, and discussion of how the complex Human Genome Project is being administered.

"Genome Issue." *Science*, October 2, 1992 and October 20, 1995. These two issues contain articles about progress on the Human Genome Project. (Look for a similar issue each fall.)

Greider, C.W., and E.H. Blackburn. "Telomere, Telomerase, and Cancer." *Scientific American*, February 1996. An enzyme that seems to keep cancer cells immortal could be a target in the war against cancer.

Hall, S.S. "James Watson and the Search for Biology's 'Holy Grail.'" *Smithsonian*, February 1990. Watson and the Human Genome Project.

HRM. *Human Genetic Disorders*. Burlington, NC: Carolina Biological Supply Company, 1989. Explores basic Mendelian principles by investigating the inheritance of specific human disorders.

Kahn, P. "Germany's Gene Law Begins to Bite." *Science*, January 31, 1992. A case study in attempts to regulate DNA technology.

Lander, E.S., and B. Budowle. "DNA Fingerprinting Dispute Laid to Rest." *Nature*, October 24, 1994. The reliability of DNA evidence in forensics.

Lawn, Richard M., and Gordon A. Vehar. "The Molecular Genetics of Hemophilia." *Scientific American*, March 1986. Details of the search for the molecular basis of this sex-linked disease and the recombination, in culture, of affected cells with the normal gene for the blood-clotting protein.

McElfresh, Kevin C., Debbie Vining-Forde, and Ivan Balazs. "DNA-Based Identity Testing in Forensic Science." *BioScience*, March 1993. A review of the admissibility of DNA data in the nation's courts over the past 5 years.

Morell, Virginia. "Huntington's Gene Finally Found." *Science*, April 2, 1993. Briefly details the history of the search and the surprising final results of the unusual search for this gene.

Mulligan, R.C. "The Basic Science of Gene Therapy." *Science*, May 14, 1993.

Mullis, Kary B. "The Unusual Origin of the Polymerase Chain Reaction." *Scientific American*, April 1990. Details the working of this method and how it was discovered.

Nova. *Decoding the Book of Life*. Princeton, NJ: Films for the Humanities and Sciences, 1992. A popular presentation on the Human Genome Project.

Patterson, David. "The Causes of Down Syndrome." *Scientific American*, August 1987. The genes thought to be responsible for many of the pathologies associated with the disorder are being identified and mapped to sites on chromosome 21.

Rennie, J. "Grading the Gene Tests." *Scientific American*, June 1994. Genetic tests and their ethical dilemmas.

Roberts, Leslie. "Two Chromosomes Down, 22 to Go." *Science*, October 2, 1992. Researchers have produced detailed physical maps of human chromosomes Y and 21, providing a boost to the Human Genome Project.

Weinberg, Robert A. "To Sequence or Not to Sequence." *Scientific American*, November 1988. An early critique of the Human Genome Project by this influential molecular biologist.

Wingerson, Lois. *Mapping Our Genes: The Genome Project and the Future of Modern Medicine*. New York: Dutton, 1990.

14

HOW POPULATIONS EVOLVE

APPROACH

This chapter begins a new unit on evolution (the central theme of biology) and diversity. Spend a few moments at the beginning of this set of lectures introducing the general topics presented in the unit: the theoretical ideas about how populations evolve, the historical development of evolutionary thought, a description of factors controlling evolution on a larger scale, and evolutionary surveys of each of the five kingdoms.

The subject of evolution might be a touchy one among nonscience majors. There may be philosophical bias as well as difficulty grasping the basic foundations of evolutionary theory (actually, evolution is a fact). Be sure to discuss the scientific basis of the material you present here, reminding students that ideas from other realms of human thought, though valid in their own context, are difficult to rationalize against the scientific development of the theory of evolution.

Evolution is the uniting theme of all of the subdisciplines that comprise biology. Take this opportunity to relate evolution to the other material covered in the course. The fact that evolution occurs in populations (not individuals) has to be emphasized and repeated. This fact makes a nice bridge from the concepts of individual genetic transmission, as discussed in Unit II, and the topics of Unit IV.

One challenging area in this chapter is the return to the probabilistic approach when discussing Hardy-Weinberg equilibrium. Be sure to review the rules for determining the outcomes of probability calculations, and explain that 0.64 is another way of saying 320 out of 500, or 64% probability.

CHAPTER OBJECTIVES

Trace the historical development of Darwin's theory of evolution, discussing the previous ideas and observations that formed his theoretical basis.

Discuss the processes of fossilization and the accumulation of fossils in sedimentary deposits.

Discuss the sources of evidence of evolution.

Define the biological species concept.

Explain the process of natural selection.

Show how the Hardy-Weinberg equation represents the stability of gene frequencies in nonevolving populations.

Explain how deviations from each of the five assumptions behind Hardy-Weinberg equilibrium provide a chance for microevolutionary change in a population.

Describe how variation is introduced into populations.

Discuss the role of reduced variation in contributing to the extinction of species.

Explain the concept of fitness and how it relates to natural selection and genetic drift.

Describe and give examples of the three modes of action natural selection can take.

Discuss the way directional selection plays a role in the evolution of resistant strains of pests and parasites.

LECTURE OUTLINE

I. Introduction.

A. *Review:* Evolution is the central theme of biology. Adaptation is a universal characteristic of living things (see Module 1.6).

NOTE: More than any other idea in biology, evolutionary theory serves to tie the discipline together: T. Dobzhansky: "Nothing in biology makes sense except in the light of evolution."

B. If you look at any organism critically, you are first struck by the differences from other organisms (*Opening Essay*).

1. Further observation often reveals that an organism's features show some relationship to where the organism lives and what it does in its environment.

2. The marine iguana of the Galapagos Islands is peculiar in several ways when compared with its nearby relative, the terrestrial iguana (compare the chapter-opening photo and Figure 14.1A).

C. A sea voyage helped Darwin frame his theory of evolution (*Module 14.1*).

1. Awareness of each organism's adaptations and how they fit the particular conditions of its environment helps us appreciate the natural world.

2. The ideas that Charles Darwin advanced, that species change over time and that living species have arisen from earlier species, were based on his keen observations of diverse life forms on the Galapagos Islands and elsewhere.

3. They are also based on previous ideas about the interrelatedness of things, which have a history dating back over 2500 years.

D. The history of evolutionary thought prior to Darwin.

1. Early Greek philosophers held various views. Anaximander (about 2500 years ago) suggested that life arose in water and that simpler forms preceded more complex forms of life. On the other hand, Aristotle, who strongly influenced later thinkers, believed that species were fixed and did not evolve.

2. This latter view was advanced by the Judeo-Christian tradition that all species were created in a single act of creation about 6000 years ago.

3. French naturalist Georges Buffon (mid-1700s) suggested that Earth was much older and raised the possibility that different species arose from common ancestors, although he later argued against this point.

NOTE: Buffon also believed in catastrophism: Following natural disasters, some species die off, while populations of others (already present in lower numbers) increase in size to become more dominant than they had been.

4. French naturalist Jean Baptist Lamarck (early 1800s) was the first to strongly support the idea of evolution, but he believed the mechanism for change was the inheritance of acquired characteristics.

E. Influential events in Charles Darwin's life.

1. Born in 1809, he joined the crew of the surveying ship *Beagle* as a naturalist for a world-encircling voyage in 1831 (Acetate 14.1C).
2. Comparisons of South American fossils with living species there and fossils elsewhere, and observations of organisms and their distributions on the Galapagos Islands, made particularly strong impressions on him.
3. Scottish geologist Charles Lyell's *Principles of Geology* promoted the idea of continual, gradual, consistent geological change.
4. After his return, Darwin began work on an essay to document his observations and his new theory of evolution.
5. English naturalist Alfred Wallace (mid-1850s) conceived of essentially the same theory, based on his observations in Indonesia. He contacted Darwin, and presentations of both their work were made to the scientific community in 1858.
6. Darwin's *On the Origin of Species by Means of Natural Selection* was published in 1859 and contains a well-constructed argument for natural selection, backed by considerable evidence. He uses the phrase "descent with modification."

F. Darwin's view of evolution: The history of life is like a tree, with multiple branching from the base of the trunk to the tips of the branches. Species on a given branch are more closely interrelated than they are to species on other branches.

G. The study of fossils provides strong evidence for evolution (*Module 14.2*).
1. Hard parts, such as skeletons and shells, remain after organic matter has decomposed. Such parts fossilize easily (Figure 14.2A).
2. Some fossils, such as those of leaves, retain remnants of organic matter with molecular fragments that can be analyzed.
3. Organisms trapped in tree resin can be fossilized intact, within the fossilized amber, protected from decomposition by bacteria and fungi (Figure 14.2B).
4. Petrified fossils form by the slow mineralization of organic materials (Figure 14.2C).
5. Fossilized molds of organisms form when a covered area decays and fills in with other sediment (Figure 14.2D).
6. The fossil record is an array of fossils appearing within the layers of sedimentary rocks. Sedimentary rocks form from accumulations of waterborne sediments. Sedimentary deposits occur in strata. Each layer contains fossils of organisms among the deposits, with younger strata on top of older strata (Figure 14.2E).
7. The fossil record shows an historical sequence of organisms from the oldest known fossils, prokaryotes, dating from ≈3.5 billion years, through the subsequent appearance of eukaryotes, on through many intermediate steps to modern forms—a sequence that has an overall pattern of change from simple to more complex forms.

    NOTE: Such a sequence, whereby links are seen between extinct organisms and species alive today, is predicted by evolutionary theory. One of the best-documented series is the evolution of the horse (see G. G. Simpson reference below).

Preview: The fossil record chronicles macroevolution (Module 16.1).

H. A mass of other evidence validates the evolutionary view of life (*Module 14.3*).

1. Biogeography: observations about the distribution of different but obviously related life forms around the world and in neighboring geographical regions. Island forms are most similar to forms found on the closest mainland, rather than those found on ecologically similar but more distant islands.

2. Comparative anatomy of homologous structures (Acetate 14.3A). For example, all mammals have the same basic limb structure.

3. Comparative embryology shows that different organisms go through similar embryonic stages. For example, evidence that all vertebrates evolved from a common ancestor is that all have an embryonic stage in which gill pouches appear in the throat region.

   NOTE: In addition to pharyngeal gill pouches, vertebrates, along with all chordates, also have in common the presence of, at some point in their life cycle, a notochord (a cartilaginous supporting rod), a dorsal hollow nerve cord (spinal cord), and a post-anal tail.

4. Molecular biology demonstrates the universality of the genetic code (see Module 10.8), the conservation of amino acid sequences in proteins such as hemoglobin, and the presence of very similar homeoboxes in very different species (see Modules 11.12 and 11.14).

   NOTE: The basics of glycolysis are also virtually the same in all species.

5. The picture shown by the fossil record is supported by other observations (biochemical, cellular) of existing life forms (Master 14.3B).

II. Darwin proposed natural selection as the mechanism of evolution (*Module 14.4*).

Review: Evolution as an explanation for the unity and diversity of life was first discussed in Module 1.6.

A. His theory was based on a number of observations:

1. Darwin's personal knowledge that species tend to produce excessive numbers of offspring, the variety inherent in populations of species, and the heritability of many of the characteristics species have.

2. English economist Thomas Malthus's essay on the inevitable human suffering resulting from populations growing faster than supplies of resources.

3. Darwin's personal knowledge of and interest in artificial selection (Figures 14.4A, 14.4B), particularly in light of the variation between closely related species (Figure 14.4C).

B. The essence of natural selection is differential reproduction.

1. Individuals in populations vary.

2. Some individuals are more suitable to a given environment and reproduce more easily and abundantly.

3. The favored characteristics are passed to the next generation and the less-favored characteristics are not.

4. Over vast amounts of time, the gradual accumulation of the favored characteristics among the individuals in a population occurs.

C. Natural selection is a prominent force in nature (*Module 14.5*).

1. Two good examples of the effects of the process can be described. In both cases, new populations have resulted, but not new species.

2. In the land snail, *Cepaea nemoralis*, shell patterns camouflage the snails in different habitats, with light, striped snails found in well-lit areas and dark snails found in shady areas (Figure 14.5A).

3. In the peppered moth, *Biston betularia*, light forms are adapted to lichen-covered tree bark and dark forms to tree bark without lichen. Increases in air pollution in England following the Industrial Revolution killed large numbers of air-pollution-sensitive lichens, exposing the light variety of peppered moth to consumption by predators when it rested on a dark substrate. The dark variety was not consumed. Recent improvements in air quality have reversed this trend.

4. Natural selection is regional and timely. Populations tend to adapt to local environments during one time period. The adaptations may be pointless in the context of other locales or times.

5. Industrial melanism in *Biston betularia* is one example of many that demonstrate that significant evolutionary change can occur in very short periods of time.

   NOTE: Other examples include the development of antibiotic-resistant bacteria and pesticide-resistant insects (see Module 14.21).

III. Background for looking at the evolution of populations.

A. Populations are the units of evolution (*Module 14.6*).

1. A population is a group of individual organisms living in the same place at the same time.

2. Evolution is measured as the change in frequency of a given characteristic within a population over a succession of generations.

3. Darwin realized this, but he did not know about the genetic mechanisms.

4. During the first half of the twentieth century, Mendel's rediscovered genetic principles have been combined with Darwin's proposal of natural selection into a comprehensive theory of evolution known as the modern synthesis.

5. Central to this synthesis is the biological species concept. A biological species is a group of populations whose individuals have the potential to interbreed. A given biological species has an overall range, with concentrations of individuals in local populations.

   NOTE: Since the biological species concept concerns actually or potentially interbreeding populations, it is difficult to apply it to the fossil record. Another species concept that is more readily applied to the fossil record is the evolutionary species concept. G. G. Simpson defined the evolutionary species as "a lineage (an ancestral-descendant sequence of populations) evolving separately from others and with its own unitary evolutionary role and tendencies."

6. Opportunities for breeding among populations of the same species vary, depending on the species and on the extent of isolation of the populations.

B. Microevolution is change in a population's gene pool (*Module 14.7*).

   Review: The basic concepts of Mendelian genetics (Chapter 9).

   1. A population's gene pool is all the alleles in all the individuals making up the population, available to be inherited by members of the next generation.

   2. Most gene loci are represented by two or more alleles across a population, and individuals (of most eukaryotes) carrying two alleles can be homozygous or heterozygous for the locus.

   3. During microevolution, the relative frequencies of the alleles governing characteristics change. For example, in the whole population of peppered moths, the allele for light color decreased and the allele for dark color increased.

C. The gene pool of an idealized, nonevolving population remains constant over the generations (*Module 14.8*).

   NOTE: This example is entirely arbitrary and will work with any numbers. Have students verify this fact by trying other examples.

   1. Use the example of a hypothetical population of 500 iguanas. Webbed feet is dominant over nonwebbed feet, and the characteristic is governed by a single gene (alleles *W* and *w*): 320 have *WW*, 160 have *Ww*, and 20 have *ww* genotypes.

   2. The Hardy-Weinberg equation shows that allele frequencies are stable in a population not undergoing microevolution: $p^2 + 2pq + q^2 = 1$. That is, the population (1) is made up of homozygous dominant genotypes (p^2) + heterozygous genotypes ($2pq$) + homozygous recessive genotypes (q^2). Also note that $p = 1 - q$.

   3. The frequencies of *WW*, *Ww*, and *ww* in the first generation are 0.64, 0.32, and 0.04. Since iguanas are diploid, there are 1000 alleles in the population, and their frequencies are $p = 0.80$ *W* and $q = 0.20$ *w* (Acetate 14.8B).

   4. If random matings occur between various members of this population, then the laws of probability will predict the genetic makeup of the next generation (see Module 9.7).

   5. On average, the next generation will have $p \times p$ *WW* individuals (= 0.64), ($p \times q$) + ($q \times p$) *Ww* individuals ($2pq = 0.32$, since there are two ways to get pq in a zygote, depending on whether the q is in the sperm or egg), and $q \times q$ *ww* individuals (= 0.04) (Acetate 14.8C). The allele frequencies remain the same.

D. The Hardy-Weinberg equation is used in public health (*Module 14.9*).

   1. Estimating frequencies of harmful alleles among the population at large helps public health scientists assess and prioritize their efforts.

   2. For example, if PKU occurs in one out of 10,000 babies ($q^2 = 0.0001$), then $q = 0.01$, $p = 0.99$, and $2pq$ (the frequency of carrier heterozygotes) = 0.0198.

   NOTE: This exercise can be used with any of the frequencies of genes from Table 13.8.

IV. **How microevolution proceeds.**

    A. Hardy-Weinberg equilibrium rarely exists in natural populations, but understanding the assumptions behind it gives us a basis for understanding how populations evolve.

    B. Five conditions are required for Hardy-Weinberg equilibrium (*Module 14.10*).

        1. The population is very large.

        2. The population is isolated (no migration of individuals, or alleles, into or out of the population).

        3. Mutations do not alter the gene pool.

        4. Mating is random.

        5. All individuals are equal in reproductive success (no natural selection).

    C. There are five potential causes of microevolution (*Module 14.11*).

        1. Genetic drift is a change in a gene pool of a small population due to chance. The effect of a loss of individuals from a population is much greater when there are fewer individuals. The bottleneck effect is genetic drift resulting from a disaster that reduces population size (such as the example of the elephant seals; Master 14.11A, Figure 14.11B). The founder effect is genetic drift resulting from colonization of a new area by a small number (even one) of individuals (likely important in the evolution of animals and plants on the Galapagos Islands).

        2. Gene flow is a gain or loss of alleles from a population due to immigration or emigration of individuals or gametes.

        3. Mutations are rare events, but they do occur constantly (as often as one per gene locus per 10^5 gametes). Mutation provides the raw material on which other mechanisms of microevolution work. Mutation is rarely, if ever, directly responsible for evolutionary change.

        4. Nonrandom mating is more often the case, particularly among animals, where choice of mates is often an important part of behavior.

        5. Differential success in reproduction is probably always the case for natural populations. The resulting natural selection is the factor that is likely to result in adaptive changes to a gene pool.

        6. Each of the above mechanisms is a deviation from the conditions required for Hardy-Weinberg equilibrium, as listed in Module 14.10.

V. **Adaptive change results when natural selection upsets genetic equilibrium (*Module 14.12*).**

    A. The degree of adaptation that can occur is limited by the amount and kind of genetic variation in the population.

    B. Variation is extensive in most populations (*Module 14.13*).

        1. Variation in a single characteristic can be caused by the effect of one or more genes or from the action of the environment inducing phenotypic change.

        2. A population is polymorphic for a characteristic if two or more morphs (contrasting forms) are noticeably present; these may be visible or biochemical characteristics.

    Review: Polygenic inheritance is discussed in Module 9.13.

3. Most populations exhibit geographic variation in the distribution of characteristics; this variation may show stratification or be clinal, varying smoothly across the population.

C. Two random processes generate variation (*Module 14.14*).

1. *Review:* Mutation (Module 10.16) and meiosis (Modules 8.15 and 8.16).

2. Mutations normally are harmful, but they may improve an organism's adaptation to an environment that is changing. Organisms with very short generation spans can evolve by mutation alone (Master 14.14A).

3. Sexual recombination shuffles the mixture of alleles in diploid organisms. Independent assortment, crossing over, and random fertilization of sperm and egg all play a role. Organisms that reproduce sexually tend to have longer life spans, and sexual recombination is necessary to increase the variation stemming from single mutations (Master 14.14B).

D. Overview: How natural selection affects variation (*Module 14.15;* Acetate 14.15).

1. An ancestral population is varied, with individuals having characteristics suited for many types of environments.

2. Over successive generations, those individuals with the characteristics best suited for the environment leave more offspring. These characteristics increase in the subsequent generations.

3. Those individuals with characteristics not suited for the environment leave fewer offspring. These characteristics decrease in subsequent generations.

4. The effects of recessive alleles are not often displayed in diploid organisms. Recessive alleles may be "hidden" from natural selection when they are found in combination with a dominant allele. Thus, variation is retained in a population subject to selection.

    Review: For example, cystic fibrosis is only expressed in individuals who are homozygous recessive for this disorder (Module 13.8).

5. Heterozygote advantage is a situation in which the heterozygote is favored over either homozygote. As a result, variation is maintained in the population. An example of this is the sickle-cell allele where one homozygote is susceptible to malaria, the other homozygote suffers from sickle-cell anemia, and the heterozygote is resistant to malaria and does not suffer from sickle-cell anemia (Module 9.12).

E. Endangered species often have reduced variation (*Module 14.16*).

1. This is becoming more and more of a problem as human activity endangers wild populations, particularly those that are small to begin with.

2. There is about 0.04% heterozygosity in the gene loci of the South African population of the cheetah, and 1.4% heterozygosity in the East African population. These animals suffered bottlenecks due to disease, hunting, and drought.

    NOTE: This is an extremely high degree of genetic uniformity—higher than some strains of highly inbred laboratory mice.

F. Not all genetic variation may be subject to natural selection (*Module 14.17*).

1. Some characteristics showing neutral variation (such as human fingerprints) apparently provide no selective advantage.

2. The frequency of these characteristics may change as a result of genetic drift, but not by natural selection.

3. It is impossible to demonstrate that an allele brings no benefit to an organism, and it may be that some supposedly neutral variations provide benefits in some environments.

VI. The overall effects of natural selection.

A. The perpetuation of genes defines evolutionary fitness (*Module 14.18*).

1. Emphasize that it is the survival of genes, not individual organisms, that is important. It is the genes that survive in time, not the individual organism(s).

2. Today, fitness is defined as the relative contribution that an individual makes to the gene pool of the next generation.

B. Natural selection acts on whole organisms and affects genotypes as a result (*Module 14.19*; Master 14.19).

1. Natural selection acts on phenotypes.

2. Each phenotype is the sum of the effects of an organism's genotype (as well as, for some characteristics, the genotype's interaction with the phenotype).

3. There is no way for natural selection to select individual gene loci; it culls, or favors, whole genomes.

C. Natural selection has three modes of action (*Module 14.20*; Acetate 14.20).

1. Stabilizing selection tends to narrow the range in population variability toward some intermediate form. This occurs in relatively stable environments.

   Preview: An example of stabilizing selection is the birth weight of human babies, which tends to be in the 6 - 10 pound range (Module 27.17).

2. Directional selection tends to move the modal (most common) form toward one of the extremes. This is most common during times of environmental change, or when organisms find themselves in new habitats.

   Preview: A good example of the effects of directional selection on life history patterns can be found in the discussion of guppies subject to different selective pressures (Module 35.6).

3. Diversifying selection occurs when environmental conditions are varied in a way that favors both extremes over the intermediate form.

D. Directional selection has produced resistant populations of pests and parasites (*Module 14.21*; Acetate 14.21).

1. In the past few decades, there has been alarming development of pesticide resistance in insects and antibiotic and drug resistance in disease-causing bacteria and protists.

2. These life forms are particularly adaptable because they are numerous, multiply rapidly, and have short generation times.

   Preview: Such species are referred to as being *r*-selected (Module 35.7).

CLASS ACTIVITIES

1. To demonstrate natural selection and microevolution, use a predator-prey exercise. Students (the predators) attempt to collect as many colored beans (the prey) as possible during a set time period. Students have different "beaks" (forks, spoons, knives, forceps) and thus will have different success rates. If the exercise is run for several generations, some beaks will become more plentiful, while others will become extinct. The exercise takes 50–60 minutes. You can reuse the exercise when covering Chapter 37 if you record the learned behaviors that occur. Some students may choose to look for prey in areas with a low density of prey, but where there is also a low density of competing predators. Other students become aggressive in later generations and may attack another's "mouth" (the cup holding the beans they have collected). Some students with like "beaks" may even work together as a gang so that at least some of their type survive.

2. When discussing Module 14.21, point out that one of the major health-care crises humans are facing is the development of antibiotic-resistant bacteria. Ask your students how many of them, when given prescription antibiotics, finish the entire prescription. There will, almost inevitably, be at least one student who does not. Point out how this contributes to the increase in the numbers of antibiotic-resistant strains of bacteria. Further, as discussed in Modules 12.2 and 12.3, bacteria can exchange plasmids; thus, bacteria never exposed to a particular antibiotic can be resistant to that antibiotic as a result of plasmid exchange.

RESOURCES AND REFERENCES

Berra, T.M. *Evolution and the Myth of Creationism: A Basic Guide to the Facts in the Evolution Debate*. Stanford, CA: Stanford University Press, 1990. A good resource for countering creationist arguments and creationists' misrepresentations of evolution.

Darwin, C. *The Origin of Species* and *The Descent of Man*. New York: Modern Library, 1990. Two classic books in one volume for around $12.

Dawkins, R. *The Blind Watchmaker*. New York: Norton, 1986. How complexity can arise in the absence of design.

Desmond, A., and J. Moore. *Darwin*. New York: Warner, 1992. A recent biography.

Diamond, J. "Founding Fathers and Mothers." *Natural History*, June 1988. The importance of genetic drift in human evolution.

Endler, J.A. *Natural Selection in the Wild*. Princeton, NJ: Princeton University Press, 1986. An excellent discussion of natural selection.

Endler, John A., and Tracy McLellan. "The Processes of Evolution: Toward a Newer Synthesis." *Annual Review of Ecology and Systematics*, 1988. A good review of current topics of debate, with a preliminary classification of the major evolutionary processes.

The Evolution of Darwin. Princeton, NJ: Films for the Humanities and Sciences, 1993. A new videotape series containing six 26-minute illustrated lectures at the Linnean Society, reviewing the history, theory, and evidence of evolutionary processes.

Genetics Series. Seattle, WA: Laser Learning Technologies, 1991. A four-disc videodisc series that gives considerable examples covering the patterns of development, diversity, evolution, and inheritance.

Gibbons, A. "The Mystery of Humanity's Missing Mutations." *Science*, January 6, 1995. Did human evolution include a bottleneck?

Gould, Stephen Jay. "'Red in Tooth and Claw.'" *Natural History*, November 1992. Traces the thinking and pain behind the most overused one-liner in biology.

Grimaldi, D.A. "Captured in Amber." *Scientific American*, April 1996. Perfectly preserved fossils and insect DNA trapped in sap.

Kauffman, Stuart A. "Antichaos and Adaptation." *Scientific American*, August 1991. Biological evolution may have been shaped by more than just natural selection. Computer models suggest that certain complex systems tend toward self-organization.

McDonald, John F. "Macroevolution and Retroviral Elements." *Bioscience*, March 1990. The insertion of viral-like DNA segments may bring about rapid and dramatic changes in gene regulation and development.

McGowen, C. *In the Beginning . . . : A Scientist Shows Why the Creationists Are Wrong.* Buffalo, NY: Prometheus Books, 1984. Debunking creationist arguments.

Milner, R. *The Encyclopedia of Evolution*. New York: Facts on File, 1990. Many interesting anecdotes.

Morey, D.F. "The Early Evolution of the Domestic Dog." *American Scientist*, July-August 1994. Perhaps the oldest cases of artificial selection.

Noonan, D. "Dr. Doolittle's Question." *Discover*, February 1990. How biochemist Russell Doolittle learns about evolution from the study of blood-clotting proteins.

Nova. *God, Darwin, and Dinosaurs*. Princeton, NJ: Films for the Humanities and Sciences, 1992. A 58-minute videotape covering all aspects of the heated debate between creationists and evolutionists.

Organic Evolution: Factoring Mendel and *The Popular Picture*. Princeton, NJ: Films for the Humanities and Sciences, 1993. Part of a videotape series tracing the developments of various theories of evolution. These two videotapes discuss the role of Mendelian genetics in the overall process and introduce and demonstrate how the Hardy-Weinberg equation model is used.

Palmiter, Michael. *Simulated Evolution*. Bayport, NY: Life Science Associates, 1989. A very simple demonstration of the process of natural selection, using electronic protozoans controlled by six "genes" that chase and eat bacteria that appear. Over the course of an hour, the protozoans will evolve unique characteristics that make them better feeders. IBM PC version only.

Ridley, M. *Evolution*. Boston: Blackwell Scientific Publications, 1993. Covers Mendelian and molecular genetics and microevolution to macroevolution.

Simpson, G.G. *Tempo and Mode in Evolution*. Reprint 1984. Originally published 1944. New York: Columbia University Press. A classic of evolution. Especially impressive when you realize that noted developments in evolutionary theory in the 1970s and 1980s were already discussed by Simpson.

Steinhart, Peter. "In the Blood of Cheetahs." *Audubon Magazine*, March/April 1992. Details some of the more recent findings about these endangered mammals.

Weiner, J. "Evolution Made Visible." *Science*, January 6, 1995. Observing evolution.

INTERNET RESOURCES FOR UNIT III

The Dinosauria

http://umcp1.berkeley.edu/diapsids/dinosaur.html

A good UC Museum of Paleontology site; can be linked to via URL shown above.

The Electronic Prehistoric Shark Museum

> http://207.67.198.22/C/celestial/epsm.htm

Field Museum of Natural History

> http://rs6000.bvis.uic.edu/museum/exhibits/Exhibits.html
>
> Look at "Life Over Time."

Human/Mouse Homology Relationships

> http://www3.ncbi.nlm.nih.gov/Homology

Mazon Creek Fossils

> http://www.museum.state.il.us/exhibits/mazon_creek
>
> Plants and animals of Illinois 300 million years ago.

Midwestern U.S. 16,000 Years Ago Content—Illinois State Museum

> http://www.museum.state.il.us/exhibits/larson
>
> Pleistocene plants and animals. Includes a discussion of causes of extinction (human versus environmental causation).

National Center for Science Education

> http://www.NatCenSciEd.org
>
> The National Center for Science Education is an organization that keeps track of creationist publications and efforts to bring creationism into public school science classes. Provides intelligent responses to creationist arguments.

Welcome to UCMP

> http://www.ucmp.berkeley.edu/index.html
>
> UC Museum of Paleontology. A resource for evolution and diversity; including phylogeny, diversity through time, and evolutionary theory.

The WWW Virtual Library: Evolution (Biosciences)

> http://golgi.harvard.edu/biopages/evolution.html

15

THE ORIGIN OF SPECIES

APPROACH

This short chapter is critical to understanding how the principles of microevolution introduced in Chapter 14 result in actual macroevolutionary change, discussed in Chapter 16. It is the mechanisms detailed here that actually lead to speciation. The microevolutionary material in Chapter 14 may be more acceptable to students who hold certain religious beliefs than the evolutionary processes discussed here because well-studied examples of evolutionarily related species are discussed.

CHAPTER OBJECTIVES

Define the terms *population* and *biological species*, and explain why it is not always possible to recognize biological species in nature.

Distinguish between *allopatric speciation* and *sympatric speciation*.

Show how various kinds of reproductive barriers can lead to speciation.

Show how various kinds of geographical barriers can lead to speciation, explaining why isolated islands are often areas rich in adaptive radiation.

Explain how new species arise by polyploidy, and trace the evolution of bread wheat as an example.

Contrast the gradual and punctuated equilibrium models of evolution.

LECTURE OUTLINE

I. Introduction.

    A. Microevolutionary changes, as discussed in Chapter 14, show us how populations change over time. When do we know that distinctly new species have evolved?

    B. Critical to determining the limits of a species is understanding if two populations are truly reproductively isolated. For example, some populations of western and eastern spotted skunks overlap in the center of the United States (*Opening Essay*).

        1. In eastern spotted skunks, mating occurs in late winter, and the young are born between April and July.

        2. In western spotted skunks, mating occurs in late summer; development is arrested at an early stage during winter and begins again in spring, so that the young are born in May or June.

        3. Thus, their individual reproductive behavior is such that it would be impossible for gene flow to occur between the two species.

    C. What is a species? (*Module 15.1*)

1. A biological species is defined as a population or group of populations whose members have the potential to interbreed and produce fertile offspring.

   Review: Since the biological species concept (Module 14.6) concerns actually or potentially interbreeding populations, it is difficult to apply it to the fossil record. Another species concept that is more readily applied to the fossil record is the evolutionary species concept. G. G. Simpson defined the evolutionary species as "a lineage (an ancestral-descendant sequence of populations) evolving separately from others and with its own unitary evolutionary role and tendencies."

2. Species are given distinct scientific names by taxonomists. These names are in the form of binomials, the forms first used by Swedish physician and botanist Carolus Linnaeus in the 1700s. The binomial for the human species is *Homo sapiens.*

3. Two different species can appear to be almost identical (for example, the skunks in the opening essay or the western and eastern meadowlarks in Figure 15.1A).

4. A single species can exhibit considerable diversity of form (for example, humans, Figure 15.1B).

5. It is not always possible to determine if species are true biological species (reproductively isolated), in which case species are distinguished by their appearance and fossil evidence.

   NOTE: In addition to appearance and the fossil record, comparative anatomy, comparative embryology, and the molecular record are also used to distinguish species.

D. The distinction between populations and biological species often blurs (*Module 15.2*).

   1. When gene flow between populations is definitely impossible, two species exist.

   2. When there may be a chance for the flow of genes between two quite different populations, such as species of deer mice, it is difficult to apply the biological species concept to distinguish the populations (Master 15.2).

   3. In this case, subspecies designations are used for a situation that appears to show a large population in the middle of evolution to two isolated species (present subspecies *Peromyscus maniculatus artemisiae* and *P. m. nebrascensis*).

II. Barriers that lead to the origin of new species.

A. Reproductive barriers keep species separate (*Module 15.3*; Table 15.3).

   1. Prezygotic barriers prevent mating or fertilization: (a) differential timing of mating (temporal isolation), such as the skunk example, or the case with many plants that reproduce at different times; (b) reproductive habitat differences (habitat isolation), such as the differences between two related species of toad; (c) behavioral isolation may involve differences in display (an example are the fireflies discussed in the opening essay of Chapter 5) or pheromones; (d) mechanical or chemical isolation of the adult sexual structures or of the gametes (Master 15.3A; Figure 15.3B).

2. Postzygotic barriers prevent the development of fertile adults: (a) the hybrids do not live (hybrid inviability); (b) they are not fertile (hybrid sterility); (c) there is progressive weakening of successive generations of interbreeding hybrids (hybrid breakdown). (Figure 15.3C).

B. Geographical isolation can lead to speciation (*Module 15.4*).

1. Allopatric speciation involves changes in allele frequencies in two or more geographically isolated populations stemming from one initial population, and is most likely in a small isolated population.

2. Changes occur by microevolutionary processes (mutation, genetic drift, and natural selection).

3. Many factors can produce geographical isolation: mountain building, deep canyons, the removal of land bridges between continents, continental drift (Figure 15.4).

4. The effectiveness of barriers depends on how effective dispersal is in the organisms that might speciate. Large mammals may find it easy to cross mountain ranges, while small mammals may be stopped by a wide river.

5. Geographical isolation does not necessarily lead to speciation. Speciation occurs only after barriers to reproduction are established.

C. Islands are living laboratories of speciation (*Module 15.5*).

1. *Review:* The case of the increased incidence of hereditary deafness on the island of Martha's Vineyard shows how allele frequencies can change over relatively short periods of time on an island (Chapter 13, opening essay).

2. Such islands must be close enough together or to the mainland to allow for occasional dispersion but far enough apart to provide isolation most of the time.

3. Darwin's finches (13 closely related species, distinguished by morphology and habitat) of the Galapagos Island chain are excellent examples of the results of island speciation (Acetate 15.5B, C).

4. Evidence from further study suggests that the progression of speciation of four of the species occurred as indicated in Figure 15.5C.

5. A contrasting situation occurs on a more isolated island (Cocos), which has just a single, unique species of finch.

6. The emergence of numerous species from a common ancestor in one diverse environment (such as the Galapagos) is known as adaptive radiation.

III. **Other factors relating to speciation.**

A. New species can also arise within the same geographic area as the parent species (*Module 15.6*).

1. Such sympatric speciation seems to be rare among animal species, but has played an important role in plant evolution.

2. The most common type of sympatric speciation occurs when an accident during cell division results in an extra set of chromosomes (polyploidy).

3. Tetraploid plants can form by self-fertilization of diploid gametes in a flower where meiosis has not reduced the chromosome set (Acetate 15.6A).

4. The plant that grows from the tetraploid zygote can reproduce by self-fertilization but cannot produce fertile offspring by mating with its diploid ancestors because these offspring would be triploid.

5. Sympatric speciation by polyploidy was first discovered in evening primroses by Dutch botanist Hugo de Vries in the early 1900s (Figure 15.6B).

B. Polyploid plants clothe and feed us (*Module 15.7*).

1. Polyploid plants most often arise by the coupling of two events: the hybridization of two parent species followed by a failure of meiosis in the hybrid.

2. Polyploid hybrids often combine the best features of the two parents into the new form.

3. Scientists estimate that 25–50% of all plants are polyploid. Most commercially grown food and fiber plants are polyploid hybrids: oats, potatoes, bananas, peanuts, barley, plums, apples, sugarcane, coffee, wheat, and cotton.

   NOTE: It has been natural for humans to select, for practical purposes, the better strains (from the human perspective) from among the offspring of chance matings.

4. The recent evolutionary history of bread wheat (*Triticum aestivum*) is believed to have occurred by a series of steps involving hybridization, a failure of meiosis, an additional hybridization, and nondisjunction (Master 15.7B).

C. The tempo of speciation can appear steady or jumpy (*Module 15.8*; Acetate 15.8A, B).

1. In the gradualist model, populations isolated from the ancestral stock change slowly as their allele frequencies shift during adaptation by natural selection. Darwin's proposals incorporated this model.

   NOTE: In Figures 15.8A and B, the sizes of the arrows' bases represent the population sizes.

2. The fossil record does not support this model because new species seem to appear suddenly in rock strata, without intermediary transitional forms.

3. In the punctuated equilibrium model, periods of rapid evolutionary change (punctuation) and speciation are interrupted by long periods of little or no change (equilibrium; stasis).

4. If species last 5 million years, on average, then a period of 50,000 years (the age of Death Valley) represents only 1% of the total, a geological instant. The unique species of pupfish in each isolated spring in Death Valley are thought to have evolved over this time period (Figure 15.8C).

5. The fact that there is debate over which model best describes the process of speciation should not be taken as disagreement about the reality of evolution, but only about the relative commonness of two models of evolution with one model being more appropriate in certain instances and the other model being more appropriate in other instances.

6. At least some aspects of this debate arise from the human perception of time. For example, is a speciation event that takes place over several thousand years abrupt? Perhaps not, in our view. Yet, if a species survives

5 million years and for the last 4.95 million years of its existence it was in equilibrium, then the speciation event took up only 1% of the overall history of the species.

IV. Talking About Science: Evolutionary biologist Ernst Mayr connects Darwinism with the modern era (*Module 15.9*).

　A. Ernst Mayr is one of the modern biologists who helped synthesize a modern understanding of the microevolutionary principles proposed by population geneticists with Darwin's theory of evolution by natural selection.

　B. Like Darwin, he spent his first professional years as a naturalist, studying the distributions of species in New Guinea and the Solomon Islands.

　C. He first developed the idea of the founder effect in populations, a process that explains why the fossil record seems inconsistent with the gradualist view of evolution.

CLASS ACTIVITIES

1. Bring in examples of domesticated plants that are polyploid (Module 15.7). The following additional plants are polyploid: cotton (the long-fiber species grown most commonly is a polyploid originally from South America; other, diploid species are grown locally elsewhere), triticale (a man-made, polyploid hybrid between wheat and rye), tobacco, chrysanthemums, pansies, and daylilies. The evolutionary origins of domesticated animals and plants (in addition to wheat) make interesting stories. Botany textbooks often include details on the evolution of corn, cotton, and potatoes (and see resources below). Often these histories involve origins from organisms that are known in the wild today and involve hybridization and polyploidy (intentionally caused or not).

RESOURCES AND REFERENCES

Culotta, E. "How Many Genes Had to Change to Produce Corn?" *Science*, June 28, 1991.

Gould, Stephen Jay. "Opus 200." *Natural History*, August 1991. Traces the origin and impact of the theory of punctuated equilibrium.

Grant, Peter R. "Natural Selection and Darwin's Finches." *Scientific American*, October 1991. How a single drought can change a population.

Hancock, James F., and James J. Luby. "Genetic Resources at Our Doorstep: The Wild Strawberries." *BioScience*, March 1993. A review of current cultivars and a description of attempts to expand these cultivars' genetic backgrounds.

Hoffman, A. *Arguments On Evolution: A Paleontologist's Perspective*. New York: Oxford University Press, 1989. A gradualist's critique of the punctuated equilibrium model.

Kluger, K. "Go Fish." *Discover*, March 1992. Rapid speciation in Lake Victoria.

Knowlton, N. "A Tale of Two Seas." *Natural History*, June 1994. Allopatric speciation resulting from the Panamanian land bridge.

Marshall, Larry G. "Land Mammals and the Great American Interchange." *American Scientist*, July/August 1988. The emergence of the Panamanian land bridge 3 million years ago permitted the mingling of the long-separated faunas of North and South America.

National Center for Science Education. The NCSE is a clearinghouse for scientific information on the evolution-creation controversy. See Appendix B for further information.

Raloff, Janet. "Corn's Slow Path to Stardom." *Science News*, April 17, 1993. Recent advances in knowledge about early varieties of cultivated corn. Mostly from an archeological perspective, but includes some details of the search for genetic ancestors.

Rennie, J. "Are Species Specious?" *Scientific American*, November 1991. Problems with the concept of biological species.

Rennie, J. "Darwin's Current Bulldog." *Scientific American*, August 1994. Ernst Mayr, Darwin scholar and a codeveloper of the modern synthesis.

SimLife—The Genetic Playground. Moraga, CA: Maxis, 1993. Available from Education Express. A learning version of the popular computer software simulation that enables the user to design organisms and manipulate pseudogenes and environmental parameters. IBM PC and Macintosh versions.

Storch, Gerhard. "The Mammals of Island Europe." *Scientific American*, February 1992. A mine at Messel in Germany has yielded magnificently preserved fossils of mammals that roamed Europe when it was an island. They clarify a key phase in evolutionary history.

Wayne, R.K., and J.L. Gittleman. "The Problematic Red Wolf." *Scientific American*, July 1995.

16

TRACING EVOLUTIONARY HISTORY

APPROACH

Tracing the evolution of life across the face of the Earth makes a fascinating story, one that ties together various details of the last two chapters and new material on geological timing and forces presented here. Macroevolution, though it takes place over inconceivably long periods of time, involves slowly dynamic changes. If possible, use video, simulations, or animation software to demonstrate some of these events, particularly the features of continental drift.

The material in this chapter shows how the separate sciences of biology and geology and their hybridization, paleontology, build on each other. In addition, students should begin to see how numerous details about fossils, knowledge of present organisms, and other information are all assembled into coherent pictures of ancient life and the likely course of its evolution. Be sure to spend some time discussing the story in the opening essay, as it emphasizes an appreciation of the ecological context of ancient life.

A review of systematics can be enhanced by concrete examples of organisms (see the Class Activities). You might want to present systematics in the context of the practical need for a system of organization. Interesting parallels can be made to other systems of categorizing life forms. For instance, systems developed by indigenous, agrarian cultures often include very fine variety-level categories for wild and cultivated plants.

Students with certain religious convictions may object that evolution is just a theory. You might explain that evolution is indeed a fact. However, like scientific theories in general, the mechanisms by which evolution occurs are explanatory ideas that are broad in scope and supported by a large body of experimentation and evidence. You might add that evolution describes aspects of the natural world but does not attempt to answer philosophical or religious questions about meaning.

CHAPTER OBJECTIVES

Distinguish between the terms *macroevolution* and *microevolution*.

Name the eras in the geological timeline, and indicate the major groups of organisms that thrived during each of these eras.

Explain how major geological and astronomical events have likely influenced the course of macroevolution, stressing the extinction of the dinosaurs at the end of the Cretaceous period and the hypotheses proposed for explaining this extinction.

Examine some of the general trends viewed in evolutionary history, such as the importance of preadapted traits, paedomorphosis, and the gradual changes of traits in one direction.

Show how phylogenetic trees, hierarchical classification systems, and data from molecular comparisons help systematists organize data describing evolutionary history.

Define the terms *homologous* and *analogous*, and explain the cladistic and classical interpretations of phylogenetic data.

Outline the distinguishing characteristics of the five kingdoms.

LECTURE OUTLINE

I. Introduction.

A. Piecing together the details of the life of extinct forms, like the duck-billed dinosaur, involves integrating knowledge about many prehistoric aspects such as geology, climate, and fossils, as well as knowledge about comparable present-day species (*Opening Essay*).

B. Fossils are particularly important. Sometimes, as in the case described, a particularly important fossil deposit can provide special insight into how ancient life forms behaved and otherwise lived (chapter-opening photo).

1. Tracks of hadrosaurs indicate they lived in large herds on outwash plains.

2. Details of skull anatomy show large ears and unusual beaks, suggesting these animals were quite vocal.

3. Fossilized nests, egg fragments, and young show that hadrosaur parents remained with their nests and cared for their young.

II. The fossil record.

A. The fossil record chronicles macroevolution (*Module 16.1*).

Review: The fossil record (Module 14.2).

1. Macroevolution, the main events in the evolutionary history of life on Earth, is determined by comparing the fossil records in strata representing various ages, from various parts of the Earth's surface.

2. The geological timeline is a standardized, hierarchical system of age categories (Table 16.1).

3. The oldest fossils are of microorganisms from 3.5 billion years ago during the early Precambrian era.

4. Late Precambrian fossils show that animal life had diversified by 670 million years ago (mya).

NOTE: The beginning of the Paleozoic, 590 mya, was originally defined in the 1800s by the first presence of macrofossils in strata of that age. Later study has shown fossilized animals, mostly with soft body parts in late Precambrian strata, and microfossils in much older rocks.

5. By 400 mya, during the middle Paleozoic ("ancient animal") era, life had moved out of water and onto dry land.

6. The Mesozoic ("middle animal") era began 248 mya and is the age of dinosaurs and cone-bearing plants.

7. The Cenozoic ("recent animal") era began 65 mya and is the age of mammals and flowering plants.

B. The actual ages of rocks and fossils mark geological time (*Module 16.2*).

1. The record of fossils in rock strata chronicles the relative ages of life.

2. The actual ages of fossils can be obtained by radioactive dating. Radioactive isotopes "decay" at a known rate relative to other isotopes. For instance, half of an amount of ^{14}C decays to ^{12}C in 5600 years. Measuring the relative amounts of the two isotopes in a sample (and comparing this ratio to the ratio known to have been in the original organism, that is, the ratio of ^{14}C to ^{12}C in the atmosphere) gives the actual

age of the sample, with an error factor of about 10%. Elements with longer half-lives are used to date older fossils.

III. Geological and astronomical influences in macroevolution.

  A. Continental drift has played a major role in macroevolution (*Module 16.3*).

   1. Continental drift was a mechanism first proposed in 1912 by German meteorologist Alfred Wegener to explain the similarities of coastal outlines of present-day continents. The proposal was not accepted because geologists knew of no method that would cause continents to move.

   2. Continents are the above-water parts of crustal plates that "float" on the lower fluid mantle. New crust is formed along ocean ridges, and old crust is destroyed at the leading margins of the plates (Acetate 16.3A).

   3. During the past 250 million years, the present continents have drifted apart from a previous supercontinent called Pangaea, first the north (Laurasia) and south (Gondwana) areas, and later the east and west (Acetate 16.3B).

   NOTE: During the Precambrian and Paleozoic eras, there were other, smaller plates that drifted widely, finally coming together as Pangaea.

   4. Recent continental drift during two time periods strongly influenced macroevolution: (a) As Pangaea coalesced, coastlines and their marine flora and fauna disappeared, and continental climates changed in ways that affected terrestrial organisms; (b) about 180 mya, the splitting up of Pangaea formed large, population-isolating landmasses.

   5. Relative distributions of present-day life forms and their fossilized ancestors are explained by the known course of recent continental drift (Master 16.3D).

   6. Continental drift is an ongoing process. For example, the ongoing collision in the Himalayan region is creating forces that are splitting the Indo-Australian plate, resulting in Australia moving independently of India.

   NOTE: The example that will be most obvious to your students are the California earthquakes (Module 16.4).

  B. Tectonic trauma imperils local life (*Module 16.4*).

   1. At shorter time and spatial scales, the sudden earth movements resulting from continental drift can affect populations.

   2. Earthquakes along faults between plates can disrupt drainages (Master 16.4A).

   3. Volcanic eruptions occur along plate margins or mid-ocean ridges and can build mountains or islands, such as the Galapagos, but can also pose a threat to local populations (Figure 16.4B).

  C. Mass extinctions were followed by diversification of life forms (*Module 16.5*).

   1. At the end of the Cretaceous period (65 mya), many lineages of terrestrial plants and animals, and about half the marine animals, became extinct.

   2. Particularly noteworthy was the demise of the dinosaurs, which had dominated the land and air for 150 million years during the Mesozoic (Master 16.5).

3. Several, not necessarily mutually exclusive, explanations have been proposed to account for this change: asteroid impact (now confirmed to have happened at the right time in the Caribbean); slow changes in climate due to continental drift; massive volcanic activity in India that contributed to cooling.

4. Another major extinction occurred at the end of the Permian period.

NOTE: The ends of the major eras (Paleozoic and Mesozoic) discussed in Module 16.3 correspond to the major extinction events. The dividing lines between the periods correspond to other major changes in fossil assemblages.

IV. Some important macroevolutionary trends.

A. Key adaptations may allow species to survive and proliferate after mass extinctions (*Module 16.6*).

1. Each of the six periods of mass extinction in the past 600 million years has been followed by an "explosion" in evolution of certain groups of organisms.

2. Chance can play a role; an organism just happens to "make it" in the right place.

3. Preadapted features arise in populations in one context but may cause organisms to be particularly well adapted to changed environments.

4. For example, the spread of bromeliads into aerial environments depended on the ancestral forms having developed catch-basins, formed from leaf bases, and water-absorbing trichomes on leaf margins (Figures 16.6A, B).

5. Changes in how organisms develop, which may involve only a few regulatory genes, can also lead to adult features that are very different and that may offer advantages in certain environments.

Review: These regulatory mechanisms are discussed in Modules 11.12–11.14.

B. Delayed maturity was a key novelty in human evolution (*Module 16.7*).

1. Paedomorphosis is the retention of juvenile body features in the adult (Figure 16.7A).

2. Paedomorphosis has been important in the evolution of humans and chimpanzees from a common ancestor. The large, paedomorphic human skull and the long period of time as a nonreproductive child provide the human with both space for a larger brain and time to learn from adults (Master 16.7B).

Preview: Human evolution is discussed in greater detail in Modules 38.2–38.11.

3. Evolutionary biologist Stephen Jay Gould contends that youthful characteristics in children elicit parental affection and care. He uses Mickey Mouse's early evolution as a cartoon character to illustrate this (Figure 16.7C).

C. What accounts for evolutionary trends? (*Module 16.8*)

1. Evolutionary trends show gradual, one-directional changes in morphology over long periods of evolutionary time, such as the increase in brain complexity among human ancestors and the increase in size and gradual development of horns among extinct titanotheres (Master 16.8A).

2. Unequal survival of new species can explain this (Master 16.8B).

3. Unequal speciation with equal survival of all new species can also explain the data. Current debate exists over the relative importance of each of these mechanisms.

4. Evolutionary trends are not preordained or unchangeable. Such trends can stop or reverse (extinction of the titanotheres, for example).

V. The biological science of systematics.

A. Phylogenetic trees symbolize evolutionary history (*Module 16.9*).

1. Phylogeny is the evolutionary history of a group of organisms.

2. Phylogenetic trees represent the most likely phylogeny of a group, based on certain evidence. For instance, the phylogeny of Galapagos finches in Figure 16.9 is based on body structures, especially beak structure, and field studies of reproductive isolation and feeding behavior.

3. Each branch axis represents the evolution of subsequent groups based on some important feature.

4. Organisms on branches close to the base of the tree are more primitive (only in the sense of having appeared earlier in time), showing more ancestral features than organisms on later branches.

B. Systematists classify organisms by phylogeny (*Module 16.10*).

1. Reconstructing phylogenies, assigning scientific names, and classifying the names are all aspects of the biological science of systematics.

2. Common names can be ambiguous because there are so many species and because different people use different names for the same species.

3. Linnaeus devised the binomial form for a species' scientific name (genus name plus species name—for example, *Canis lupus*, the wolf) and a hierarchical system of progressively broader categories (Table 16.10).

NOTE: The species name is always italicized or underlined.

4. Ideally, the species taxon is based on a real group in nature (biological species), but all the other, larger taxa are determined according to systematists' understanding of relationships between and among species.

5. Although identifying species often requires judgment calls, classifying species into higher taxa always does.

6. It has always been a goal of taxonomists to have their taxonomic systems reflect the evolutionary relationships and phylogeny of whatever groups are in the system (Acetate 16.10).

7. Systematics reflects the hierarchical nature of biology. The major levels of classification, from most to least inclusive, from a lesser degree of relationship to a greater degree of relationship, are Kingdom, Phylum, Class, Order, Family, Genus, and Species.

NOTE: A good mnemonic for this hierarchy is "King Phillip Come Out For Goodness' Sake."

C. Homology indicates common ancestry, but analogy does not (*Module 16.11*).

1. Homologous structures may function differently, such as the wing of a bird, the arm of a human, and the flipper of a whale, but they exhibit

fundamental similarities—in this example, the bones supporting these appendages.

Review: The concept of homology is introduced in Module 14.3.

2. The greater the number of homologies between two species, the more closely related they likely are.

3. Analogous structures may look and function the same, but they do not exhibit the fundamental similarities that reflect common ancestry (Figure 16.11). Instead, they more likely reflect the results of convergent evolution.

4. Convergent evolution is a process by which different evolutionary lineages evolve similarities as a result of similar regimes of natural selection.

NOTE: A good example of convergent evolution as a result of common selective pressures in a common environment is the evolution of torpedo-shaped bodies in sharks (a fish), penguins (a bird), and dolphins (a mammal).

D. Molecular biology is a powerful tool in systematics (*Module 16.12*).

1. Bears and raccoons have long been recognized to be closely related mammals based on morphology, but such relationships can be clarified by analyzing DNA and proteins. This example is based on DNA-DNA hybridization and blood protein amino acid sequencing data. The data suggest that lesser pandas are more closely related to raccoons than to bears (Master 16.12A). Another method used to clarify such relationship is to directly compare the DNA of the two different species by means of DNA sequencing.

2. Comparing the amino acid sequence in human cytochrome *c* and that of other animals: humans differ from chimpanzees by 0 amino acids, from rhesus monkeys by 1 amino acid, from dogs by 13 amino acids, from rattlesnakes by 20, and from tuna by 31.

3. Other sequence comparisons can be made of DNA nucleotides from nuclei and from mitochondria.

4. DNA-DNA hybridization measures the extent of hydrogen bonding between fragments extracted from two or more comparable species (Master 16.12B).

E. Systematists debate how to interpret phylogenetic data (*Module 16.13*).

1. Cladistic taxonomy is concerned only with the order of branching in phylogenetic lineages. Each branch on a cladogram represents the most recent ancestor common to all the taxa beyond that point. All the taxa above a branch share some one or more homologous features (Master 16.13A).

2. Homologous characters shared by a group of species and their common ancestor are called primitive characters. For example, the common ancestor of all vertebrates had five toes; therefore, the presence of five toes is a primitive character.

3. Homologous characters unique to each lineage are called derived characters.

4. Derived characters are used to identify the branch points of a cladogram. For example, hair and mammary glands are derived characters that distinguish the mammalian lineage.

5. Cladistics is particularly suited for analysis of the similarities and differences of molecular data, which may be done entirely objectively.

6. Classical evolutionary taxonomy also takes into account the apparent degree of divergence of taxa (Acetate 16.13B).

VI. Arranging life into kingdoms is a work in progress (*Module 16.14*).

A. Linnaeus used a two-kingdom system to categorize life at the most inclusive level of classification.

B. In 1969, American ecologist Robert Whittaker proposed a five-kingdom system, largely to split up the kingdom Plantae, which in the two-kingdom system contained organisms such as fungi and bacteria that were definitely not closely related to plants (Acetate 16.14).

1. Monera: Prokaryotes.

2. Protista: All eukaryotes that do not fit the definition of plants, fungi, or animals. Usually unicellular, but contains relatively simple multicellular organisms that are thought to be direct descendants of the unicellular protists.

   NOTE: The multicellular protists are highly variable in the degree to which their cells are cooperative/specialized. A sequence can be set up that reflects hypothesized intermediates from unicellularity to true multicellularity.

3. Plantae: Plants are complex eukaryotes that make their own food by photosynthesis.

4. Fungi: Mushrooms and molds are complex eukaryotes that have threadlike cells. Fungi decompose the remains of other organisms and absorb the resulting small organic molecules; formerly in Plantae.

5. Animalia: Animals are complex eukaryotes that ingest the food they require.

6. Other systems have been proposed, and the current five-kingdom system is likely to soon be replaced. The five-kingdom system is based on the distinction between two basic cell types, the prokaryotes and the eukaryotes. Figure 16.4B illustrates a scheme based on three main groups: two prokaryotic, the Eubacteria and the Archaebacteria, and one eukaryotic, the Eukarya. Distinctions between the two prokaryotic groups are discussed in Chapter 17. Another system proposes splitting the kingdom Protista into at least three separate kingdoms.

CLASS ACTIVITIES

1. Introduce and follow up your discussions of species and systematics with this demonstration. (It can be used in classes as large as 100.) Bring in the following varieties of *Brassica oleracea* (Latin *Brassic*, cabbage, and *oler*, greens) and other vegetables and ask students to categorize them into genus and species: kale and Napa cabbage (leaves versus *acephala*, Greek for "no head"); cabbage (leaves versus *capitata*, Latin for head); broccoli, cauliflower, and broccoflower (immature flowers versus *botrytis*, Greek for bunch of grapes); brussels sprouts (leaves on lateral stems versus *gemmifera*, Latin for bud-bearing); and kohlrabi (thickened stem versus *Caulo-Rapa*, Latin for stem turnip). These are all artificial varieties, bred for the traits indicated. Wild-type *Brassica oleracea* grows on the sea cliffs of Europe. Compare it with turnip (*Brassica rapa*), bok choy (*Brassica chinensis*), wild mustard (*Brassica arvensis* and *Brassica campestris*, both from the Latin for field), and lettuce (*Lactuca sativa*). Students will probably group species

according to the varieties and their look-alikes. Once everyone is certain of the species arrangement, show them the actual groupings. Obvious traits are not always important in defining species (in this case, details of flowers define genera, and details of leaf arrangement on stalks define species of *Brassica*). If you have time, discuss the scientific names. This particular set is good for reinforcing the idea that scientific names do mean something (if you know Latin or Greek). Remind students of the role that artificially bred varieties of plants and animals played in the development of Darwin's ideas (see Module 14.4).

2. Have the students imagine what the dominant life form might now be if the Cretaceous mass extinction had not taken place (Module 16.5). Would a dinosauran lineage have evolved humanlike intelligence? What would have happened to that insectivore lineage whose ancestors evolved into modern humans?

RESOURCES AND REFERENCES

Alvarez, W., F. Asaro, and V. Courtillot. "What Caused the Mass Extinction?" *Scientific American*, October 1990. Two articles on asteroids versus volcanoes as the main cause of Cretaceous extinctions.

Bambach, Richard K., Christopher R. Scotese, and Alfred M. Ziegler. "Before Pangaea: The Geographies of the Paleozoic World." *American Scientist*, January/February 1980. Pre-Pangaean (from the late Cambrian) configurations of continents and oceans reconstructed according to knowledge of geological and biological processes; a very useful resource.

Barnes, R.S.K., ed. *A Synoptic Classification of Living Organisms*. Sunderland, MA: Sinauer Associates, 1984. A comprehensive collection of brief, diagnostic descriptions down to the level of order for all phyla in each of the five kingdoms; the work of several authors attempting a consensus; and an up-to-date view of the interrelationships among the organisms in the phyla.

Bishop, Beth A., and Charles W. Anderson. "Student Conceptions of Natural Selection and Its Role in Evolution." *Journal of Research in Science Teaching*, May 1990. Results from tests administered to nonscience majors describing the effects of instruction and belief in evolution on student conceptions.

Carroll, S.B. "Homeotic Genes and the Evolution of Arthropods and Chordates." *Nature*, August 10, 1995. Duplication of homeotic genes may have played a key role in the evolution of animals.

Cherfas, J. "Ancient DNA: Still Busy After Death." *Science*, September 20, 1991. Molecular biologists go to work on fossils.

Cloud, Preston. *Oasis in Space. Earth History from the Beginning*. New York: Norton, 1988. A wonderful amalgamation of geological and biological history. Includes considerable maps, charts, and illustrations of changing paleogeographies and other physical, chemical, and geological influences on the evolution of life.

Dalziel, I.W.D. "Earth Before Pangaea." *Scientific American*, January 1995. Overview of continental movements.

Dawkins, R. "God's Utility Function." *Scientific American*, November 1995. An argument that the diversity of life is the result of a survival contest among selfish genes.

Disc-O-Saurus. Ottawa, Canada: Timebox, 1994. A CD-ROM that provides innovative access to a large database on dinosaurs, including descriptions and illustrations for several hundred species, animated sequences of feeding activity, and animated maps of geological forces involved in their evolution. Macintosh version only.

Ezzell, C. "Conserving a Coyote in Wolf's Clothing? Molecular Systematics and Endangered Species." *Science News*, June 15, 1991.

Gore, Rick. "Extinctions." *National Geographic*, June 1989. Nicely illustrated with timelines and photographs; a discussion of current ideas about what caused the major extinctions, including the recent, human-caused increase in extinction rate.

Gould, Stephen Jay. "We Are All Monkey's Uncles." *Natural History*, June 1992. Cladists and human classification.

Lipscomb, Diana. "Broad Classification: The Kingdoms and the Protozoa." *Parasitic Protozoa*, Volume 1, 2nd ed. New York: Academic Press, 1991. A review of classical and modern approaches to classifying all the kingdoms, with emphasis on the protozoan protists.

Margulis, Lynn, and Dorian Sagan. *Microcosmos: Four Billion Years of Microbial Evolution*. New York: Simon & Schuster, 1986. Evolution from the perspective of microorganisms.

Margulis, Lynn, and Karen V. Schwartz. *Five Kingdoms: An Illustrated Guide to the Phyla of Life on Earth*, 2nd ed. New York: W. H. Freeman, 1987. Capsule summaries of the characteristics and roles, with illustrations of important members in each phylum, plus a discussion about each kingdom.

17

THE ORIGIN AND EVOLUTION OF MICROBIAL LIFE: PROKARYOTES AND PROTISTS

APPROACH

In addition to tracing the evolution of the first cells, this chapter introduces and begins the kingdom survey that continues through the end of this unit. The approach emphasizes major evolutionary steps and major ecological aspects within each of the five kingdoms. By way of example, representative species are discussed. The lecture outlines follow the modules closely, surveying all groups mentioned.

It is important to tell students how much and what kind of detail from these chapters you expect them to assimilate: how much subgroup taxonomy, how many representative organisms, and so on. It is very easy for students to lose sight of the forest for all the trees in a survey. You may wish to omit certain details in the text and add some of your own, depending on the time available, your own knowledge of the groups, and their importance in your region.

Exhaustive lists of different groups and their characteristics will be intimidating to nonscience majors. If you are going to cover this material in one lecture, plan on discussing only a few groups of bacteria and protists, emphasizing their uniqueness, their distinguishing characteristics, their evolution, and their ecological importance.

Particularly for this chapter and the next two chapters, where students have less direct experience with the organisms, provide as many practical connections and examples as possible. The life forms discussed in this chapter often are poorly known by people, yet they play crucial, often pivotal roles in environments. To spark interest in these intriguing organisms, make use of videos and computer software that show dynamic aspects of processes and species.

CHAPTER OBJECTIVES

Describe the abiotic environment in which life probably began on the early Earth, and trace a likely scenario for the evolution of the first (prokaryotic) cells, including some of the experiments that support this evolutionary process.

Outline the characteristics of the kingdom Monera and its overall impact on environments.

Define the terms *autotrophic, chemoautotrophic, photosynthetic, heterotrophic,* and *chemoheterotrophic* as nutritional modes.

Describe the archaebacteria, explaining why they may better be classified in a sixth kingdom.

Name some eubacteria, and describe their unique characteristics and activities.

Trace a likely scenario for the evolution of the first eukaryotic cells.

Outline the characteristics of the kingdom Protista, its general subcategorization, and its overall impact on environments.

Name some protists, and describe their unique characteristics and activities.

Trace a likely scenario for the evolution of the first multicellular eukaryotes.

LECTURE OUTLINE

I. Introduction.

A. The evolution of life has had a profound effect on the Earth (*Opening Essay*).

1. Photosynthetic cyanobacteria evolved very early in the history of life and left unique fossilized communities as stromatolites.

2. Modern-day cyanobacteria of this type, less common because of predation, are virtually indistinguishable from the early forms (chapter-opening photo).

3. In addition to being the ancestors of today's cyanobacteria, these first cyanobacteria produced the Earth's first oxygen-rich atmosphere.

B. This chapter begins a survey of all of Earth's life forms in an evolutionary context, beginning with the very big steps of the evolution of life itself.

NOTE: The earliest organisms were ancestors of modern prokaryotes. They and the first eukaryotes (early protists) have inhabited our planet for a much longer time than members of any other kingdom. Through the evolutionary events that resulted in their great diversity, all existing metabolic reactions evolved, at least in an early form.

C. Life began on a young Earth (*Module 17.1*; Master 17.1C).

1. The age of the universe is estimated to be between 10 and 20 billion years old, while Earth coalesced from gathering interstellar matter about 4.5 billion years ago (bya).

2. Volcanic gases released water vapor, carbon monoxide, carbon dioxide, nitrogen, methane, and ammonia into the early atmosphere.

3. Earth's crust cooled and solidified about 4.1 bya, condensing water vapor into early seas.

4. Fossil evidence shows early cyanobacteria by 3.5 bya (Figures 17.1B, D).

NOTE: The immensity of geological time and the very early events discussed can be made more meaningful by putting them in perspective. Borrowing an idea used by many, use a geologic time scale divided into a "life-on-Earth year." On such a scale, prokaryotic life evolves in mid-March, eukaryotes first appeared around September 1, dinosaurs flourished around Christmas, and the typical human life span of 70 years is represented by the last half-second on December 31.

5. Because cyanobacterial photosynthesis is complex and advanced, the first cells likely evolved earlier, perhaps as early as 4.0 bya.

II. How did life originate? (*Module 17.2*)

A. Some early writers believed life arose by spontaneous generation.

1. Experiments in the 1600s showed that larger organisms cannot arise spontaneously from nonliving matter.

2. In the 1860s, French scientist Louis Pasteur confirmed that all life today, including microbes, arises only from preexisting life.

B. Most biologists subscribe to the hypothesis that the earliest life forms were simpler than any that exist today, and that they evolved from nonliving matter.

    1. Although extraterrestrial organic molecules could have seeded Earth's early environment, most scientists think these molecules arose from nonorganic molecules present in the early oceans and atmosphere.

    2. A possible scenario: Monomers evolve first, then polymers, then aggregates that eventually formed in the particular arrangement that allowed simple metabolism and self-replication. Data supporting the likelihood of many of these steps exist from a number of experiments.

    NOTE: The following modules detail some of the experimental evidence and theory supporting the steps in this scenario. No one has completed all the steps in order, nor are they likely to soon. Today's environment (even in laboratories) is very different from the environment of the early Earth. Huge amounts of time are needed for these complex developments to occur.

C. Talking About Science: Stanley Miller's experiments showed that organic molecules could have arisen on a lifeless Earth (*Module 17.3*).

    1. In the 1920s, Russian biochemist A. I. Oparin and English geneticist B. S. Haldane proposed that organic chemistry could have evolved in the early Earth's environment because it contained no oxygen and was reducing.

    2. An oxidizing environment (like Earth's O_2-rich environment today) is corrosive, tending to break molecular bonds. Thus, life could not spontaneously arise today on Earth.

    3. A reducing environment tends to add electrons to molecules, building more complex forms from simple ones.

    4. In 1953, American chemist Stanley Miller tested this hypothesis using an artificial mixture of inorganic molecules (H_2O, H_2, CH_4, and NH_3) in a laboratory environment that simulated conditions on the early Earth (Acetate 17.3B).

    5. Within days, the mixture produced amino acids, some of the 20 amino acids that are found in organisms today (see Module 3.12).

    6. More recent experiments, using modifications of Miller's setup to more closely mimic the early Earth's environment, have produced all 20 naturally occurring amino acids, sugars, lipids, and nitrogenous bases of nucleotides.

D. The first polymers may have formed on hot rocks or clay (*Module 17.4*).

    1. *Review:* Polymerization occurs by dehydration synthesis (Module 3.3).

    2. Although biological polymerization occurs enzymatically in organisms today, the reactions can also occur when dilute solutions of monomers are dripped on hot mineral surfaces (heat forces the dehydration synthesis) or on clays (electric charges concentrate monomers, and metallic atoms act as catalysts).

    3. American biochemist Sidney Fox has made polypeptides ("proteinoids") from mixtures of amino acids dripped on hot mineral surfaces.

E. The first genetic material and enzymes may both have been RNA (*Module 17.5*).

    1. *Review:* The flow of genetic information from DNA to RNA to protein is intricate and probably did not evolve as such (see Module 10.15).

The Origin and Evolution of Microbial Life: Prokaryotes and Protists

2. The essential difference between cells and nonliving matter is replication.

3. A number of lines of reasoning and some experiments support the hypothesis that the first genes may have been made of RNA. Short RNA molecules have been created in test tubes without cells or enzymes from precursor nucleotides. Some of these sequences will self-replicate if placed with additional monomer nucleotides. Further, some RNAs can act as enzymes (ribozymes), even one that catalyzes RNA polymerization (Acetate 17.5; Module 11.9).

4. The hypothetical period in the evolution of life, when RNA played the role of both genetic material and enzyme, is termed the RNA world.

F. Molecular cooperatives enclosed by membranes probably preceded the first real cells (*Module 17.6*).

1. Life requires the close and intricate cooperation of many different polymers.

2. Experimental evidence shows that proteinoids and lipids self-assemble into microspheres, fluid-filled droplets with semipermeable, membranelike coatings. Though not alive, these microspheres grow by the attraction of additional proteinoids and divide when they reach a certain maximum size (Figure 17.6B).

3. Early molecular cooperation may have involved a primitive form of translation of polypeptides directly from genes in RNA. If these cooperating molecules were incorporated into a microsphere, the basic structures for self-replicating cells would be present (Acetate 17.6A, C).

4. At this point, a primitive form of natural selection would favor those molecular co-ops that were most efficient at growing and replicating.

III. Characteristics of prokaryotes.

A. Prokaryotes have inhabited the Earth for billions of years (*Module 17.7*).

Review: Prokaryotic cells (Module 4.4).

1. Fossil evidence shows that prokaryotes were the first living things on Earth 3.5 bya, and they evolved alone for the following 2 billion years.

2. Prokaryotes are ubiquitous, numerous, and small, surviving in environments that are too hot, cold, acidic, salty, or alkaline for any eukaryote (Figure 17.7).

3. Despite being small, prokaryotes influence all other life—as the cause of disease and other problems, as benign inhabitants of all environments, and, more commonly, in beneficial relationships with all other living things.

4. Probably the most essential activities carried out by prokaryotes are the numerous ways they function in the decomposition of the dead remains (cellular and molecular) of other organisms.

B. Archaebacteria and eubacteria are the two main branches of prokaryotic evolution (*Module 17.8*).

Review: Classification of prokaryotes is first discussed in Module 16.4.

1. The main differences between these two groups concern their rRNA, RNA polymerase, introns, antibiotic sensitivity, cell wall composition, and membrane lipid structure (see Table 17.8).

2. In some of these differences (some rRNA sequences, RNA polymerase) the archaebacteria are more similar to eukaryotes than to eubacteria.

3. The differences between the eubacteria and archaebacteria are so great that many biologists feel that they should be placed in separate kingdoms.

   NOTE: When viewed through a microscope, these two groups look similar.

C. Prokaryotes come in a variety of shapes (*Module 17.9*).

1. Cocci are spherical and often occur in defined groups of two or more (Figure 17.9A).

2. Bacilli are rod-shaped and usually occur unaggregated (Figure 17.9B).

3. Vibrios resemble commas, and spirilla and spirochetes (Figure 17.9C) are spiral-shaped.

D. Prokaryotes obtain nourishment in a variety of ways (*Module 17.10*; Table 17.10).

1. *Review:* Cellular respiration, fermentation (Chapter 6), and photosynthesis (Chapter 7).

2. Modes of nutrition refer to how organisms obtain energy and carbon.

3. Autotrophs are "self-feeders" that make carbon compounds from the carbon in CO_2 and the energy in sunlight (photoautotrophs) or inorganic compounds (chemoautotrophs).

   Preview: Chemoautotrophic bacteria living in hydrothermal vents are discussed in the introduction to Chapter 34.

4. Examples of photosynthetic prokaryotes include cyanobacteria that use H_2O as a source of electrons and release O_2 as a waste product, and several other groups that use other electron sources, such as H_2S (Figure 17.10A).

   NOTE: These bacteria do not release O_2 as a waste product (those using H_2S release S) and are probably like the first photosynthetic bacteria.

5. Heterotrophs are "other-feeders" that make carbon compounds from the carbon in existing organic compounds and obtain energy from those same compounds (chemoheterotrophs) or from sunlight (photoheterotrophs).

6. *E. coli* is an important chemoheterotroph that lives in the human intestine, that can live on simple sugars alone (Figure 17.10).

   NOTE: Although chemoheterotrophs are the most common nutritional types of organism, an important aspect of the prokaryotes is that, taken as a whole, they exhibit more kinds of metabolic activities (nutritional and otherwise) than are found in organisms in all other kingdoms combined.

7. With generation times as short as several hours or less, prokaryotic populations can multiply exponentially as long as there is a ready supply of nutrients.

E. The first cells probably used chemicals for both carbon and energy (*Module 17.11*).

1. Absorbing nutrients from the environment is chemically simpler than synthesizing them anew; thus the first organisms are most likely to have been heterotrophs.

2. A possible scenario for the evolution of an early form of chemoheterotrophic nutritional metabolism: (a) free ATP is used as a source of energy for early cells; (b) ATP is used up faster than it can be regenerated by

abiotic processes; (c) the evolution of proteins (enzymes) that catalyzed the synthesis of ATP occurs, using energy stripped from other organic compounds; (d) the step-by-step evolution of glycolysis occurs (Master 17.11).

    3. Another, more popular, hypothesis assumes that little ATP was available in the primordial organic soup. In this scenario, the organism would have used CO_2 as a carbon source. Energy would have been obtained by chemical reactions involving sulfate and iron compounds. Such an organism would have been a chemoautotroph that generated its ATP through a primitive form of chemiosmosis (Module 6.12).

F. Archaebacteria thrive in extreme environments (*Module 17.12*).

    1. Archaebacteria and eubacteria are likely to have diverged from each other early in the history of life.

    2. The halophiles thrive in salt flats and salty lakes.

    3. The thermoacidophiles thrive in hot springs at temperatures close to boiling and at low pH.

    4. The methanogens are a group of anaerobic, methane-producing bacteria that thrive in some vertebrate intestines and in the mud of swamps.

    5. Methanogens are the organisms responsible for the production of marsh gas and are a major contributor to flatulence in humans. Methanogens also digest cellulose in the gut of animals such as cattle and deer.

IV. Diverse structural features help eubacteria to thrive almost everywhere (*Module 17.13*).

A. Flagella.

    1. Flagella can be either scattered over a cell or in bunches at one or both ends.

    2. Size, structure, and function differ from those aspects of eukaryotic flagella.

    3. They are composed of protein in two parts: external, nonmembrane-bounded filaments and rotating rings embedded in the plasma membrane and cell wall (Master 17.13A).

    4. Motion is produced as they spin on their axes like propellers.

B. Pili.

    1. *Review:* The role of sex pili in conjugation (Module 12.2).

    2. Pili are protein filaments thinner than bacterial flagella (Figure 17.13B).

    3. Pili help bacteria stick to each other or to surfaces in their environments.

C. Endospores.

NOTE: Because this is the first time the term *spore* is used, you might want to define spores as single-celled dispersive structures, similar in function to seeds.

    1. Thick-walled spores are formed inside the parent cell walls around a replicated copy of DNA (Figure 17.13C).

    2. Endospores are extremely resistant to decomposition and disintegration.

3. *Clostridium botulinum* is a bacterium that grows in anaerobic, low-acid environments, such as poorly canned vegetables. The toxin released by colonies of this bacterium causes botulism when consumed by humans.

D. Hyphae.

NOTE: The term *hypha* is not introduced until Module 18.17, but it is entirely suitable here.

1. Hyphae are tubular filaments from which actinomycetes are constructed. They may be branched and/or divided into many compartments by cross walls (Figure 17.13D).

2. Actinomycetes are important chemoheterotrophic soil eubacteria. Some are commercial sources of antibiotics (for example, streptomycin from *Streptomyces* eubacteria).

V. Additional roles played by prokaryotes.

A. Cyanobacteria sometimes "bloom" in aquatic environments (*Module 17.14*).

1. Blooms are population explosions of microorganisms (usually photosynthetic) in freshwater lakes and marine bays.

2. A bloom of a red species of cyanobacterium gives the Red Sea its name.

3. Large blooms of cyanobacteria indicate that a lake is polluted (Figures 17.14A, B). Such blooms often indicate pollution by organic wastes, such as phosphates and nitrates, from agricultural runoff.

NOTE: Blooms of cyanobacteria can release large amounts of toxins that kill fish. Blooms of all sorts of microorganisms can use up O_2 during nighttime respiration and suffocate other O_2-requiring organisms. Relate the information in this Module to the material presented in Chapter 38 concerning the human impact on the biosphere.

B. Some bacteria cause disease (*Module 17.15*).

1. Eubacteria cause about half of all known human diseases and are responsible for diseases in all other eukaryotes.

2. Diseases can be caused by eubacterial growth on, and destruction of, tissues, but they are more likely to be caused by the release of exotoxins out of growing bacteria or the presence of endotoxins on the surfaces of these bacteria.

3. *Staphylococcus aureus* is a normal skin eubacterium, but when it grows inside a person, the exotoxins it produces can cause serious disease (Figure 17.15A).

4. Species of *Salmonella* produce endotoxins that cause food poisoning and typhoid fever.

NOTE: The recent cases of fatal poisonings by a toxic strain of *E. coli* are due to endotoxins.

5. The cause of Lyme disease, *Borrelia burgdorferi*, is carried by a tick and elicits a distinctive set of symptoms and potential disorders (Figure 17.15B).

6. Sanitation, the use of antibiotics, and education are three of our defenses against eubacterial diseases.

7. However, antibiotic-resistant bacteria have evolved and are now a major health issue (Module 14.21).

C. Koch's postulates are used to identify disease-causing bacteria (*Module 17.16*).

1. Discovering the cause of disease is the first step in preventing or curing the disease.

2. In 1876, German physician Robert Koch presented diagnostic criteria proving *Bacillus anthracis* to be the cause of anthrax. (a) The same pathogen must be found in each host. (b) The pathogen must be isolated into pure culture. (c) The original disease must be produced in new hosts inoculated with the culture. (d) The same pathogen must be reisolated from the new host (Master 17.16B).

3. For some pathogens, Koch's postulates cannot be used because the organism cannot be cultured outside the host. The cause of syphilis, the spirochete *Treponema pallidum*, is such an organism, but the first postulate is true, and other evidence leaves no doubt that this bacterium is the cause of the disease.

D. Chemical cycles in our environment depend on prokaryotes (*Module 17.17*).

1. Because of the variety of metabolic capabilities, prokaryotes play many beneficial roles in cycling elements among living and nonliving components of environments.

2. *Preview*: Chemical cycles are discussed more fully in Chapter 36.

3. Only prokaryotes are capable of nitrogen fixation, the conversion of N_2 gas to nitrogen in amino acids. Important nitrogen fixers include many cyanobacteria and many chemoheterotrophs in the soil.

    Preview: Many plants depend on bacteria for nitrogen (Modules 32.13 and 32.14).

4. The breakdown of organic wastes by decomposers is one of the most common beneficial roles of prokaryotes.

5. Prokaryotic decomposers are part of the aerobic and anaerobic communities of organisms functioning in sewage-treatment plants (Master 17.17).

E. Bacteria may help us solve some environmental problems (*Module 17.18*).

1. Natural bacteria are encouraged, or recombinant strains are used, to decompose away the remains of oil spills on beaches (Figure 17.18A).

2. Species of *Thiobacillus*, autotrophs that obtain energy from oxidizing ions in minerals, can be used to help remove toxic metals from old mine and industrial waste sites (Figure 17.18B). However, their use in this role is limited since their metabolism adds sulfuric acid to the water.

F. Summary: Prokaryotes are at the foundation of life on Earth (*Module 17.19*).

1. All life depends on prokaryotes, in both an evolutionary and an ecological sense.

2. Prokaryotes were not only the first producers of O_2 but the first organisms tolerant of increased levels of O_2 in the atmosphere.

3. Prokaryotes are extremely important in nutrient cycles of all kinds.

VI. The eukaryotic cell probably originated as a community of prokaryotes (*Module 17.20*).

    A. The fossil record indicates the first eukaryotes evolved between 1.5 and 1 bya.

    B. Two hypotheses have been proposed to explain how this happened.

        1. The infolding of prokaryotic plasma membranes and the specialization of internal membranes into membrane-bounded organelles (Acetate 17.20A), except mitochondria and chloroplasts.

        2. Endosymbiosis is the likely basis of the origin of mitochondria and chloroplasts (Acetate 17.20B). The ancestral mitochondria may have been small heterotrophic prokaryotes and, similarly, the ancestral chloroplasts may have been small photosynthetic prokaryotes.

    C. Considerable evidence supports the endosymbiotic hypothesis.

        1. Mitochondria and chloroplasts are similar in size and shape to eubacteria and include circular DNA and eubacterial-type ribosomes.

        2. These organelles replicate in eukaryotic cytoplasm in a manner resembling binary fission.

        3. The inner, but not the outer, membranes of these organelles contain enzymes and electron transport molecules characteristic of prokaryotes, not eukaryotes.

        NOTE: Endosymbiosis is common today between protists and/or prokaryotes.

    D. The two hypotheses are not mutually exclusive. Some organelles, such as the nucleus and endomembrane system, may have evolved by plasma membrane invagination.

    E. Unicellular eukaryotes and their direct multicellular descendants are called protists (*Module 17.21*).

        1. Protists are diverse, united in one kingdom by their relative simplicity and their eukaryotic cells.

        2. As a group, protists are nutritionally diverse, being heterotrophic or autotrophic, but not as diverse as the whole Kingdom Monera.

        3. Protists are found in all habitats but are most common in aquatic ones (Figure 17.21).

        4. As eukaryotes, their cells are more complex, with many kinds of organelles.

        5. Life's most complex cells belong to the ciliates in this kingdom.

        6. Evolutionarily, protists were pivotal because it was ancestral protists that evolved into ancestral plants, fungi, and animals.

        7. Studies of protistan rRNA provide evidence that suggests that these eukaryotic cells evolved from several different prokaryotic ancestors. Thus, the currently used classification probably does not reflect the evolutionary relationship among the protistans.

VII. Heterotrophic protists (protozoans and relatives).

    A. Protozoans are protists that ingest their food (*Module 17.22*).

        1. Most species are free-living inhabitants of watery environments. A few are causes of dangerous diseases of humans and other animals.

NOTE: Most protozoans have cells that lack cell walls, although some have relatively rigid protein skeletons below their plasma membranes.

2. Flagellates move by one or more flagella. *Giardia* is a flagellate that can cause abdominal cramps and severe diarrhea. What makes *Giardia* particularly interesting is its lack of mitochondria. This suggests that its lineage may have arisen before mitochondria evolved by endosymbiosis.

3. Another interesting flagellate is a species of *Trypanosoma* that is spread by tsetse flies and causes African sleeping sickness when it grows among blood cells. Trypanosomes escape a host's immune response by changing the molecular appearance of the proteins in the membranes (Figure 17.22A.2).

4. Amoebas move and feed by means of pseudopodia (Figure 17.22B).

5. Apicomplexans are all parasites, some causing serious human disease. *Plasmodium* species are spread by mosquitoes and cause malaria when they reproduce inside red blood cells (Figure 17.22C).

Review: The relationship between the sickle-cell allele and malaria (Module 9.12)

6. Ciliates are common, free-living forms that use cilia to move and feed. Daily activity is controlled by a polyploid macronucleus, and sexual recombination involves as many as 80 diploid nuclei (Figure 17.22D).

NOTE: Cilia are defined as numerous, short, flagella-like structures arranged in more complex patterns than flagella.

B. Cellular slime molds have multicellular stages (*Module 17.23*; Figure 17.23).

1. The unicellular stage exists as individual, amoeboid cells that feed on bacteria in rotting vegetation, increasing their populations by mitotic cell division.

2. When their food supply runs out, the individual cells mass into a sluglike, multicellular colony.

3. The slugs wander about, moving to an advantageous location to reproduce. Some cells form a stalk below, and others form reproductive spores above.

4. Because they are eukaryotes with a simple developmental sequence, cellular slime molds have played a role in research on cellular differentiation.

C. Plasmodial slime molds have brightly colored stages with many nuclei (*Module 17.24*; Figure 17.24).

1. These protists exist in several different forms, including single cells, multinucleate feeding webs, resistant bodies, and multicellular reproductive structures.

2. They are common inhabitants of moist, rotting leaves and dead logs.

3. Life starts as individual amoeboid or flagellated cells (that can change back and forth, depending on the availability of water). As these cells ingest bacteria, they grow into an amoeboid plasmodium (a single undivided mass of protoplasm containing many nuclei).

4. When food runs out, the plasmodium mounds up into reproductive structures.

NOTE: Because the nuclei in the feeding stage all divide by mitosis at exactly the same time, some plasmodial slime molds were used in early research on the chemistry of mitosis.

VIII. Autotrophic protists.

    A. Photosynthetic protists are called algae (*Module 17.25*).

        1. All algae have chloroplasts with chlorophyll *a*.

           NOTE: The variety of accessory photosynthetic pigments causes many algae to be other colors than the typical grass-green color of chlorophyll *a*.

        2. Most algae have cell walls composed of cellulose or silica.

        3. The morphology of species varies considerably, from single cells to colonial filaments to plantlike bodies (seaweeds).

        4. Dinoflagellates are uniquely shaped and move by two flagella in perpendicular grooves. Some dinoflagellates are responsible for toxin-releasing blooms in marine water that are known as red tides (Figure 17.25A). Nutritionally, some dinoflagellates are photoautotrophs, others chemoautotrophs, and others chemoheterotrophs.

        5. Dinoflagellates are responsible for the toxic red tides that sometimes appear in warm coastal waters.

        6. Diatoms are unicellular, with uniquely shaped and sculptured silica walls. They are common components of watery environments (Figure 17.25B). In terms of being a food source, diatoms are to marine animals what plants are to land animals.

        7. Fossilized diatoms make up thick sediments of diatomaceous earth, which can be used either for filtering or as an abrasive.

        8. The green algae are common inhabitants of fresh water and include a large variety of forms. Green algae share some features with higher plants and are considered to be the plant kingdom's ancestors (Figure 17.25C).

    B. Seaweeds are multicellular marine algae (*Module 17.26*).

        1. Seaweeds are the most complex of the photosynthetic protists. Some have complex bodies with leaflike, stemlike, and rootlike structures, all analogous rather than homologous to similar structures in higher plants.

        2. As seems to be the case with nearly every group, the seaweeds are also apparently due for taxonomic revision. Molecular and cellular studies suggest that there are actually three groups of seaweeds. One group (brown algae and diatoms) may belong in a separate kingdom, whereas the other two groups (red and green algae) may actually be plants.

        3. Brown algae include the most complex seaweeds. Some can grow to lengths of 100 m, forming kelp "forests" that are rich with other life (Figure 17.26A).

        4. Red algae are most common in tropical marine waters (Figure 17.26B).

        5. Some green algae are seaweeds, such as *Ulva*. The reproductive pattern of this alga involves alternation of generations, alternating between haploid gametophytes that give rise to gametes directly by mitosis and diploid sporophytes that give rise to spores by meiosis (Acetate 17.26C).

        6. *Preview:* This pattern is found among many, but not all, algae and all plants and is important in understanding plant evolution (Module 18.4).

IX. The origin of the three multicellular kingdoms of life.

    A. Multicellular life may have evolved from colonial protists (*Module 17.27*).

        1. Most multicellular organisms, including seaweeds, slime molds, fungi, plants, and animals, are characterized by the differentiation of cells that perform different activities within one organism.

        2. Multicellularity undoubtedly evolved several times within the kingdom Protista. Some of these organisms evolved further into ancestors of the plant, fungus, and animal kingdoms.

        3. A hypothetical scenario for the evolution of a multicellular plant or animal from an early protist: (a) formation of ancestral colonies, with all cells the same; (b) specialization and cooperation among different cells within the colony; (c) delimitation of specialized sexual cells from the somatic cells (Master 17.27).

    B. Multicellular life has diversified over hundreds of millions of years (*Module 17.28*; Master 17.28).

        1. The oldest fossils of multicellular organisms (red algae and invertebrate animals) date from 700 mya. These organisms were red algae and animals resembling corals, jellyfish, and worms. Other kinds of multicellular algae probably existed as well, but their remains are yet to be found in the fossil record.

        2. A mass extinction occurred between the Precambrian and Paleozoic eras.

        3. Up until 500 mya, life was aquatic and represented by diverse animals and multicellular algae, along with ancestral protists and prokaryotes.

CLASS ACTIVITIES

1. If you do not have access to videotapes or videodiscs demonstrating microbes but do have a microscope-mounted video camera, be sure to feature lecture demonstrations of Monera and Protista. A scraping of cheek epithelium from inside your mouth and stained with methylene blue will always include countless bacteria (human cell nuclei and the bacteria stain blue), and the contrast in size is dramatic. Drops of cultured protists—or better, of diverse communities of pond microbes—will also provide material for demonstrating the diversity of the Protista (numerous algal and protozoan groups are usually represented, and bacteria are always present but often overlooked because of their small size). Highly motile forms can be slowed down using thick, aqueous solutions of methyl cellulose. Another impressive source of protozoa is the hind gut of North American termites.

RESOURCES AND REFERENCES

Amabile-Cuvas, C.F., M. Cardenas-Garcia, and M. Ludgar. "Antibiotic Resistance." *American Scientist*, July-August 1995. The battle to keep up with rapid prokayotic evolution.

Attenborough, David. *Life on Earth. A Natural History*, Boston: Little, Brown, 1979. A natural historian's well-delivered views on the ecological and geological forces that shaped the evolution of life on Earth. The book is organized around the original video series, which includes thirteen 1-hour programs produced by BBC in the same year. Videotapes available in libraries or in an abridged version from Carolina Biological Supply Company.

Bacteria and Viruses. Princeton, NJ: Films for the Humanities and Sciences, 1987. A 20-minute videotape that reviews the biology and importance of bacteria and contrasts the differences between bacteria and viruses, particularly regarding how they can be controlled.

BioShow. Redwood City, CA: Benjamin/Cummings, 1993. This videodisc contains several video sequences of living bacteria and protists, including a demonstration of the action of the bacterial flagellum, phagocytosis by an amoeba, various other freshwater protozoans and algae, and cellular slime mold development.

Brock, T.D., and M.T. Madigan. *Biology of Microorganisms*, 7th ed. Englewood Cliffs, NJ: Prentice-Hall 1994.

Currie, Philip J. "Long-Distance Dinosaurs." *Natural History*, June 1989. Evidence that a group of dinosaurs migrated between environments in the Arctic and what is now Alberta, Canada.

de Duve, C. "The Beginnings of Life on Earth." *American Scientist*, September 1995. Early molecular evolution.

Hidden Kingdoms: The World of Microbes. Burlington, NC: Carolina Biological Supply Company, 1990. An 11-minute videotape showing the dynamic activities of a variety of protists.

Horgan, J. "In the Beginning . . ." *Scientific American*, February 1991. A discussion of new controversies about the origin of life.

Mann, C. "Lynn Margulis: Science's Unruly Earth Mother." *Science*, April 19, 1991. A profile of the chief advocate of the endosymbiotic theory.

Margulis, Lynn. *Early Life*. Boston: Jones & Bartlett, 1984. The best introduction to the forces (chemical and then biological) involved in the evolution of the simplest life forms: bacteria, protists, and fungi.

Margulis, Lynn, and Dorian Sagan. *Microcosmos: Four Billion Years of Microbial Evolution*. New York: Simon & Schuster, 1986. Evolution from the perspective of microorganisms.

McDermott, Jeanne. "A Biologist Whose Heresy Redraws Earth's Tree of Life." *Smithsonian*, August 1989. A biographical sketch that includes Lynn Margulis's contributions to our understanding of the endosymbiotic theory.

Microbial Engine: Algae and Protozoa—Ecology to Biotechnology. Burlington, NC: Carolina Biological Supply Company, 1990. A 36-minute videotape that documents diversity in form and function, ecological importance, and uses in research and biotechnology.

Nanney, D.L., D.O. Mobley, R.M. Preparata, E.B. Meyer, and E.M. Simon. "Eukaryotic Origins: String Analysis of 5S Ribosomal RNA Sequences from Some Relevant Organisms." *Journal of Molecular Evolution*, 32, 1991.

National Geographic Society. *Bacteria*. Washington, DC: National Geographic Society. A 23-minute videotape that reviews the forms of bacteria, the structure of bacterial cells, their reproduction, and their importance.

Origin of Life. Paradise, CA: Projected Learning Programs, 1989. Simulation software for IBM PC and Apple II. Experiment with atmospheric composition, temperature, UV light, and electrical energy and their interacting effects on macromolecular formation.

Radetsky, P. "How Did Life Start?" *Discover*, November 1992.

Sagan, Dorion, and Lynn Margulis. *Garden of Microbial Delights: A Practical Guide to the Subvisible World*. New York: Harcourt Brace Jovanovich, 1988. A good resource for details on the biology of, and practical information about, bacteria, protists, and fungi.

Sarkar, Sahotra. "Sex, Disease, and Evolution—Variations on a Theme from J.B.S. Haldane." *BioScience*, June 1992. How did sex evolve? This article reviews Haldane's original hypothesis concerning the role of immunity in evolution, and some current hypotheses about the evolution of sexual reproduction.

Schwartzman, David, Mark McMenamin, and Tyler Volk. "Did Surface Temperatures Constrain Microbial Evolution?" *BioScience*, June 1993. Changes in the interpretation of geological deposits suggest Earth was much warmer during the early evolution of microbial life.

Sonea, Sorin, and Maurice Panisset. *A New Bacteriology*. Boston: Jones & Bartlett, 1983. The authors believe that bacteria function in superorganisms, not as species in the sense of eukaryotes; they propose that, in nature, bacteria form a unified global entity in which they are all linked, both genetically and by specific high-level functions.

Tortora, G.J., B.R. Funke, and C.L. Case. *Microbiology: An Introduction*, 5th ed. Redwood City, CA: Benjamin/Cummings, 1995. A general text with an emphasis on disease-causing microbes.

Videodiscovery. *Cell Biology I Videodisc*. Seattle, WA: Videodiscovery, 1987. Includes a demonstration of flagellar movement in prokaryotes and protists, feeding in *Paramecium*, and cell division in unicellular algae.

Woodworth, Robert H. *Non-Cellular Slime Molds* and *Cellular Slime Molds*. Burlington, NC: Carolina Biological Supply Company, 1993. Two 13-minute videotapes that document the ecology, activity, and reproduction of these interesting groups.

INTERNET RESOURCES FOR UNIT IV

The Basking Spot

http://www.contrib.andrew.cmu.edu/~iguana/herp.html

Herpetological resource.

The Cephalopod Page

http://is.dal.ca/~ceph/wood.html

The Froggy Place

http://www.cs.yale.edu/HTML/YALE/CS/HyPlans/loosemore-sandra/froggy.html

Very playful and silly, but the "Scientific Amphibian" section has useful links.

Images

gopher://muse.bio.cornell.edu:70/11/images

Many images of plants, animals, and so on. Read "About This Biology Image Archive" before searching for images.

NetVet/Electronic Zoo Animal Information & Archives Home Page

http://netvet.wustl.edu/ssi.htm

Welcome to Casiano Zoo

http://sparky.cs.nyu.edu:19234/welco.html

Welcome to Cockroach World

http://www.nj.com/yucky/roaches/index.html

Juvenile, but worth looking at. Cockroach ecology and diversity.

18

PLANTS, FUNGI, AND THE COLONIZATION OF LAND

APPROACH

Some of the subtleties of plants and fungi are missed by nonscience majors who tend to focus on animal biology. Spend some time checking that students fully comprehend the events outlined in Module 18.4. The development of adult gametophytes by mitosis and the production of gametes by mitosis are usually novel concepts for nonscience majors. Introduce these concepts in the context of the human life cycle: What would happen if our zygotes developed into parasitic life forms that we carried around, and those structures produced spores that developed into humans? Likewise, the idea of a fungal body being a diffuse network or mycelium (and, if you choose to go into it, the fungal life cycle in Module 18.18) is difficult to grasp because these structures cannot be seen.

Using an evolutionary context works well for both kingdoms. On the one hand, the progression from alga to bryophyte to flowering plant shows the increasingly complex adaptations to life on land. On the other hand, the coevolution of symbiotic and decomposer relationships of fungi to plants focuses on the distinctive characteristics of the fungi.

The stepwise progression of evolutionary advances is used as the underlying structure in the modules for both plants and (in Chapter 19) animals. The evolution of plants is previewed in Module 18.3 with a phylogenetic tree. Otherwise, the message is spread out with some modules detailing these steps, and other modules providing ancillary material. The lecture outline is structured to emphasize the evolutionary steps.

If you only have time for one lecture on each group, you can cover just the bryophytes and conifers here; leave the flowering plants until Chapter 31. An abridged treatment of the fungi might include a brief survey of the groups; a mention of the economic uses of yeasts, molds, and mushrooms; and an emphasis on the three important ecological roles of all fungi: decomposition of plants, mycorrhizae, and plant parasitism.

CHAPTER OBJECTIVES

Compare aquatic and terrestrial environments as habitats for algae and plants, outlining the major kinds of adaptations to the terrestrial environment most land plants have.

Distinguish between the terms in the following pairs: *xylem* and *phloem*, *gametangium* and *sporangium*, *gametophyte* and *sporophyte*, *gymnosperm* and *angiosperm*, and *diploid* and *dikaryotic*.

Outline the major evolutionary steps exhibited by several lineages of plants: gametangia, vascular tissue, seeds, and flowers and fruits.

Name the major features of the bryophytes, ferns, conifers, and flowering plants.

Compare the life cycles of primitive, nonvascular plants (mosses) and higher, vascular plants (conifers). If time permits, also compare the fern and flowering plant life cycles.

Discuss the coevolution and interdependence of plants and animals.

Explain the ways in which flowering plants are related to animals and fungi.

List the characteristics of fungi, and discuss the ecological roles. If time permits, describe the life cycle of a mushroom.

Describe some parasitic fungi and their important economic implications.

LECTURE OUTLINE

I. **Introduction.**

    A. The lives of modern plants and fungi are intertwined (*Opening Essay*).

        1. We depend on plants and, indirectly, fungi for much of our food.

        2. Plants are harmed most by fungi.

        3. On the other hand, nearly all plants in the wild are aided by mycorrhizal fungi. Mycorrhizae are rootlike structures made of both fungi and plants. The fungi help plants obtain nutrients and water, and protect plant roots from parasites, in exchange for food the plants make by photosynthesis (Figure 18.0).

        4. Modern agricultural practices, such as killing the parasitic fungi with fungicides, may disrupt mycorrhizal fungi, forcing the need for fertilizer.

        5. *Preview:* This kind of dilemma often results when humans try to manipulate complex natural systems (Chapter 38).

    B. Plants and fungi evolved together as life moved onto land over 400 million years ago (mya). This is supported by the earliest plant fossils having mycorrhizae.

II. **What is a plant?** (*Module 18.1*).

    A. *Review:* In the two-kingdom system, Linnaeus classified algae as plants. In the five-kingdom system, algae are protists (Modules 16.14 and 17.25).

    B. The definition of plants as multicellular, eukaryotic photosynthesizers also describes multicellular algae (Modules 17.25 and 17.26).

    C. Multicellular seaweeds, the most complex algae, are adapted for life in water, while plants are adapted for life on land (Acetate 18.1A).

    D. Seaweeds live in water.

        1. All the resources, including water, carbon dioxide, and minerals, are in direct contact, and waste products can be washed away.

        2. Water supports and suspends the body of the alga.

        3. Holdfasts anchor the alga, and all other parts can be photosynthetic.

        4. Water provides a means of dispersal for gametes and offspring.

    E. Life on land imposes problems.

        1. Water and nutrients are concentrated in the ground, while carbon dioxide and light are most abundant above the ground.

        2. Air provides no support against the force of gravity.

        3. Air will dry out reproductive cells.

    F. Adaptations to a terrestrial environment in plants include the following:

        1. Discrete organs: roots, stems, leaves, and gametangia specialized for anchorage and absorption, support, photosynthesis, and reproduction, respectively.

2. Mycorrhizal fungi to increase the efficiency of absorption of their roots.

3. Vascular tissue to move food (phloem) and water (xylem) among the parts (Figure 18.1B).

4. A cuticle on leaves and stems to help retain water.

5. Dependency on wind or animal carriers to disperse gametes and offspring.

G. Plants probably evolved from green algae (*Module 18.2*).

1. Plants and green algae have a number of homologous features, such as identical photosynthetic pigments, food storage molecules, cell walls, and mechanisms of cell division (Modules 17.25 and 17.26).

2. Plants may have evolved from green algae about 500 mya. The ancestor is not known, but analysis of nucleic acid sequences, cell structure, and biochemistry indicate that a group of green algae, the charophytes, may be the closest (extant) relatives of plants (Figure 18.2A).

3. Early plants thrived on moist shorelines where space was essentially limitless and there was abundant sunlight.

4. Among the earliest plant fossils is *Cooksonia*, which lived 415 mya. This plant had branched, upright, photosynthetic stems and spore sacs (sporangia) but lacked leaves (Master 18.2B).

III. Plant diversity provides clues to the evolutionary history of the plant kingdom (*Module 18.3*).

A. Plant phylogeny shows branching of several plant lineages, reflecting major evolutionary steps. These are previewed here, and details are added in the modules that follow (Acetate 18.3A).

1. The origin of early, bryophytelike plants (500 mya). They had gametangia and cuticles but lacked vascular tissue. Like their algal ancestors, they had flagellated sperm and depended on water for reproduction (Figure 18.3B).

    NOTE: Some botanists think that modern bryophytes are not directly related to vascular plants but separately derived from green algal ancestors. The bryophytes include the mosses, liverworts, and hornworts. The bryophytes that resemble the first plants the most are the strap-shaped liverworts.

2. Vascular plants from early bryophytes (425 mya). Vascular tissue transports water and nutrients and provides support (Figure 18.3C). In addition, these plants have embryonic development within gametangia.

3. Seed plants from the early vascular plants (360 mya). Seeds provide the embryonic plant with a protective coating and a food supply. These plants also produce pollen, vehicles that transfer nonflagellated sperm to the female parts of plants. The earliest seed plants to appear were the gymnosperms (naked seed).

4. Flowering plants (angiosperms) evolved from the gymnosperm line (130 mya). Flowers increase the efficiency of pollination and develop into fruits that protect seeds and increase the efficiency of seed dispersal.

    Preview: Angiosperms are discussed in detail in Chapter 31.

B. Haploid and diploid generations alternate in plant life cycles (*Module 18.4*).

1. *Review: Ulva* life cycle (Acetate 17.26C); human life cycle (Acetate 8.14).

2. Plant life cycles alternate between multicellular, diploid ($2n$) and haploid ($1n$) adults, unlike animals that are diploid as adults and produce single-celled haploid gametes (Acetate 18.4).

3. Sporophytes grow from diploid zygotes and produce spores by meiosis.

4. Gametophytes grow from haploid spores and produce gametes by mitosis.

5. Alternation of generations provides two chances to produce large numbers of offspring, zygotes, and spores.

6. Major changes from the algal life cycle that occur during the evolution of plants are the change in dominance from the gametophyte to the sporophyte, the change in dependency of sporophytes on gametophytes to the opposite dependency, and the loss of flagellated sperm.

   NOTE: Compare the mosses (dominant gametophytes, semiparasitic sporophytes, flagellated sperm) to the conifers or flowering plants (dominant sporophytes, semiparasitic, reduced gametophytes, pollen).

IV. Lower plants.

A. Mosses have a dominant gametophyte (*Module 18.5*; Acetate 18.5).

   1. The green growth we see consists mostly of gametophytes (Figure 18.3B).

      NOTE: Adult gametophytes consist of nonvascular stems and leaves and threadlike holdfasts. For water and minerals, they depend on rain or on water flowing along the stem from the ground surface.

   2. Gametophytes produce sperm and eggs in gametangia.

   3. Flagellated sperm require a film of water in which to swim from male gametangium to female gametangium.

   4. The zygote remains in the female gametangium and develops into the sporophyte.

   5. Meiosis occurs in the sporangium and produces spores.

   6. Spores are dispersed by wind and later develop into gametophytes by mitosis.

      NOTE: Adult sporophytes lose their chlorophyll and become more dependent on the female gametophytes. The sporangia are wonderfully adapted to dispersing spores by means of movable teeth around the sporangium opening, on wet/dry cycles.

B. Ferns, like most plants, have a dominant sporophyte (*Module 18.6*; Figure 18.6).

   1. The fern fronds we see are sporophytes.

      NOTE: Ferns have underground stems with small roots, and all parts are vascularized. The fronds are the principal organs of photosynthesis.

   2. Gametophytes produce flagellated sperm and eggs.

   3. Like mosses, fern zygotes remain in the female gametangium and develop into adult sporophytes. Eventually the sporophytes take over.

   4. Sporangia in clusters on the frond's underside produce spores by meiosis.

      NOTE: Fern sporangia have a unique mechanism to release their spores. As the sporangia dry out, thick-walled cells along the "back" of the sporangium pull the thin-walled "front" cells into tension. In a fraction of a second, the front cells rupture, scattering the spores into the air.

    5. Spores are dispersed by wind and grow into small, heart-shaped gametophytes by mitosis.

    6. Today, about 95% of all plants have a dominant sporophyte generation.

  C. Seedless plants formed vast "coal forests" (*Module 18.7*).

    1. During the Carboniferous period (285–360 mya), vast forests grew in tropical, swampy areas over what is now Eurasia and North America.

    2. The forests included dominant lycopods, horsetails (both seedless plants more primitive than ferns), and tree ferns (Figure 18.7).

    3. The compressed remains of these forests survive as deposits of coal and other fossil fuels.

    4. At the end of the Carboniferous period, the climate turned colder and drier, providing the conditions for further plant evolution (of seeds).

V. Gymnosperms called conifers replaced the swamp forests (*Module 18.8*).

  A. Conifers are the most successful of several groups of gymnosperms because they are adapted to harsh winter climates. They readily shed snow, and their needles resist drying and remain photosynthetic throughout the year.

  B. A pine tree is a sporophyte with tiny gametophytes in its cones (*Module 18.9*; Figure 18.9).

    1. Cones are a significant adaptation to land, protecting all reproductive structures: sporangia, microscopic gametophytes, and zygotes.

    2. Female cones are larger. Scales bear a pair of ovules (sporangia with a covering, or integument). Within the ovule, one of the four products of meiosis develops into a tiny, multicellular, "female" gametophyte.

    3. Male cones are smaller, and scales bear many sporangia that make spores by meiosis. Spores develop into tiny male gametophytes (pollen grains), each consisting of a few cells. Pollen grains house the cells that will develop into sperm.

    NOTE: The particular pollen grains in Figure 18.9 are typical of the pine family: The central cell and cells within it compose the gametophyte. The two lateral "cells" are outgrowths of the cell wall that function as floats.

    4. Pollen is dispersed by wind. Pollination occurs when a pollen grain lands near an ovule and grows into it.

    5. Meiosis occurs in the ovule where a haploid spore cell beings to develop into the female gametophyte. Months later, a few of the female gametophyte's cells inside the ovule function as eggs. The pollen tube delivers a male gametophyte's nucleus (functioning as a nonflagellated sperm) to fertilize an egg.

    6. The zygote develops into the embryonic sporophyte, and the remaining ovule develops into the other parts of the seed (the integument becomes the seed coat, and the female gametophyte becomes a food supply).

    7. Seed dispersal is by wind or animal.

    8. Under favorable conditions, the seed germinates and the embryo grows into a tree.

C. Coniferous forests are threatened by our demand for wood and paper (*Module 18.10*).

1. A dilemma has developed between environmentalists who want to conserve the remaining U.S. coniferous forests (for future use and to protect the plant and animal species they contain) and those who depend on employment in the timber and paper-pulp industries.

2. Two possible ways to make the forest industry sustainable: increase the recycling of forest products, and reduce paper consumption.

Preview: The material is this Module should be related to the impact of humans on the biosphere (Chapter 38).

VI. Angiosperms dominate most landscapes today (*Module 18.11*).

A. Although coniferous forests dominate northern areas, flowering plants dominate most other land areas. About 80% of plant species are angiosperms.

B. Angiosperms supply nearly all our food and much of our natural fibers.

C. Several adaptations account for the success of the angiosperms: flat, efficiently photosynthetic leaves; vascular tissue with thicker, stronger cell walls; and flowers and fruits.

D. The flower is the centerpiece of angiosperm reproduction (*Module 18.12*).

Preview: The life cycle of the flowering plant (Modules 31.9–31.13).

1. Flowers expose an angiosperm's sexual parts and are the sites for pollination and fertilization (Figure 18.12A).

2. Flowers are short stems with the following modified leaves: sepals, to protect flower buds; petals, to attract animal pollinators; stamens, the male parts (holding the anthers, in which pollen develops); and carpels, the female parts (consisting of sticky stigma to trap pollen, and the ovary, which bears ovules and later develops into fruit) (Acetate 18.12B).

E. The angiosperm plant is a sporophyte with gametophytes in its flowers (*Module 18.13*; Figure 18.13).

1. *Review:* Gymnosperm life cycle (Figure 18.9).

2. Flowers protect all microscopic reproductive structures: sporangia, male and female gametophytes, and zygotes.

3. Meiosis in the anthers leads to spores that develop into the male gametophytes (pollen grains).

4. Meiosis in the ovules leads to spores that develop into the female gametophytes, each of which produces an egg, inside the ovules.

5. Pollen is dispersed by either wind or insects. Pollination occurs when a pollen grain lands on the stigma.

6. The pollen grows into the ovary and delivers one nucleus (again functioning as a nonflagellated sperm) to the egg nucleus in the ovule. Fertilization usually occurs within 12 hours after pollination.

7. Following fertilization, the zygote develops into the embryonic sporophyte in the seed, and the ovary develops into the fruit.

8. Seed dispersal is also by wind or animal and is aided by the enveloping fruit.

9. Under favorable conditions, the seed germinates and the embryo develops into a mature sporophyte.

10. *Preview:* Further details about the flowering plant life cycle, including the unique adaptation of double fertilization, are covered in Modules 31.10–31.14.

F. The structure of a fruit reflects its function in seed dispersal (*Module 18.14*).

1. Fruits develop and ripen quickly, within one growing season.

2. Some fruits are adapted for wind dispersal of seeds (Figure 18.14A).

3. Others are adapted to hitch a ride on animals (Figure 18.14B).

4. Fleshy, edible fruits are attractive to animals as food. Seeds of these fruits usually pass unharmed through the animal's digestive tract and are deposited with fertilizer far from the parent plant (Figure 18.14C).

G. Interactions with animals have profoundly influenced angiosperm evolution (*Module 18.15*).

1. Most flowering plants depend on insects, birds, or mammals for pollination and seed dispersal.

2. Most land animals depend on flowering plants for food.

3. Pollen and nectar provide food for pollinators. Bees are one of the most highly coevolved groups of pollinators (Figure 18.15A).

4. Flowers that are pollinated by birds are usually pink or red (Figure 18.15B).

5. Flowers that are pollinated at night by bats or moths are usually large and light-colored (Figure 18.15C).

6. These mutual dependencies have been favored by natural selection.

VII. Fungi.

A. Fungi and plants moved onto land together (*Module 18.16*).

1. *Review:* Mycorrhizae helped make colonization of land possible (Opening Essay, Figure 18.0).

2. Fungi are heterotrophic and require external sources of food molecules. They likely evolved to terrestrial habitats from aquatic ancestors, joining their food supply, which was early plants on land.

3. Fungi are found in both terrestrial and aquatic environments.

4. Many fungi are parasites of plants (Figure 18.16A).

   Preview: Some fungi and plants have a mutualistic relationship (Module 32.11).

5. A major role of fungi is the decomposition of dead organisms, particularly plants, and their organic molecular remains. Some fungi can break down other organic materials, including residues of pesticides and cancer-causing chemicals (Figure 18.16C).

6. A few fungi have unique adaptations that enable them to trap animals for food (Figure 18.16B).

B. Fungi absorb food after digesting it outside their bodies (*Module 18.17*).

1. The kingdom Fungi includes heterotrophic eukaryotes that digest their food externally and absorb the resulting small molecules as nutrients. Other characteristics include: cell walls of chitin, spore production, and the absence of motile stages. Fungi are grouped in their own kingdom. They differ from plants, with which they were once classified, because they cannot make food by photosynthesis, and in the details of their cellular and molecular structure.

2. Fungi are usually multicellular, but their bodies (mycelia) are composed of extensive networks of hyphae (filamentous cell-like units). Hyphae may be branched, may be a syncytium, or may be divided into cell-like compartments (Master 18.17A, Figure 18.17D).

3. The hyphae of the mycelium extend rapidly into the food source, developing a huge surface area from which digestive enzymes are secreted and through which the digested food is absorbed. A large mycelium can add as much as a kilometer of hyphae each day (Figure 18.17B).

4. Some fungi produce mushrooms, external reproductive bodies composed of packed hyphae (Figure 18.17C).

C. Many fungi have three distinct phases in their life cycle (*Module 18.18*).

1. The phases of mushroom-producing fungi and their relatives are characterized by a nuclear condition (Acetate 18.18).

2. Fertilization of one nucleus in a dikaryotic pair with another in the mushroom (fruiting body) produces diploid zygotes. These are held in terminal cells of the hyphae in the mushroom. Meiosis of this zygote produces the haploid nuclei that form spores.

3. Haploid spores are shed from the fruiting body, are wind-dispersed (or dispersed by water or animals), and develop into haploid mycelia. Two compatible, haploid mycelia can fuse to form a dikaryotic mycelium.

4. The dikaryotic mycelium maintains the compatible (and genetically different) nuclei, paired but separate. This phase is dominant in the life of the fungus, often lasting for many years as it continues to grow into its food source. Dikaryotic hyphae also makes up the tissues of the mushrooms produced by this mycelium periodically.

NOTE: The "humongous fungus" in Michigan, a species of *Armillaria*, is estimated to occupy 30 acres and be 1500 years old. However, the mycelium of *Armillaria* is a unique exception among mushroom-producing fungi. It is diploid for most of its life.

5. Some fungi are very simple. Yeasts are single-celled and mostly reproduce by mitotic cell division.

NOTE: The term *yeast* refers to a growth form and ecological habitat preference (plant saps), not to a taxonomic category. There are many types of yeasts, related by reduction in form to several not-closely-related fungi. Many yeasts exhibit sexual stages.

D. Lichens consist of fungi living mutualistically with photosynthetic organisms (*Module 18.19*; Master 18.19B).

1. Lichens are associations of millions of green algae or cyanobacteria held in a tangled network of fungal hyphae.

2. The fungus receives food from the photosynthesis of its partner.

3. The alga or cyanobacterium receives housing, water, and the minerals trapped by the hyphal network.

4. All lichen fungi and most lichen algae and cyanobacteria cannot grow independently.

5. Asexual reproduction of the fungus and algae together is advantageous; dispersal of both partners occurs, and they can immediately reestablish the lichen.

6. Lichens are able to survive in habitats where neither partner (nor any other multicellular organism) could grow alone (Figure 18.19A). However, they are highly susceptible to airborne pollutants.

    Preview: This is relevant to species loss being indicative of environmental problems (Module 38.15).

7. Lichens play important ecological roles in soil formation on rock surfaces and as food for animals (Figure 18.19C).

    NOTE: Cyanobacterium-containing lichens fix nitrogen and play important roles in the nitrogen cycle.

E. Parasitic fungi harm plants and animals (*Module 18.20*).

1. Of the 100,000 species of known fungi, about one-third are mutualistic in mycorrhizae and lichens, one-third are decomposers, and one-third are parasites.

2. Parasitic fungi are the most serious plant pests. Particularly dangerous are nonnative parasites, such as the fungus that causes Dutch elm disease (Figure 18.20A).

3. Some parasitic fungi attack developing seeds and fruits of grains (Figure 18.20B).

4. Fungi cause a few diseases in humans. Some infections of lung tissue can be fatal, particularly in people weakened by other diseases. Less serious are the fungal infections of the outer layers of the skin known as ringworm of the scalp and athlete's foot. Fortunately, most of these fungal parasites can be controlled by fungicidal ointments.

F. Fungi have many commercial uses (*Module 18.21*).

1. Mushrooms and other fungi are eaten (Figure 18.21).

2. Certain molds lend distinctive flavors and textures to foods such as Roquefort and blue cheese, and underground truffles (which also are important mycorrhizal fungi).

3. Different strains of one kind of yeast are used in baking, brewing, and winemaking.

4. Some fungi are important commercial sources for antibiotics.

5. There is evidence that fungi evolved from the same protistan lineage that gave rise to animals.

CLASS ACTIVITIES

1. The lively activities of spore dispersal in moss and fern sporangia can be demonstrated to small classes using an overhead projector, or a video camera attached to a dissecting microscope. As the heat from the light source dries out, the moss sporangium teeth, or whole fern sporangia, will release the spores. Mosses have quite a variety of

mechanisms, depending on the genus. Most release spores on a drying cycle, and protect the spores from release on a wetting cycle. As you watch the mechanism in action, discuss how it would be adaptive to life on land.

2. A considerable number of wild, cultivated, and cultured fungi are now available in many markets in the United States, both in fresh and dried condition, depending on the season and local availability of mushroom farms or proper habitat. Bring in examples to stimulate discussion and interest in the topic of fungi. Point out that the mushroom you see is like the apple of an apple tree: the "plant" (mycelium) that produces the mushroom is invisible to the naked eye. The mycelium of the most commonly available mushroom, *Agaricus campestris,* is a litter decomposer of composted straw. Oyster mushrooms and the Japanese mushroom, shiitake, are quite common in better markets and are both litter decomposers of wood (but can be grown on artificial logs made of sawdust). In season you may be able to find golden chanterelles (the mycelium of which forms mycorrhizae, so these are not domesticated), truffles (very expensive and also mycorrhizal), or morels (a recent, secret process allows these to be cultured). The moldy cheeses (Brie, Camembert, and blue); wine, beer, and soy sauce bottles; and a loaf of bread can also be used as props to stress the diversity of our dependency on fungi for food. In health-food stores you may also be able to find tempeh, a soybean product fermented with *Rhizopus mold.*

3. Almost every region of the United States has at least one serious plant-pathogenic fungus. Contact your local agricultural extension office to obtain details about the identity of these fungi, and their classification, hosts, and controls. Featuring a specific representative will help clarify the general characteristics of the kingdom as a whole, particularly the reproductive advantage of producing spores and the ease with which their hyphae enter plant tissues. In presenting special information on local parasitic fungi, be sure to remind your students that the majority of fungal species are beneficial to humans, either in their roles as recyclers or mycorrhiza-formers.

RESOURCES AND REFERENCES

Alexopoulos, C.J., and C.W. Mims. *Introduction to Mycology,* 3rd ed. New York: Wiley, 1979. A classic general text, presented in a taxonomic framework.

Ancient Forests. Washington, DC: National Geographic Society, 1992. A 25-minute videotape that provides a general survey of the old-growth forests along the coast of the U.S. Pacific Northwest, with a discussion of the conflicting forces on management.

Bakker, Robert T. "How Dinosaurs Invented Flowers." *Natural History,* November 1986.

Barron, George. "Jekyll-Hyde Mushrooms." *Natural History,* March 1992. Documents the connection between nematode-trapping fungi and some wood-rotting mushrooms.

Clay, Keith. "Trespassers Will Be Poisoned." *Natural History,* September 1989. Fungi help grasses ward off their nibbling enemies.

Denison, William C. "Life in Tall Trees." *Scientific American,* June 1973. The high forest canopy consists of more than branches and leaves. Entire communities of lichens, bryophytes, and animals dwell here and provide the trees with nitrogen.

Dilcher, David, and Peter R. Crane. "In Pursuit of the First Flower." *Natural History,* March 1984. In the last decade, studies of fossil flowers, once dismissed as too rare to further the search for angiosperm origins, have begun to bear fruit.

Doyle, Jeff J. "DNA, Phylogeny, and the Flowering of Plant Systematics." *BioScience,* June 1993. Molecular technology now produces data easily, but analysis is fraught with theoretical and technical problems.

Gould, Stephen Jay. "A Humongous Fungus Among Us." *Natural History*, July 1992. Is a 30-acre fungus the world's largest organism?

Gray, Jane, and William Shear. "Early Life on Land." *American Scientist*, September/October 1992. Minute fossils offer evidence that life invaded the land millions of years earlier than previously thought.

Inheritance in a Fungus. Princeton, NJ: Films for the Humanities and Sciences, 1986. A 15-minute videotape that illustrates the experimental techniques involved in the genetic analysis of an ascospore color mutant in the ascomycete *Sordaria*.

Jamerette, Carol. "An Introduction to Mycorrhizae." *American Biology Teacher*, January 1991. A nicely illustrated guide that reviews structure, importance, and economic applications.

Lewington, A. *Plants for People*. New York: Oxford University Press, 1990. The many uses of plant products.

Masueth, J. *Botany*. Philadelphia: Saunders, 1991. A beautifully illustrated introduction to plants.

Moore-Landecker, E. *Fundamentals of the Fungi*, 3rd ed. Englewood Cliffs, NJ: Prentice-Hall, 1990. An introduction to the Fungi.

Mushrooms and Fungi. Princeton, NJ: Films for the Humanities and Sciences, 1987. A 28-minute videotape that describes the structure, habitat, reproductive habits, and taxonomic status of a variety of fungi.

Newhouse, Joseph R. "Chestnut Blight." *Scientific American*, July 1990. A fungus that has ravaged the American chestnut is coming under biological attack. Plant pathologists hope the new parasite will eventually control the old one.

Pirozynski, K.A., and D.W. Malloch. "The Origin of Land Plants: A Matter of Mycotrophism." *BioSystems*, June 1975. The authors argue that early land plants evolved from previous unions of algae and aquatic fungi.

Sharnoff, Sylvia Duran, and Stephen Sharnoff. "Lowly Lichens Offer Beauty—and Food, Drugs and Perfume." *Smithsonian*, April 1984. This lovely photographic essay includes many details about the biology and importance of lichens.

Streberg, S. "The Emerging Fungal Threat." *Science*, December 9, 1994. The potential rise in fungal diseases in humans.

Strobel, Gary A., and Gerald N. Lanier. "Dutch Elm Disease." *Scientific American*, August 1981. This deadly fungus may be brought under control with the aid of new biological techniques.

19

THE EVOLUTION OF ANIMAL DIVERSITY

APPROACH

This chapter on animal diversity emphasizes the numerous and successful invertebrates. Chapters in the next unit (Unit V) cover vertebrate anatomy and function in considerable detail.

The evolution of the animal kingdom is an illuminating story. This framework facilitates a student's assimilation of the details because each phylum builds on the last. Stepwise evolutionary changes are used as the underlying structure for the animal kingdom. They are reviewed in Module 19.23 with a phylogenetic tree. See the Approach section of Chapter 18 for a discussion of how the structure of the lecture outline relates to that of the text.

With all the material in this unit, the challenge is deciding how much detail to cover in one lecture. Try to balance the number of groups you cover, and the general features and concepts concerning them, with details about actual species (one or two per phylum). The text follows this guideline, and you may wish to reemphasize those examples in your lectures. On the other hand, if you know of examples from local habitats, try to work with them in place of the examples in the text and lecture outline below.

CHAPTER OBJECTIVES

Distinguish the animal kingdom from other kingdoms, and describe additional characteristics of animals.

Distinguish between the terms in the following sets: *radial symmetry* and *bilateral symmetry*; *gastrovascular*, *pseudocoelomic*, and *coelomic cavities*; *tissue*, *organ*, and *organ system*; and *invertebrate* and *vertebrate*.

Name the characteristics of, and selected individuals in, the various phyla illustrated in this chapter.

Explain the advantages to evolving animal groups of the following features: bilateral symmetry, coelomic cavities, segmentation, endoskeletons, jaws, amniotic eggs, feathers, and placentas.

Trace the phylogenetic relationships among the animal phyla.

LECTURE OUTLINE

I. **Introduction.**

    A. Most known organisms are animals (*Opening Essay*).

        1. Of the 1.5 million known species, one-third are animals.

        2. Most animal species live on land, but the greatest diversity of animal phyla are marine. The most diverse communities of animals are tropical coral reefs; 27 of the 30 animal phyla may be found here (chapter-opening photo).

3. These reefs are composed of lime from the skeletons of corals and the leftover cell walls of coralline red algae.

   Review: Red algae (Module 17.26).

4. Corals feed on small food particles that they trap on their tentacles (Figure 19.UN1).

5. Corals also receive food from the photosynthesis of mutualistic dinoflagellates (Module 17.25), in exchange for compounds the corals give off as wastes (CO_2 and nitrogenous compounds).

6. *Preview:* Zoologists distinguish between animals with internal skeletons (vertebrates) and those without internal skeletons (invertebrates). All but one animal phylum (our own, phylum Chordata) are invertebrates (Module 19.15).

B. What is an animal? (*Module 19.1*)

1. Animals are eukaryotic, multicellular heterotrophs that lack cell walls.

2. The life cycle of most animals includes a dominant, diploid adult that produces eggs or sperm by meiosis. These gametes fuse to form a zygote. The zygote develops into the adult animal, passing through a series of embryonic stages, many of which are shared by most members of the animal kingdom (Acetate 19.1).

   Review: This is an example of homology (Module 16.11).

3. In all animals the embryonic stages include the blastula (hollow ball of cells) and, in most, a gastrula (a saclike embryo with one opening and two layers of cells). Most further develop an additional layer of cells.

4. In many animals, the gastrula develops into one or more immature stages, larvae, that develop into the sexually mature adults only after metamorphosis.

   NOTE: Larvae are dispersive stages that help an animal's offspring find suitable habitats before continuing to grow. They are also extremely important sources of food for many other animals in aquatic habitats.

5. Animals have unique types of cell junctions (Module 4.20).

6. Most animals digest food inside their multicellular bodies.

7. Animals are, with the exception of the gametes, composed of diploid cells.

C. The animal kingdom probably originated from colonial protists (*Module 19.2*).

1. Fossils of the oldest known animals date from the late Precambrian era, 700 mya.

2. The modern animal phyla evolved during the Cambrian period, 600 mya.

3. A hypothetical evolutionary scenario leading to the first animal proceeds as follows: (a) colonial protists of a few, identical, flagellated cells; (b) larger, hollow, spherical colonies that ingested organic materials suspended in the water around the colony; (c) colonies with cells specialized for somatic (movement, digestion, etc.) and reproductive functions; (d) differentiated entities with an infolded, temporary digestive region; and (e) "protoanimals," completely infolded, with two-layered body walls (Acetate 19.2).

4. These protoanimals probably "crawled," feeding on the ocean bottom.

II. Primitive, radially symmetrical invertebrates.

    A. Sponges have relatively simple, porous bodies (*Module 19.3*).

        1. Sponges are classified in the phylum Porifera. Most are marine and live singly attached to a substrate, and range in height from 1 cm to 2 m (Figure 19.3A).

        2. Many sponges are built on a body plan having radial symmetry, that is, similar shapes as mirror images around a central axis (Acetate 19.3B).

        3. The cell layers include an outer layer of flattened, pore-containing cells, a middle layer of noncellular skeletal elements and amoebocytes, and an inner layer of choanocytes, surrounding an inner chamber (Acetate 19.3C).

        4. Sponges feed when the choanocyte flagella beat, pulling water in through the pores, through the collars, trapping bacteria in mucus, and out through the large, upper opening. The choanocytes then phagocytize the food and package it in vacuoles. Amoebocytes pick up the food vacuoles, digest the food, and carry the nutrients to other cells.

        Preview: Contrast this with the structure of the digestive system of other animals (Module 21.3).

        5. In addition to digesting and distributing food, amoebocytes also transport oxygen, dispose of waste, and manufacture skeletal elements. Further, amoebocytes can change into other cell types.

        NOTE: Sponges function as complex colonies of differentiated but interchangeable cells, the amoebocytes being the central cell type that forms most of the others. If a sponge is pressed through a sieve and all the cells and skeletal elements are separated, they will reorganize themselves back into the same layers and similar shape.

        6. Sponges are likely to have been a very early offshoot from the multicellular organisms that gave rise to the animals. Sponges retain several protistan characteristics, including not having a digestive track and having intracellular digestion. Developmentally, sponges do not go through a gastrula stage and thus their three cell layers are not homologous to those of other animals. Sponges also lack nerves and muscles.

        7. Sponges likely descended from choanoflagellates (Figure 19.3D) in the late Precambrian. Choanoflagellates are colonial protists composed entirely of collar cells (choanocytes), but as protists, they do not show cellular differentiation.

    B. Cnidarians are radial animals with stinging threads (*Module 19.4*).

        1. Most cnidarians are marine.

        2. These animals may be in the form of a polyp (relatively fixed in position) or a medusa (swimming), or they may alternate between polyp and medusa forms. Both body plans have a central tubular body (surrounding a gastrovascular cavity), one opening ("mouth") into this cavity, and tentacles arranged around the mouth (Figures 19.4A, B).

        3. Along the tentacles are cnidocytes (stinger cells) that function in defense and the capturing of food. The coiled thread in each cell is discharged, stinging or entangling the prey or predator as it brushes against the cnidocyte (Master 19.4C).

4. Cnidarians trap food with their tentacles, then maneuver it into a gastrovascular cavity where it is digested and distributed throughout the body. Undigested food is eliminated through the mouth.

5. Cnidarians are built at the tissue level of construction, with several different cell types arranged in layers and having common functions. Muscle cells are arranged in groups that allow the body to extend, move tentacles, and contract. Nerve tissue coordinates this movement.

6. Development includes a gastrula stage, like all the remaining animal phyla.

III. Most animals are bilaterally symmetrical (*Module 19.5*).

A. Bilateral symmetry means that an animal can be divided into mirror images (left and right sides) by a single plane (Acetate 19.5).

1. Associated with bilateral symmetry is the division of the animal into head (anterior), tail (posterior), back (dorsal), bottom (ventral), and side (lateral) surfaces.

2. The head houses sensory structures and the brain.

3. Bilateral animals are fundamentally different from radial animals. In contrast to radial animals that sit or drift passively, bilateral animals actively move through their environments headfirst.

B. Flatworms are the simplest bilateral animals (*Module 19.6*).

1. Flatworms are classified in the phylum Platyhelminthes. There are three major groups of flatworms: free-living planarians that live on rocks in marine and fresh water, parasitic flukes, and tapeworms.

2. Like cnidarians, planarians and most flukes have a gastrovascular cavity and no other body cavities. The body normally has a head end with a concentration of sensory nerves. The mouth opens from the ventral surface, and the gastrovascular cavity runs and branches through the entire length of the body (Master 19.6A).

NOTE: Flatworms are not quite at the organ level of construction.

3. Most flukes have a complex life cycle, including reproduction in more than one host and one or more larval stages. *Schistosoma* (blood fluke) adults live and permanently mate inside blood vessels, where they feed on blood. As many as 1000 fertilized eggs a day are produced, and they leave through the host's intestine. The eggs grow into larvae that infect snails. Asexual reproduction in the snail produces different larvae that infect humans (Master 19.6B).

4. Tapeworms are highly adapted parasites that inhabit the digestive tracts of their hosts. Unlike other flatworms, tapeworms are segmented and lack a gastrovascular cavity, absorbing their predigested food directly. The anterior end attaches to the host with hooks and suckers, and a region behind generates segments. Sexual reproduction occurs in each segment; the oldest segments break off and leave the host via the feces. Many tapeworms produce larvae that infect the prey animal, while the adult tapeworms infect that prey animal's predator (Master 19.6C).

NOTE: Although drugs are available to treat many parasitic flatworm infections (such as niclosamide for the human parasite *Taeniarhynchus*, as mentioned in the text), it is sometimes very difficult to kill these parasites without harming the host. The treatment for schistosomiasis is long and not always successful. These parasites, after all, are more closely related to their human hosts than are bacterial, fungal, and protozoan parasites.

IV. Most animals have a body cavity (*Module 19.7*).

    A. *Review:* Animal development (Acetate 19.1).

    B. Depending on how tissue regions develop from the gastrula stage, three body plans exist among animal phyla (Acetate 19.7A, B, C).

        1. The body is solid, except for the gastrovascular cavity, as in flatworms and cnidaria.

        2. The body contains a pseudocoelom, an internal space in direct contact with the inner layer of the digestive tract, as in roundworms.

        3. The body contains a coelom, an internal space completely lined by tissue of the middle layer, as in all other animals.

        4. For each body plan, the skin and other outer layers develop from the outside of the gastrula, the inner lining of the digestive tract or gastrovascular cavity (yellow) develops from the inside of the gastrula, and the middle layer (pink) develops from outgrowths of the inner layer.

        5. *Preview:* Two different paths of development of the middle layer distinguish two lines of animal evolution (Module 19.23).

    C. Advantages of having a body cavity:

        1. Provides greater flexibility.

        2. Helps cushion internal organs and provides room for internal organ growth.

        3. In some animals (like earthworms), body cavities under pressure function as a skeleton.

        4. The fluid in the cavity circulates nutrients and oxygen and aids in waste collection and disposal.

    D. Roundworms have a pseudocoelom and a complete digestive tract (*Module 19.8*).

        1. Roundworms, classified in the phylum Nematoda, are numerous and diverse in most environments. Most are important decomposers or parasites of plants or animals.

        2. The roundworm body is cylindrical, includes a complete intestinal tract, and is covered by a tough, nonliving cuticle (Figure 19.8A).

        3. Food passes in one direction along a digestive tract that includes regions specialized for certain functions: food intake, breakup, digestion, absorption, and waste elimination.

           NOTE: Roundworms are tending toward the organ level of construction.

        4. One free-living species, *Caenorhabditis elegans*, is an important organism for genetic research and one of the best-understood organisms. Researchers have been able to trace the developmental lineage of each of an adult's 1000 cells.

        5. Trichinosis is a disease caused by the roundworm *Trichinella spiralis*. Humans get this parasite by eating raw pork (Figure 19.8B).

    E. Diverse mollusks are variations on a common body plan (*Module 19.9*).

        1. Mollusks are classified in the phylum Mollusca.

2. The basic body plan of a mollusk includes bilateral symmetry, a complete digestive tract, a coelom, and many internal organs.

3. Two distinctive characteristics of the phylum are a muscular "foot" and a mantle, an outgrowth of the body surface that drapes over the animal, functions in sensory reception, often secretes a shell, and usually houses gills that function in gas exchange and waste removal.

4. Mollusks also have a true circulatory system (in contrast to the circulatory function of the gastrovascular cavity of cnidarians and flatworms, and the pseudocoelom of roundworms).

5. Most have a radula, an organ used to scrape food, such as algae, off surfaces in the environment (Acetate 19.9A).

6. Evolution has modified the basic body plan in different groups of mollusks. Gastropods include snails and slugs. Bivalves include clams, scallops, and oysters. Cephalopods include octopuses and squids (Figures 19.9B–F).

7. Cephalopods have large brains and sophisticated sense organs.

   NOTE: The cephalopod eye and the vertebrate eye are an example of convergent evolution (Module 16.11).

V. Most animals have segmented bodies (*Module 19.10*).

A. Segmentation is the subdivision of the body into repeated parts.

B. In an earthworm, the segments are clearly visible from the outside, outlining the repeating pattern of organs inside. The nervous, circulatory, and excretory systems all have repeating, mostly identical parts in each segment (Acetate 19.10A).

C. In other animals, segmentation may involve fewer segments or be less obvious (Figures 19.10B, C).

D. Advantages of segmentation:

   1. Greater body flexibility and mobility.

   2. Aids movement: In the earthworm, rhythmic alternating contractions and elongations of segments propel the worm into or along the ground. In many animals, the segments are the sites of insertion of walking legs or muscles.

      NOTE: There are two important evolutionary advantages. Genetically speaking, it is easier to build a large, complex animal by repeating a single developmental sequence in many smaller units rather than following a longer sequence of developmental steps for the whole region. Also, during evolution, animal groups (polychaete annelids, the insects and other arthropods, and the chordates) have segments specialized into different functional regions, working from a basic pattern; segments are often fused.

E. Earthworms and other segmented worms are called annelids (*Module 19.11*).

   1. These worms are classified in the phylum Annelida. They live in the sea, in most freshwater habitats, and in damp soil. Annelids usually have one or more anterior segments specially modified into a head region.

   2. Earthworms are one group of annelids adapted to life in soil. They consume the soil, digest the organic parts, and eliminate undigested soil

and other waste products in their feces, improving soil texture in the process.

3. Polychaete worms are mostly marine. They are characterized by segmental appendages with broad, paddlelike appendages and bristles. Some polychaetes show modification of anterior segments (Figures 19.11A, B).

4. Leeches are free-living carnivores of aquatic animals or blood-sucking parasites on vertebrates (Figure 19.11C). A blood-sucking leech cuts the skin with razor-sharp jaws and secretes an anesthetic and an anticoagulant. Leeches are still used in medicine, and their anticoagulants have recently been produced by genetic engineering.

F. Arthropods are the most numerous and widespread of all animals (*Module 19.12*).

1. In terms of diversity, geographical distribution, and sheer numbers, the phylum Arthropoda is the most successful that has ever lived.

2. Arthropods probably evolved from annelids or segmented ancestors of annelids. Fossils of animals with intermediate characteristics have been found in Cambrian rocks laid down 550 mya.

3. Arthropods are segmented (often fused), have jointed appendages, and have an exoskeleton composed of chitin and proteins (Acetate 19.12A).

Preview: To facilitate movement, muscle tissue is attached to the inside of the exoskeleton (Module 30.2).

4. To grow, arthropods molt their exoskeleton, swell in size, and thicken a new, developing exoskeleton.

5. Horseshoe crabs are "living fossil" life forms that have survived for hundreds of millions of years with little change (see Module 15.8: equilibrium). A very close relative of the modern genus was abundant 300 mya (Figure 19.12B).

6. Arachnids include scorpions, spiders, ticks, and mites. Their ancestors were among the first terrestrial carnivores. Except for mites, arachnids are carnivores (Figure 19.12C).

7. Crustaceans include crabs, shrimps, lobsters, crayfish, and barnacles; they are mostly aquatic (Figure 19.12D).

8. Millipedes have segments with two pairs of appendages each and feed on decaying plants. Centipedes have segments with one pair of appendages each and are carnivorous. Both groups are terrestrial (Figure 19.12E).

G. Insects are the most diverse group of organisms (*Module 19.13*).

1. About one million insect species are known to biologists, perhaps half of those that exist. They have been important aspects of terrestrial life for 400 million years, but less so in aquatic and, especially, marine habitats.

2. Insects are united as a group in having a three-part body plan with head, thorax, and abdomen. The head has sensory appendages and mouthparts specialized for a particular diet. The thorax contains three pairs of walking legs and, usually, one or two pairs of wings (Acetate 19.13A). The abdomen houses digestive and reproductive organs.

3. Grasshoppers (order Orthoptera) have biting and chewing mouthparts. Most species are herbivorous (mantids are an exception). They have two pairs of wings.

4. Damselflies and dragonflies (order Odonata) have biting mouthparts and are carnivorous. They have two identical pairs of wings (Master 19.13B).

5. True bugs (order Hemiptera) have piercing, sucking mouthparts, and most species feed on plant sap (bedbugs feed on blood). They have two pairs of wings (Master 19.13C).

6. Beetles make up the largest order (Coleoptera) in the animal kingdom, with some 500,000 known worldwide from all types of habitats. They have biting and chewing mouthparts, and most species are either carnivorous or herbivorous. They have two pairs of wings, and the forewings serve as protective covering for the hindwings (Master 19.13D).

7. Moths and butterflies (order Lepidoptera) are the second most numerous insects. They have drinking-tube mouthparts for sipping nectar or other liquids, and have two pairs of scale-covered wings (Master 19.13E).

8. Flies, gnats, and mosquitoes (order Diptera) have lapping mouthparts and feed on nectar or other liquids (mosquitoes have piercing, sucking mouthparts and suck blood). They have a single pair of functional wings, with the hindwings reduced to halteres to maintain balance (Master 19.13F).

9. Ants, bees, and wasps (order Hymenoptera) are the third most numerous insects. They have chewing and sucking mouthparts, and many are herbivorous. They have two pairs of wings. Many in this group display complex behavior and social organization (Master 19.13G).

NOTE: These social groups function as "superorganisms." Some aspects of their social behavior are covered in Chapter 37.

VI. A second branch of animal evolution.

A. Echinoderms have spiny skin, an endoskeleton, and a water vascular system for movement (*Module 19.14*).

1. The phylum Echinodermata includes sea stars and sea urchins (all marine) and represents a second branch of evolution in the animal kingdom. Similarities in embryonic development suggest that this phylum is closely related to our own, the phylum Chordata.

2. Echinoderms lack segmentation and bilateral symmetry as adults, but larvae are bilaterally symmetrical. Members of this phylum are noted for their regenerative capacity. Most have tubular endoskeletons composed of fused plates lying just under the skin. Unique to the phylum is the presence of a water vascular system. This is a network of water-filled canals that branch into extensions called tube feet (Acetate 19.14A).

3. Sea stars have flexible "arms" that bear the tube feet. They wrap these arms around a bivalve prey, pull the valves apart, extrude their stomach out their mouth and into the opening, and digest the soft parts (Figure 19.14B).

4. Sea urchins are spherical, with five double rows of tube feet running radially, with which they pull themselves along. They eat seaweeds (Figure 19.14C).

B. Our own phylum, Chordata, is distinguished by four features (*Module 19.15*).

1. A dorsal, hollow nerve cord.

2. A notochord: a flexible, longitudinal rod located between the digestive tract and the nerve cord.

3. Gill structures (slits and supports) in the pharynx region behind the mouth.

4. A post-anal tail.

5. The most diverse chordates are vertebrates. However, there are several invertebrate groups in this phylum.

6. Lancelets are small, bladelike chordates that live anchored by their tails in marine sands and expose their heads and mouths, filtering and trapping organic particles in mucus around the gill slits. These animals show the clearest presentation of the chordate body plan (Master 19.15A, B). Molecular evidence indicates that lancelets are the closest living relatives of vertebrates.

7. Tunicates (sea squirts) are another group of marine chordates. As adults, they do not exhibit the chordate pattern of notochord and nerve cord. As stationary filter feeders, they use gill slits much like lancelets. Larval tunicates do exhibit the complete chordate pattern and look very much like adult lancelets. It is likely that during the Cambrian (500 mya), by paedomorphosis (Module 16.7), vertebrates evolved from chordates similar to larval tunicates.

VII. Vertebrates.

A. A skull and a backbone are hallmarks of vertebrates (*Module 19.16*; Master 19.16).

1. A skull forms a case for the brain.

2. A segmented backbone composed of vertebrae encloses the nerve cord.

3. Most vertebrates have skeletal support for paired appendages (fins, legs, arms, wings).

4. The vertebrate skeleton is an endoskeleton of cartilage or bone. These nonliving materials contain living cells and grow as the animal grows, unlike the exoskeletons of arthropods, which must be molted prior to growth.

B. Most vertebrates have hinged jaws (*Module 19.17*).

1. One group of vertebrates, the lampreys (class Agnatha), lack jaws but have skeletal supports between their gill slits, and they lack paired appendages. Otherwise, they are superficially similar to fishes. Most are parasites on fish, boring a hole in the host and sucking its blood (Figure 19.17A).

2. Jaws evolved by modification of the first two pairs of skeletal supports of the gill slits. Similar events occur during the embryonic development of all fishes today (Acetate 19.17B).

3. The oldest fossils of jawed vertebrates appear in rocks that were formed 450 mya. Jaws enabled vertebrates to catch and consume a wider variety of foods than were available to filter feeders.

C. Fishes are jawed vertebrates with gills and paired fins (*Module 19.18*).

1. Fish extract oxygen from water with their gills. Their paired fins help stabilize their bodies (Acetate 19.18C).

2. The cartilaginous fishes (class Chondrichthyes) include the sharks and rays and have skeletons of flexible cartilage. Sharks have a keen sense of smell and can sense minute vibrations with a pressure-sensitive lateral line system. Because they cannot pump water through the gills, they must swim in order to move the water (Figure 19.18A).

3. Bony fishes (class Osteichthyes) are more common and diverse and have a stiff skeleton of bone reinforced by hard calcium salts. Like sharks, they have a keen sense of smell and a lateral line system. In addition, they have a keen sense of sight, and a bony flap over the gills helps move water through the gills when the fish is stationary (Figure 19.18B).

4. Bony fish also have a swim bladder that can act as a buoyant counterbalance to their heavier bones. Sometimes this is connected to the digestive tract, enabling certain species to gulp air to increase oxygen intake.

5. Bony fish, the largest class of vertebrates, are divided into two groups, the ray-finned fish and the lobe-fined fish. The fins of ray-finned fishes are supported by thin, flexible skeletal rays. The fins of lobe-finned fishes are muscular and supported by bones.

D. Amphibians were the first land vertebrates (*Module 19.19*).

1. Amphibians include frogs, toads, and salamanders.

2. Most amphibians are tied to water because their eggs and larvae (tadpoles) develop there.

3. Tadpoles are aquatic, legless scavengers with gills, a tail, and a lateral line system (Figure 19.19B).

4. They undergo a radical metamorphosis to change into an adult that is often a terrestrial hunter, with paired legs, external eardrums, air-breathing lungs, and no lateral line (Figure 19.19A).

5. Amphibians were the first land vertebrates and evolved about 400 mya from lobed-finned fishes. These fish had muscular fins and saclike lungs in their digestive tracts, both adaptations to living in habitats that often dried up or became stagnant (Master 19.19C).

6. Early amphibians thrived on insects in the coal-producing forests of the Carboniferous period. This period is known as the Age of Amphibians.

E. Reptiles have more terrestrial adaptations than amphibians (*Module 19.20*).

1. Reptiles include snakes, lizards, turtles, and crocodilians.

2. Reptiles have skin protected with protein (keratin), eggs with coatings that retain water, an internal fluid-filled sac (amnion) that bathes the embryo, and a food supply (yolk). The young hatch as juveniles, bypassing the need for free-living larvae (Figure 19.20A).

3. Most modern reptiles are ectothermic. They warm up by absorbing external heat rather than generating much of their own metabolic heat.

4. Following the decline of prehistoric amphibians, reptilian lineages expanded and dominated Earth during the Age of Reptiles (200–65 mya). These dinosaurs may have been endothermic (Figure 19.20C).

5. When the dinosaurs died off 65 mya (Module 16.5), one line survived and evolved into birds.

The Evolution of Animal Diversity

F. Birds share many features with their reptilian ancestors (*Module 19.21*).
    1. Birds (class Aves) have amniotic eggs, scales on their legs, and a reptilian body form. They evolved from one line of dinosaurs 150–200 mya.
    2. Feathers, the most distinctive characteristic of birds, are derived from scales. Feathers shape bird wings into airfoils that create lift and enable birds to maneuver in the air.
    3. *Archaeopteryx* was an early birdlike animal with feathers and a body much like a two-legged dinosaur. This body had few of the modern adaptations to lighten it (Figure 19.21A). *Archaeopteryx* is unlikely to be ancestral to modern birds.
    4. Many bird groups became extinct about 65 mya, along with the rest of the dinosaurs. Thus, the bird lineage appears to have gone through a bottleneck (Module 14.11) about 65 mya (the 65 mya mass extinction) and to have evolved from a very few surviving groups.
    5. Modern birds have additional adaptations for flight, including an absence of teeth, no claws on their wings, very short tailbones, hollow bones, large breast muscles, efficient lungs, and an extremely high rate of metabolism to provide energy.

G. Mammals also evolved from reptiles (*Module 19.22*).
    1. Mammals evolved from very early reptiles, about 225 mya, but were minor parts of habitats during the Age of Reptiles. Once the dinosaurs died off, the mammalian lineages underwent adaptive radiation.
    2. Most mammals are terrestrial, with a number of winged and totally aquatic species. The largest animal that has ever existed is the blue whale, a species that can reach 30 m in length.
    3. Mammals are endothermic, and they have hair and mammary glands.
    4. The monotremes, such as the platypus, are egg-laying mammals that live in Australia and Tasmania (Figure 19.22A).
    5. The marsupials, such as the American opossum and kangaroo, give birth to embryonic young that complete development in their mother's pouch. Most marsupials live in Australia, on neighboring islands, and in Central and South America.

        NOTE: Both groups evolved in isolation from other mammals in areas that were part of the Pangaean supercontinent, Gondwana (see Module 16.3).

    6. The placentals include all other mammals. Their embryos are nurtured inside the mother by a placenta, an organ that includes both maternal and embryonic vascular tissue.
    7. Humans are placental mammals belonging to the order Primates. The relationship of humans to other primates was discussed in Module 16.13.
    8. *Preview:* Unit V will cover many other details of the lives of animals, emphasizing humans, mammals, and vertebrates, in that order. Chapter 38 will examine further details of the evolutionary history of primates.

VIII. **A phylogenetic tree gives animal diversity an evolutionary perspective (*Module 19.23*).**

    A. *Review:* The names of the phyla and representative animals in this chapter (Table 19.23).

B. The major evolutionary steps of the evolution of the animal phyla are conveniently reviewed in a phylogenetic tree (Acetate 19.23).

1. Like most phylogenetic trees, this one is hypothetical and serves mainly to stimulate research and focus discussion.
2. Colonial protists were the ancestors of the animal kingdom.
3. Sponges may be a separate line of evolution, or a very early side lineage.
4. Two lines of tissue-level animals evolved: radially symmetrical animals, represented by cnidarians, and bilaterally symmetrical animals, the most primitive of which are represented by the flatworms.
5. Among the bilateral animals, two lines evolved: those with pseudocoeloms, represented by the roundworms, and those with coeloms, the most primitive of which are represented by the mollusks.
6. Among the coelomate animals, two lines evolved: those whose coelom develops from hollow outgrowths of the digestive tube, the echinoderms and chordates (deuterostomes); and those whose coelom develops from solid masses of cells, the mollusks, annelids, and arthropods (protostomes).
7. Segmentation in the annelids and arthropods on the one hand, and in the chordates, on the other, are secondarily derived characteristics of those lines.

IX. **Change is a fundamental feature of life and the environment (*Module 19.24*).**

A. Species change by adaptation through natural selection to a changing environment.

B. In turn, the presence of species changes the environment for others.

C. Chance has also affected the evolutionary process by providing variation through mutation and the events involved in sexual reproduction. Chance also played a role in the origin of living things in the first place. And later, chance played a role during many extinction events, such as when a collision from an extraterrestrial body caused extinction of the dinosaurs 65 mya.

D. *Preview:* We will return to the idea of change, over shorter periods of time, when we look at succession in Module 36.6.

CLASS ACTIVITIES

1. Bring in a sample of local pond water, and have the students key out the organisms found in the sample. Then have the students develop their own dichotomous key to the members of the class. This will illustrate both the shared (primitive) characters and the unique (derived) characters of species (Module 16.13).

RESOURCES AND REFERENCES

Brusca, R.G., and G.J. Brusca. *Invertebrates*. Sunderland, MA: Sinauer Associates, 1990. An evolutionary approach to invertebrate animals.

Carroll, R.C. *Vertebrate Paleontology and Evolution*. New York: W. H. Freeman, 1987. An authoritative and accessible text.

Fischman, J. "One That Got Away." *Discover*, January 1992. Using molecular systematics to trace our vertebrate roots to fish.

Forey, P., and P. Janveir. "Evolution of the Early Vertebrates." *American Scientist*, November-December 1994. The earliest fishes.

Gore, Rick, and O. Louis Mazzatenta. "The Cambrian Period. Explosion of Life." *National Geographic*, October 1993. An illustrated discussion of the fauna of the Burgess Shale and a recently documented similar fauna from China.

Gould, Stephen Jay. *Wonderful Life: The Burgess Shale and the Nature of History*. New York: Norton, 1989. Gould's best-seller about animal evolution, the role of fate in evolution, and the unique paleofauna in an isolated Cambrian deposit.

Gregory, Ed. "Tuned-In, Turned-On Platypus." *Natural History*, May 1991. How platypuses use electricity to feed.

Knoll, Andrew H. "End of the Proterozoic Eon." *Scientific American*, October 1991. An increase in oxygen levels and major tectonic change some 800 million years ago may have opened the door to large animals.

Lenhoff, Howard M., and Sylvia G. Lenhoff. "Trembley's Polyps." *Scientific American*, April 1988. Describes and illustrates elegant experiments done on hydras in the 1740s, marking the dawn of experimental zoology.

Levinton, Jeffrey S. "The Big Bang of Animal Evolution." *Scientific American*, November 1992. Almost 600 million years ago, animal evolution demonstrated an unmatched burst of creativity. Has the mechanism of evolution altered in ways that prevent fundamental changes in the body plans of animals?

Mitchell, L.G., J.A. Mutchmor, and W.D. Dolphin. *Zoology*. Menlo Park, CA: Benjamin/Cummings, 1988. An accessible introductory text.

Morris, S. Conway. "Burgess Shale Faunas and the Cambrian Explosion." *Science*, October 20, 1989. Among the components of these faunas appear to be survivors of the preceding evolutionary assemblages and a suite of bizarre forms that give unexpected insights into morphological diversification.

Nowak, R.M. *Walker's Mammals of the World*, 5th ed. Baltimore, MD: Johns Hopkins University Press, 1991. A comprehensive account.

Rice, J.A. (ed). "The Marvelous Mammalian Parade." *Natural History*, April 1994. Sixteen articles on extinct mammals.

Thomson, K.S. *Living Fossil: The Story of the Coelacanth*. New York: W.W. Norton & Company, 1991.

"The Thunder Lizards." *Natural History*, December 1991. Several articles about how dinosaurs lived.

Wellnhofer, Peter. "Archaeopteryx." *Scientific American*, May 1990. Although sometimes misclassified or even derided as a fraud, this prehistoric flier remains a rich source of information about the evolution of flight in birds.

Wiewandt, Thomas A. "Ancient Seas and Creatures. Images from Another, Distant Era." *Audubon*, November 1986. A lovely photographic essay about ancient animal fossils.

Within the Coral Wall. Oxford, England: Oxford Scientific Films, 1989. Available from Carolina Biological Supply Company. A videotape about Australia's Great Barrier Reef that examines the many types of coral animals, their food, and their predators.

Yoffe, E. "Silence of the Frogs." *The New York Times Magazine*, December 13, 1992. Amphibians are in global decline. What are the causes and ecological implications?

Zimmer, C. "Coming onto Land." *Discover*, June 1995. Vertebrates that first invaded land.

Zimmer, C. "Ruffled Feathers." *Discover*, May 1992. The controversy about the origin of birds.

20

UNIFYING CONCEPTS OF ANIMAL STRUCTURE AND FUNCTION

APPROACH

The unit on animal structure and function begins with this chapter. The chapter provides important conceptual and organizational details that will help students interpret the remaining chapters in the unit. Spend time here previewing the entire unit.

To many students, the different tissue types (reduced to a manageable number in the text) may seem somewhat arbitrarily categorized. The system of general tissue types does reflect developmental pathways and helps organize our view of the body's diverse cellular patterns. In this chapter and all the others in Unit V, help students decide how much anatomical terminology is appropriate to learn. This decision must take into account the length and focus of your course. In a short, one- or two-quarter course without an animal focus, plan to cover this chapter and one or a few others in detail, rather than skipping around and assigning scattered modules throughout the unit.

In Chapters 21–30, module coverage is divided between anatomical description, physiology, and medical concerns. The anatomy is well illustrated and should pose little problem to students. Using transparencies or video illustrations of cells, tissues, and organs will help make the examples more concrete. The most challenging areas will be in physiology, for example, fluid flow from blood to interstitial fluid in capillary beds (Module 23.12). In such cases, be sure students understand the underlying physical, chemical, and biological bases of the functions, reviewing previous material as necessary. Throughout this unit, the "medical connection" can be used to provide interest in a particular subject. Each chapter has at least one module focusing on medical problems involving the system being studied. These will provide a constant connection to students' own biological concerns.

CHAPTER OBJECTIVES

Distinguish between the terms *tissue*, *organ*, and *organ system*.

Define each of the major tissue types, give examples from each category, and explain the relationship between structure and function for each example.

Name and describe each of the major organ systems in a mammalian body.

LECTURE OUTLINE

I. Introduction.

    A. Each animal species is an accumulation of different structural and functional adaptations to life in its particular environment.

    B. This is particularly evident when one studies animals in extreme environments such as the open oceans of the Southern Hemisphere, which are frequented by the wandering albatross (*Opening Essay*; chapter-opening photo).

1. Only a few birds can remain at sea, completely away from land. One difficulty is obtaining water: Seawater is hypertonic to body tissue and pulls water from those tissues. An albatross has special salt-excreting glands in its nostrils to rid its body of excess salts from the water and food it consumes.

2. An albatross has extraordinary flight capabilities that are adapted to the distribution of winds at sea. It glides downwind, then coasts sharply upwind, using the winds to provide lift (Master 20.0).

3. Both these functions stem from special structural adaptations of its body.

C. This chapter introduces the unit on animals. Each succeeding chapter examines the function and structure of a different organ system (or the combination of two closely interrelated systems).

D. In this chapter, the general, overall body organization is discussed: cells to tissues to organs to organ systems.

II. Structure fits function in the animal body (*Module 20.1*).

A. The ability to fly results from special structural features of birds.

1. Feathers are dead protein (keratin) formed into complex three-dimensional structures by special pits in a bird's skin. These form airfoils (Master 20.1).

    Review: Birds' feathers are derived from scales (Module 19.21).

2. A bird's bones are reduced in number and motility, allowing the wing to function as a unit, and they are hollow but strongly reinforced, to reduce weight.

    Review: These bones follow the typical tetrapod pattern (Module 16.11)

3. Flight muscles sit below the bird, mostly off the wings, so the wings do not have to work hard to move the weight. This position also provides balance.

    NOTE: The bones of the wings of birds and bats are homologous, following the typical tetrapod pattern. Bird flight and bat flight are analogous. Contrast how bats, mammals which developed independently of birds, fly with how birds fly. The power stroke of birds, a downward stroke, uses only the pectoralis major muscle. In contrast, bats have four power stroke muscles, the pectoralis major and three other muscles (anterior serratus, cleidodeltoideus, and subscapularis).

B. Animal structure has a hierarchy (*Module 20.2*).

1. *Review:* Hierarchy of organization (Module 1.1; Master 1.1).

2. There is even a hierarchy of structure in a feather, from the molecules of keratin, to the arrangements of keratin in the feather's parts (shaft, barbs, and barbules), to the arrangement of these parts into the whole feather.

    NOTE: There are genetic advantages in building structures by repeating parts. Recall the advantages of segmentation in animal evolution (Module 19.10).

3. In the whole animal, the hierarchy of structure is as follows: cells, tissues (cooperating cells), organs (cooperating tissues), organ systems (cooperating organs), organism (cooperating organ systems) (Acetate 20.2).

III. Tissues are groups of cells with a common structure and function (*Module 20.3*).

   A. *Review:* Animal cell junctions (Module 4.19; Acetate 4.19B).

   B. The cells composing a tissue (from the Latin word for "weave") are specialized: Their particular structure enables them to perform their particular function.

   C. Cells in tissues are held together within the context of nonliving material they organize, with sticky glue that coats the cells, or with special membrane junctions.

   D. There are four major categories of tissue: epithelial, connective, muscle, and nervous.

   E. Epithelial tissue covers and lines the body and its parts (*Module 20.4*; Acetate 20.4A–D).

      1. Epithelial tissue occurs as sheets of closely packed cells. One "free" surface forms barriers or exchange surfaces; the other surface is attached to underlying tissues by a basement membrane.

         Review: The attachment to the underlying basement membrane is accomplished by a type of anchoring junction (Module 4.19).

      2. Tissues are categorized according to the number of cell layers and the shape of the individual cells.

      3. The structure of each type of epithelium fits its function.

      4. Stratified squamous epithelium regenerates rapidly by division of the cells at its attached surface; it covers surfaces that are subject to abrasion, such as the epidermis and the lining of the esophagus.

      5. Simple squamous epithelium is thin and leaky, suitable for the exchange of materials by diffusion; it lines our lungs and blood vessels.

         Preview: The lungs, a component of the respiratory system, are discussed in Chapter 22.

      6. Cuboidal epithelium and columnar epithelium have large cells that make secretory products and form large, often folded, surface areas. They line the digestive tract and air tubes, where they form a moist epithelium, a mucous membrane.

      7. The mucous membranes of air tubes are important for keeping debris out of the lungs. Particles get trapped in the mucoid secretions, and cilia beat them up and out of the air tubes.

         NOTE: Smoking paralyzes cilia and thus allows debris to reach the lungs that would otherwise be trapped and removed. Further, the paralysis of cilia in the oviducts probably contributes to the higher incidence of ectopic pregnancies among smokers.

   F. Connective tissue binds and supports other tissues (*Module 20.5*; Acetate 20.5A–F).

      1. Connective tissue consists of a sparse population of cells scattered in a nonliving matrix that is synthesized by the cells.

      2. Loose connective tissue is a loose weave of the protein collagen; it holds many other tissues and organs in place.

      3. Adipose tissue contains fat to pad and insulate the body and store energy.

4. Blood has a fluid matrix (plasma, consisting of water, salts, and proteins) and red and white blood cells. It functions in transport and immunity.

   Preview: Blood, a component of the circulatory system, is discussed in more detail in Chapter 23. Immunity is discussed in greater detail in Chapter 24.

5. Fibrous connective tissue consists of densely packed collagen fibers that form tendons (muscles to bone) and ligaments (bone to bone).

6. Cartilage is strong but flexible skeletal material with collagen fibers embedded in a rubbery matrix.

7. Bone is rigid tissue made of collagen fibers embedded in calcium salts.

   Preview: Bone tissue is discussed in more detail in Modules 30.5 and 30.6.

G. Muscle tissue functions in movement (*Module 20.6*; Acetate 20.6).

   Preview: Skeletal muscle is discussed in more detail in Modules 30.8 and 30.9.

   1. Muscle tissue consists of bundles of long muscle cells and is the most abundant tissue in most animals.

   2. Skeletal muscle is attached to bones by tendons and is responsible for voluntary movement. Its cells are multinucleate, striated, and unbranched.

   3. Cardiac muscle causes the involuntary contractions of the heart. Its cells are striated and branched.

   4. Smooth muscle is found in the walls of the digestive tract, urinary bladder, and arteries. Its cells are unstriated, spindle-shaped, and cause slow, but strong, involuntary movements.

H. Nervous tissue forms a communications network (*Module 20.7*).

   Preview: The nervous system is discussed in more detail in Chapters 28 and 29.

   1. The nervous system functions to relay information regarding the internal and external environments and to relay information from one part of the body to another.

      NOTE: The nervous system and the endocrine system (Chapter 26) are the control and communication systems of the body. The difference is that the nervous system acts more rapidly than the endocrine system, and the effects of nervous system activity are not as long-lasting as those of the endocrine system.

   2. Nervous tissue consists of interconnected neurons, cells specialized to conduct nerve signals, and other cells that support the neurons. They function in transmitting sensory signals and in coordinating internal events.

   3. Each neuron has a cell body, dendrites that transfer messages to the cell body, and axons that transfer messages away from the cell body (Acetate 28.2).

   4. Between neurons, signals are transferred by the diffusion of chemicals.

IV. Organs and organ systems.

A. Multiple tissues are arranged into organs (*Module 20.8*; Master 20.8).

   1. All animals except sponges have some organs.

2. Organs consist of several tissues adapted to perform specific functions as a group. They perform functions that none of the component tissues can perform alone.

3. The heart consists of muscle (the major proportion, providing the contractile, pumping force), epithelial tissue (providing a smooth, low-friction inner surface), connective tissue (tying all the tissues together into a strong, elastic structure), and nervous tissue (directing the contractions).

4. The stomach consists of muscle tissue (smooth muscle churns and moves the food), epithelial tissue (secretes digestive juices and protects other tissues from the juices' enzymatic functions), and connective tissue (binds all into an elastic structure that surrounds the lumen). There are a few nerves to sense and regulate its function.

B. The body is a cooperative of organ systems (*Module 20.9*).

1. An organ system is a group of several organs that work together to perform a vital body function.

2. In vertebrates, there are twelve organ systems. Each one is introduced below, followed by the chapter in which it is covered.

3. Digestive system. Organs of the digestive tract ingest food, break it down into smaller chemical units, absorb these units, and eliminate the unused parts (Chapter 21; Master 20.9A).

4. Respiratory system. The lungs and associated breathing tubes exchange gases with the environment (Chapter 22; Master 20.9B).

5. Cardiovascular system. The heart and blood vessels supply nutrients and O_2 to the body and carry away wastes and CO_2 (Chapter 23; Master 20.9C).

6. Lymphatic and immune systems. Lymph vessels and nodes supplement the work of the cardiovascular system, particularly as components of the immune system, a diffuse system of cells (including lymphocytes and macrophages, both of which are types of white blood cells) and processes that protect the body from foreign invasion (Chapter 24; Master 20.9D). Together, the lymphatic and cardiovascular systems are referred to as the circulatory system.

7. Excretory system. The kidneys, bladder, and urethra remove nitrogen-containing wastes from the blood and maintain osmotic balance (Chapter 25; Master 20.9E).

8. Endocrine system. The endocrine glands secrete into the blood hormones that regulate most other activities (Chapter 26; Master 20.9F).

9. Reproductive system. There are two separate systems, one in females and one in males. Ovaries and testes and associated organs produce female and male gametes, and help in fertilization and embryo development (Chapter 27; Master 20.G).

10. Nervous system. The brain, spinal cord, nerves, and sense organs work together with the endocrine system to sense the outside environment, affect responses, and coordinate body activities (Chapters 28 and 29; Master 20.9H).

11. Muscular system. All skeletal muscles provide movement as they work with the skeletal system (Chapter 30; Master 20.9I).

12. Skeletal system. Bones and cartilage provide support and protection, and work with the muscular system to provide movement (Chapter 30; Master 20.9J).

13. Integumentary system. Skin, hair, and nails protect the internal body parts from mechanical injury, infection, extremes of temperature, and drying out (Chapter 30; Master 20.9J).

V. New imaging technology reveals the inner body (*Module 20.10*).

A. X-rays show shadows of hard structures but fail to image soft tissues; X-rays produce flat, two-dimensional images.

B. Computerized tomography (CT) uses computers to combine the images produced by many weak X-ray sources. This technology can detect small differences between normal and abnormal tissues in many organs (Figure 20.10A, B).

C. Magnetic resonance imaging (MRI) measures changes in the magnetic signal when the hydrogen atoms in living materials are excited. MRI images soft tissues extremely well (Figure 20.10C).

D. Positron-emission tomography (PET) yields information about metabolic processes by imaging the pattern of radioactivity from isotopically labeled glucose or other metabolic precursors. PET is most valuable for measuring metabolic activity in the brain (Figure 20.10D).

VI. The relationship between an animal's outer and inner environments.

A. Structural adaptations enhance chemical exchange between animals and their environments (*Module 20.11*).

1. Animals are not closed systems; from the cellular through the organismal level of organization they must obtain materials from the outside environment and excrete metabolic wastes in that same environment.

2. In simple animals with gastrovascular cavities (cnidarians and flatworms, Modules 19.4 and 19.6), virtually every cell has plasma membrane exposed directly to the aqueous environment (Acetate 20.11A).

3. Most other animals have relatively smaller outer surfaces compared to their volumes. They rely on specialized, inner surfaces for the exchange of materials (Acetate 20.11B; Figure 20.11C).

4. Surface areas of the lungs, intestines, and kidneys provide for the exchange of materials between the outer environment and the blood. Bodies with greater numbers of cells to be serviced have correspondingly larger total surfaces of exchange.

5. The interstitial fluid mediates the exchange of materials between the blood and the body's inner cells.

B. Animals regulate their internal environment (*Module 20.12*).

1. Homeostasis is the maintenance of an organism's steady state in the face of environmental fluctuations (Acetate 20.12B).

NOTE: The term "steady state" should not be taken to mean unchanging. Homeostasis maintains the body in a dynamic equilibrium.

Preview: Module 26.8. A good example of homeostasis is the regulation of blood sugar levels. After a meal, when blood sugar levels rise, the body releases insulin to lower blood sugar levels. Between meals, when blood sugar levels have fallen, the body releases glucagon to stimulate the release of sugar into the blood.

2. An animal's homeostatic control systems maintain internal conditions within a range where life's metabolic processes can occur.

3. For example, our own bodies maintain salt and water balance and also keep our internal fluids at about 37°C.

C. Homeostasis depends on feedback control (*Module 20.13*).

1. A thermostat uses negative feedback control to keep the room temperature constant. When a sensor falls below a set temperature, the heat turns on. When the sensor rises above that point, the heat turns off (Master 20.13A).

Review: Negative feedback with regard to cell metabolism is discussed in Module 5.8.

2. Maintenance of blood temperature in mammals (and most homeostatic mechanisms) functions by negative feedback. The brain's hypothalamus senses temperature and raises and lowers body temperature by sending nervous signals to two sets of structures in the skin: sweat glands and blood vessel networks (Acetate 20.13B).

Preview: Thermoregulation (Module 25.2) and thermoreceptors (Module 29.3).

CLASS ACTIVITIES

1. Throughout this unit, the use of a color video camera and TV monitor to detail structures and functions seen under a dissecting or compound microscope can be used to great advantage in lecture. These small cameras have excellent color rendition even at very low light intensities, so lighting will not interfere with the activity demonstrated. Your own, internal cheek epithelial cells and blood cells (previously included in many laboratory exercises but now banned from most student labs because of AIDS) can be rapidly prepared and examined in front of a class. The arrangement of tissues in organs (lung, heart, striated muscle, kidney, brain, and other material from slaughterhouses) can be demonstrated under a dissecting microscope to very large classes. Combine a living surface view of human skin with the details seen under a prepared slide showing a cross section of skin. Activities such as feeding, responses to stimuli, various kinds of movement, and the circulation of blood can be demonstrated in invertebrates, or with fish, frogs, and small reptiles.

2. Homeostasis is easily demonstrated. On a hot day people sweat; on a very cold day they shiver.

RESOURCES AND REFERENCES

Allen, Frank. *Human Light Microscope Anatomy*. Seattle, WA: Health Sciences Center for Educational Resources, 1992. A videodisc with over 2500 tissue-section images. Includes accompanying software for the IBM PC.

Animated Dissection of Anatomy for Medicine. Marietta, GA: A.D.A.M. Software, 1993. An advanced, interactive multimedia reference (computer software and videodisc formats) on human anatomy. Designed for medical students but simple and flexible enough for demonstrations to introductory nonscience majors. Details of various body regions can be peeled away, showing the relationships among systems. Additional extensions include demonstrations of medical imaging technology, animations, and pathologies. Available in Macintosh and Windows platforms.

Caplan, I. "Cartilage." *Scientific American*, October 1984. The structure and function of an important tissue.

Diamond, J. "Building to Code." *Discover*, May 1993. Discussion of the safety factors built into animals as a result of natural selection.

Eckert, R., and D. Randall. *Animal Physiology: Mechanisms and Adaptations*, 3rd ed. New York: W.H. Freeman, 1988. A widely used textbook of comparative animal physiology.

Goldberger, Ary L., David R. Rigney, and Bruce J. West. "Chaos and Fractals in Human Physiology." *Scientific American*, February 1990. A photographic essay showing how chaos theory can explain self-similarity in levels of organ structure in a number of systems, help define the healthy body, and foreshadow disease.

Learning All About Dissection. Fairfield, CT: Queue, 1992. A CD-ROM introduction to animal dissection procedures, as well as an overview of dissection as a scientific tool. Vivid color photographs of actual dissections, including in-depth studies of the earthworm, crayfish, fish, frog, and fetal pig. Available in Macintosh and IBM PC versions.

The Living Body Series. Princeton, NJ: Films for the Humanities and Sciences, 1987. A 26-part series of half-hour videotape programs covering all aspects of human anatomy and function. Uses cinematography, microphotography, and animation to illustrate all the aspects of animal structure and function covered in this unit. Also available in videodisc format, with indexing and chaptering.

Marieb, E. *Human Anatomy and Physiology*, 2nd ed. Redwood City, CA: Benjamin/Cummings, 1992. A basic text, beautifully illustrated.

National Geographic Society. *Medical Technology*. Washington, DC: National Geographic Society, 1986. A 25-minute videotape that demonstrates advances in modern medical technology, including imaging systems discussed in this chapter, the use of computers to design prosthetics, and noninvasive methods of surgery.

Schmidt-Nielson, K. "How Are Control Systems Controlled?" *American Scientist*, January-February 1994. Many unanswered questions.

Schmidt-Nielsen, K. *Animal Physiology: Adaptations and Environment*, 4th ed. New York: Cambridge University Press, 1990.

Sochurek, H., and P. Miller. "Medicine's New Vision." *National Geographic*, January 1987. Methods of photographing the internal human body.

INTERNET RESOURCES FOR UNIT V

Arthritis Foundation

http://www.arthritis.org

Food and Consumer Service

http://www.usda.gov/fcs

The Heart: An Online Exploration

http://sln2.fi.edu/biosci/heart.html

An exploration of the human cardiovascular system.

Netfrog Title Page

http://teach.virginia.edu/gdfrog

An on-line frog dissection.

New Jersey Online's Yucky Site: Worm World

    http://www.nj.com/yucky/worm/index.html

    Juvenile, but worth a look. Worm anatomy and ecology.

Nutrition Analysis and Information

    http://www.orst.edu/food-resource/nutrient/index.html

The Nutrition Expert

    http://www.alaska.net/~tne

    Registered dietitians.

USDA Home Page

    http://www.usda.gov

21

NUTRITION AND DIGESTION

APPROACH

Details of the human digestive system, diet, and dietary deficiencies are normally an important part of high school biology or health courses, but these subjects intrigue all of us. Despite this, most of your students probably have poor nutritional habits and many nutritional misconceptions. For example, poll the class to see how many students eat breakfast on a regular basis. Eating breakfast has been demonstrated to reduce the risk of early morning strokes and heart attacks.

A common approach to presenting the material is to follow the course of a specific meal, as is done in the text. A nutritional handbook will help you identify the nutrients in the food in any meal.

The main nutrients covered in this chapter are carbohydrates, lipids and proteins. It would be helpful to review their characteristics as discussed in Modules 3.4–3.16.

Be sure to let students know how many of the details presented in the text's illustrations and tables they are required to memorize.

You might want to investigate and report on how RDAs are determined. This would fit nicely into a general discussion of how to interpret conflicting claims for different vitamin regimens (Module 21.17).

CHAPTER OBJECTIVES

Describe the general ways that animals feed.

Distinguish between the four stages of the digestive process: ingestion, digestion, absorption, and elimination.

Explain how chemical digestion generally involves enzyme activity, in the environments provided by specific organs, to hydrolyze macromolecules into their monomeric units.

Describe the specific digestive functions of the oral cavity, esophagus, stomach, small intestine, large intestine, and the digestive glands associated with each.

Name and explain the three basic chemical needs satisfied by digested food.

Discuss the aspects of a healthy diet and the potential problems stemming from improper diet.

LECTURE OUTLINE

I. **Introduction.**

   A. *Review:* Feeding (ingesting food) is a distinctive characteristic of the animal kingdom.

   B. The humpback whale, from an unusual habitat, shows how an animal's structure and behavior are directly tied to feeding and food processing (*Opening Essay*).

1. Humpback whales are suspension feeders that strain small fish and crustaceans from the ocean. A 72-ton whale processes as much as 2 tons of food a day.

2. These whales use "bubble nets" to help concentrate their food at the surface. The mouth has a tremendous volume when expanded and uses the comblike baleen to sift the food from the water. The stomach can hold up to half a ton of food at a time.

3. For 4 months in the summer, these whales feed in the rich, cold oceans of polar regions and store up vast fat reserves. In the winter, they migrate to warm, southern oceans to breed. They eat little for 8 months until they return to the polar regions.

C. Animals ingest their food in a variety of ways (*Module 21.1*).

1. Feeding can be by absorption (as in a few parasitic worms) or by ingestion.

2. Ingestive feeding can be categorized by type of food. Omnivores (such as humans and crows) eat both plants and animals. Herbivores (such as deer or sea urchins) eat plants or algae. Carnivores (such as lions and spiders) eat only meat.

3. Ingestive feeding can be categorized by the size and location of the food. Those ingesters that do not consume larger prey whole or in pieces (most animals) suspension-feed on small animals (such as the humpback whale); suspension-feed on microscopic protists, plants, and animals (such as clams and other bivalves); substrate-feed by burrowing into their food (such as earthworms); or fluid-feed on plant sap (aphids) or animal fluids (mosquitoes).

II. Organization of the digestive process and the system that carries it out.

A. Food processing occurs in four stages: an overview (*Module 21.2*; Acetate 21.2).

1. Ingestion is the act of eating.

2. Digestion is the breakdown of food into small enough molecules to absorb. Digestion occurs in two steps: mechanical and chemical breakdown.

3. The products of digestion are then used for either cellular respiration or biosynthesis (Modules 6.16 and 6.17).

4. Absorption is the taking up of these small nutrient molecules.

    NOTE: Food does not actually enter the body until it is absorbed. Prior to absorption, food is in a tube (in the case of animals with an alimentary canal) that runs through the body.

5. Elimination is the release of undigested material.

6. Food consists of large polymeric fats, carbohydrates, proteins, and nucleic acids that animals cannot absorb directly. All animals need the same monomeric components: fatty acids, simple sugars, amino acids, and nucleotides.

7. During digestion, larger polymers are chemically digested into these smaller components by hydrolysis (Module 3.3; Acetate 3.3A, B). Each step is catalyzed by specific enzymes.

B. Digestion occurs in specialized compartments (*Module 21.3*).

1. These compartments provide environments that favor the action of the specific digestive enzymes and ensure that the enzymes will not attack an animal's own macromolecules.

2. Even single-celled organisms, such as amoeba, have specialized compartments for digestion. Sponges, like these single-celled organisms, carry out all of their digestive functions within their cells (Module 19.3).

3. Simple animals, such as cnidarians and flatworms, have a single digestive compartment, a gastrovascular cavity in which digestion and absorption occur, with a single opening for ingestion and elimination (Acetate 21.3A).

4. Other animals have a series of compartments (organs) arranged along a tube (alimentary canal) that extends between the mouth and anus (Acetate 21.3B).

5. Ingested food passes to the first cavity via a muscular pharynx and esophagus.

6. The first cavity may be a crop (a pouchlike organ for temporary storage and food softening), a gizzard (a muscular pouch that contains teeth or grit), or a stomach (a muscular pouch without grinding structures).

7. Chemical digestion and nutrient absorption occur mainly in the intestine. Intestines typically have modifications that increase their inner surface area and thus increase the absorptive surface.

8. Undigested materials are expelled through the anus.

9. The exact nature of an animal's alimentary canal reflects its diet. An earthworm is an omnivorous substrate feeder with an intestine that has an inner, dorsal fold to increase its absorptive area. A grasshopper is a herbivore with a number of adaptations for the efficient processing of plant material. Different birds eat different foods, but most store food in a crop and use a gravel-containing gizzard to grind food swallowed whole.

C. The human digestive system consists of an alimentary canal and accessory glands (*Module 21.4*; Acetate 21.4).

1. The main parts of the alimentary canal are the mouth, oral cavity, tongue, pharynx, esophagus, stomach, small intestine, large intestine, rectum, and anus.

2. Digestive glands—the salivary glands, pancreas, and liver—secrete digestive enzymes into the cavities with which they are associated.

NOTE: These glands secrete into a duct; this makes them exocrine glands. In contrast, endocrine glands secrete into the blood (Module 20.9).

3. Food is propelled through the alimentary canal by wavelike contractions (peristalsis) of smooth muscle.

4. Sphincter muscles control the passage of food from one cavity to the next.

5. The total digestive process takes about 5–6 hours.

NOTE: Lipids take longer to digest than carbohydrates and proteins.

III. Ingestion and the beginnings of digestion.

A. Digestion begins in the oral cavity (*Module 21.5*; Master 21.5).

1. Salivary juices contain lubricants, buffers, antibacterial agents, and a digestive enzyme (salivary amylase) that hydrolyzes starch. The release of salivary juices is triggered by meal time and the sight or smell of food.

NOTE: Oral stimulation also triggers the secretion of saliva. Other functions of saliva include helping keep teeth clean, dissolving food so that it can be tasted, and aiding in the formation of the bolus.

   2. Mechanical and chemical digestion begin in the oral cavity as food is chewed.

   3. Humans have four kinds of teeth (arranged in four sets, right and left in the upper and lower jaw): two bladelike incisors for biting, one pointed canine for tearing, two premolars, and three molars for grinding and crushing food.

   NOTE: Only animals with a palate and cheeks chew their food; the palate prevents food from entering the nasal cavity while chewing, and the cheeks prevent food from falling out of the mouth. Compare how a crocodile gulps its food with how a human chews its food. Vertebrates vary greatly in their complement of teeth. A pattern of dentition is related to an animal's diet. For example, horses have incisors to shear off, and molars to grind, the grass they eat, but they have no canines. Rodents have strong, continually growing incisors for gnawing on cellulose-rich plant materials. Lions and other carnivores have prominent canines with which they tear large hunks of flesh from their prey.

   4. The tongue tastes the food, manipulates the food, and shapes the food into a bolus, which it then pushes to the back of the oral cavity and into the pharynx (swallowing).

   B. The food and breathing passages both open into the pharynx (*Module 21.6*).

   1. Most of the time, when not eating, the human pharynx opens into the windpipe for breathing and speaking (as air vibrates vocal cords in the voicebox).

   2. When a bolus of food passes into the pharynx, the swallowing reflex is triggered. The esophageal sphincter muscle relaxes, the tracheal opening is closed off by the epiglottis, and the food passes into the esophagus (Master 21.6).

   C. The esophagus squeezes food along to the stomach (*Module 21.7*).

   1. Esophageal muscles are arranged in two smooth muscle layers, one circular and the other longitudinal.

   2. Peristalsis moves the bolus down the esophagus toward the stomach (Master 21.7).

   NOTE: The esophagus itself has no digestive function. However, salivary amylase continues to act on the food during its passage through the esophagus. The digestion of carbohydrates stops in the acidic environment of the stomach and then continues in the small intestines.

IV. Digestion, absorption, and elimination.

   A. The stomach stores food and breaks it down with acid and enzymes (*Module 21.8*).

   1. The stomach can store up to 2 liters of food. It empties its contents slowly (after 2–6 hours) by opening the pyloric sphincter.

   NOTE: Fats remain in the stomach longer than do carbohydrates and proteins. Thus, one feels fuller after a high-fat meal than after a low-fat meal. This is the phenomenon that is responsible for the cliché about being hungry one hour after eating a meal at a Chinese restaurant.

2. The inner surface of the stomach is highly folded and has pits that terminate in gastric glands (Master 21.8).

3. Chemical digestion continues in the stomach and is aided by contractions of smooth muscle in the stomach wall. The digestion of proteins into smaller polypeptides occurs by the action of the enzyme pepsin.

4. Gastric juice also includes mucus, which protects the stomach lining and also lubricates, and hydrochloric acid, which converts pepsinogen to pepsin and provides the proper pH for the action of pepsin.

   NOTE: The HCl secreted by the stomach has a pH of ≈1. HCl is also important in denaturing proteins so as to allow greater exposure of peptide bonds to pepsin, deactivating hormones present in food, and killing bacteria. Further, HCl is important for the absorption of nutrients such as vitamin B_{12} and iron.

5. Gastric activity is initiated by a nervous signal from the brain (after seeing, tasting, or smelling the food) and is continued by the secretion of gastrin, a gastric gland hormone, when food is actually present in the stomach.

6. The release of gastric juice by the gastric glands, under the control of gastrin, is a negative-feedback mechanism.

7. Occasional backflow of the stomach contents (acid chyme) into the esophagus causes heartburn (what the commercials like to call acid reflux). During vomiting, the contents are eliminated by reverse peristalsis back up the esophagus and out of the oral cavity.

   NOTE: The lower esophagus and region around the heart both signal the same nerve to the brain. Therefore, the presence of acid in the esophagus is perceived as coming from the region near the heart. A not uncommon problem is a hiatal hernia; this occurs when a portion of the stomach protrudes upwards through the esophageal hiatus (the opening in the diaphragm through which the esophagus passes). An individual with a hiatal hernia may be asymptomatic, or may experience problems with acid reflux (heartburn).

B. Bacterial infections can cause ulcers (*Module 21.9*).

1. A gastric ulcer is an open sore on the stomach lining. The major symptom is pain in the upper abdomen associated with eating.

2. At one time ulcers were thought to be due to the overproduction of pepsin and/or acid. However, evidence now indicates that a major cause of ulcers is the bacterium *Helicobacter pylori*.

   NOTE: This applies to the small intestines as well as the stomach. In addition to *H. pylori*, the other major cause of ulcers is the (over)use of nonsteroidal anti-inflammatory drugs (NSAIDS) such as aspirin and ibuprofen. NSAIDS reduce inflammation by inhibiting prostaglandin synthesis. Prostaglandins play a major role in cytoprotection (protecting the cells lining the stomach from damage by HCl).

3. The body's response to an *H. pylori* infection results in stomach inflammation (gastritis). Gastritis may progress to an ulcer.

   NOTE: *H. pylori* infection is also associated with an increased cancer risk.

4. Gastric ulcers usually respond to an antibiotic regime in combination with drugs such as bismuth (the active ingredient in Pepto Bismol).

Nutrition and Digestion

NOTE: Not only is this treatment regime more effective than the older regime; it is also less expensive.

    5. Duodenal ulcers occur in the first portion of the small intestine.

C. The small intestine is the major organ of chemical digestion and nutrient absorption (*Module 21.10*).

    1. All remaining chemical digestion and most absorption of nutrients occur in the small intestine. This organ is about 6 meters long and 2.5 cm in diameter. Peristalsis moves the mixture.

    NOTE: The chyme that enters the small intestines from the stomach has a pH of ≈2–3.

    2. Digestion continues in the first 25 cm (the duodenum).

    3. Glandular secretions are released into the duodenum from the liver and gallbladder (the liver produces bile that contains salts to make fats more soluble; bile is stored in the gallbladder until it is needed in the small intestine) and from the duodenum wall and the pancreas (produce enzymes and bicarbonate ions to neutralize the acid chyme and raise its pH) (Master 21.10A).

    Preview: The role of the liver in homeostasis is discussed in more detail in Module 25.13.

    NOTE: Folate (a B vitamin that is of great importance during pregnancy) is secreted along with bile and is reabsorbed in the small intestines. Anything that inhibits this reabsorption can result in a folate deficiency.

    4. Each type of macromolecule (carbohydrates, proteins, fats, and nucleic acids) is digested sequentially by specific enzymes. The digestion of carbohydrates and proteins continues on fragments produced by previous chemical breakdown. The digestion of fats and nucleic acids starts here (Table 21.10).

    5. The surface area of the lower part of the small intestine is huge, with several levels of folding. The wall is folded into circular pleats. These pleats contain projections (villi) of cells, and the cells have further projections (microvilli). The total surface area is about 600 m^2 (Master 21.10B).

    6. The core of each villus contains capillaries and lymph ducts. Nutrients diffuse from intestine chamber to blood, or they are moved across microvillous membranes by an ATP-requiring transport mechanism.

    7. Nutrient-laden blood from the small intestine passes to the liver, which gets the first chance to process or store the nutrients, particularly storing excess glucose as animal starch (glycogen).

D. The large intestine reclaims water (*Module 21.11*).

    1. The large intestine is about 1.5 m long and 5 cm in diameter (Master 21.11).

    2. About 7 liters of digestive contents pass into the large intestine each day. About 90% of its water is absorbed back into the blood.

    3. Colon bacteria, including *E. coli*, live in the undigested material. They produce and release important vitamins (biotin, folic acid, B vitamins, and vitamin K) that humans cannot make themselves.

NOTE: The amount of a vitamin synthesized by the intestinal fauna that is available for absorption is not the same for each vitamin. For example, intestinal bacteria can meet ≈50% of an adult's need for vitamin K; whereas it is not yet known how much of the biotin synthesized by the intestinal bacteria is absorbed.

    4. The remaining, undigested material is compacted by peristalsis and stored in the rectum until it is defecated.

    5. The appendix is a gland at the top of the large intestine that has a minor immune system function. Appendicitis occurs if the appendix becomes infected following irritation, or when its opening is jammed by undigested food.

V. Adaptations of vertebrate digestive systems reflect diet (*Module 21.12*).

  A. Herbivores and omnivores have longer alimentary canals than carnivores, to allow more time and surface area for digesting plant material (Figure 21.12A).

  B. Dietary needs change from the larval stage to the adult stage in amphibians. The alimentary canal of the larva is proportionally longer relative to body size than that of the adult.

  C. Herbivores have a variety of mechanisms to aid in the digestion of cellulose.

    1. Most rely on the cellulose-digesting enzymes of protozoans and bacteria, and populations of these organisms are housed in parts of the animals' alimentary canals.

    2. Rabbits, and some rodents, produce soft fecal pellets first, which include microorganisms that have digested the cellulose in the cecum (a pouchlike region where small and large intestine meet). They reingest these pellets and absorb the digested cellulose (glucose molecules) through their small intestines, and then defecate hard fecal pellets.

    3. Ruminant mammals, such as cattle and deer, have an elaborate, four-chambered stomach, part of which houses the microorganisms. Ingested grass enters the rumen and reticulum, where bacteria begin to digest the cellulose. Periodically, a cow regurgitates some of this material and helps mechanically digest it by "chewing the cud." It is then swallowed into the omasum, where water is absorbed, passing to the abomasum, where the cow's own enzymes complete the digestion process (Master 21.12B).

VI. Digestion and dietary needs.

  A. A healthful diet satisfies three needs: an overview (*Module 21.13*).

    1. Fuel to power all body activities.

    2. Raw materials needed to make an animal's own molecules.

    3. Essential nutrients (substances the animal cannot make itself).

  B. Chemical energy powers the body (*Module 21.14*).

  Review: Cellular respiration in Chapter 6.

    1. The energy content of food (carbohydrates, fats, and, when these are in short supply, proteins) is measured in kilocalories (kcal), the accurate form of the popular word "calories."

  NOTE: A kilocalorie is the amount of energy required to raise one kilogram of water one degree Celsius.

2. The basal metabolic rate (BMR) is the amount of energy required to maintain all cellular metabolism in a resting animal. The average BMR for adult humans is 1300–1800 kcal per day.

3. Various levels of activity add to a human's caloric requirements, and various foods supply these requirements (Table 21.14).

4. Metabolic rates are measured by measuring the rate of oxygen consumption. Each liter of O_2 consumed liberates 4.83 kcal from food (Figure 21.14).

C. Body fat and fad diets (*Module 21.15*).

1. Fat is an essential component of the human body. It insulates the body against cold, and a moderate amount is correlated with a healthy immune system. Below-normal amounts of fat interfere with vitamin A formation and may make people susceptible to some cancers.

2. Those fatty acids that our body cannot make for itself and that are required in the diet are called essential fatty acids.

3. Ideally, fat should be 20–25% and 15–19% of the body weights of women and men, respectively. "Overfat" levels should be regarded to be about 20% higher than ideal levels.

4. Too much body fat increases the chance of developing certain diseases, such as heart disease, and decreases life span.

5. Fad diets not only are usually ineffective but may be harmful (Table 21.15).

6. Following the guidelines for the Recommended Dictary Allowances (RDAs) for nutrients, and getting regular aerobic exercise, can keep body fat at normal levels.

NOTE: One good way to add body fat is to consume fat. When carbohydrate and protein consumption increases, the body's metabolism of these nutrients increases (however, overfeeding/force-feeding of either carbohydrates or proteins will result in their conversion to body fat). In contrast, when fat consumption increases, metabolism of fat does not increase; thus body fat is added. Another good way to add body fat is to consume alcohol. Alcohol blocks the entry of acetyl CoA into the Krebs cycle (Module 6.11), and instead acetyl CoA is converted into fatty acids.

D. Nine amino acids are essential nutrients (*Module 21.16*).

1. The human body can make a great variety of organic molecules (including 11 amino acids) from basic sources of organic carbon and nitrogen provided in digested food.

2. Some substances (essential nutrients) cannot be made and must be obtained directly from food.

3. Meat and egg products provide all nine essential amino acids.

4. A combination of plant proteins can also provide all nine essential amino acids (Master 21.16).

E. A healthful diet includes 13 vitamins (*Module 21.17*; Table 21.17).

1. A vitamin is an organic nutrient that is essential but required in much smaller quantities than the essential amino acids.

2. Most vitamins serve as coenzymes, or parts thereof, that are reused in metabolic reactions or in a variety of roles in maintaining cellular health.

Review: Enzyme function is discussed in Modules 5.5–5.9.

3. Vitamins are grouped into those that are water-soluble and those that are fat-soluble. Unlike water-soluble vitamins, excess fat-soluble vitamins are not easily eliminated from the body and build up in body fat, where they may have toxic effects.

4. RDAs have been established for all vitamins, but some people recommend higher levels of vitamins C and E.

5. Extreme deficiencies of each vitamin cause specific sets of symptoms.

F. Essential minerals are required for many body functions (*Module 21.18*; Table 21.18).

1. Minerals are chemical nutrients other than carbon, hydrogen, oxygen, and nitrogen.

2. All minerals are essential. Depending on their roles in structure and function, essential minerals are required in various amounts.

3. Too much of some minerals (such as the sodium in salt) can cause high blood pressure in humans.

NOTE: Recent studies are showing that only a subset of the population is sodium-sensitive and that those individuals who are not sodium-sensitive have less need to be concerned about their sodium intake (but shouldn't overdo it). Further, it appears that African Americans with hypertension can lower their blood pressure by lowering their salt intake and by taking drugs that increase salt excretion by the kidneys. On the other hand, medications that reduce cardiac output seem more effective for Caucasians with high blood pressure.

VII. What do food labels tell us? (*Module 21.19*; Figure 21.19)

A. A list of ingredients arranged according to weight, from the greatest amount to the least amount.

B. The number of kilocalories, carbohydrates (total and dietary fiber and sugars), proteins, fats (total and saturated fats and cholesterol), and selected vitamins and minerals supplied in one serving appear on the label and are expressed as percentages of a daily value.

VIII. Diet can influence cardiovascular disease and cancer (*Module 21.20*).

A. Linked to cardiovascular disease are diets low in fruits and vegetables and rich in saturated fats, which in turn are correlated to high levels of blood cholesterol. Cardiovascular disease is linked to high levels of low-density lipoproteins (LDLs), while high-density lipoproteins (HDLs) are correlated with lower risk of cardiovascular disease. Exercise tends to increase—and smoking to decrease—HDLs (Figure 21.10).

NOTE: Inherited (familial) hypercholesterolemia is discussed in Modules 5.20 and 9.10. However, as discussed here, lifestyle (lack of exercise, a high-fat diet) will also result in hypercholesterolemia.

B. Linked to some forms of cancer are high levels of dietary fat and low levels of dietary fiber (Table 21.20).

Review: See Module 8.11 for a discussion of the cellular basis of cancer. Also see Module 13.18 for a discussion of lifestyle and cancer risk.

NOTE: The benefits of fiber include slowing glucose absorption and lowering blood cholesterol levels. In addition, foods high in fiber tend to be lower in fats. Until the body adapts to a high-fiber diet, the result can be diarrhea or constipation, gas, and abdominal discomfort. A diet can also be too high in fiber; the result can be insufficient consumption of energy or nutrients, inhibition of nutrient absorption, and formation of phytobezoars (fiber balls that can obstruct the GI tract).

CLASS ACTIVITIES

1. Demonstrate the role of antacids by adding about 1 mL of vinegar to 400 mL of water containing Universal Indicator. The solution will start out orange, but after adding one antacid tablet and shaking the bottle, the pH will rise and the color will change to green or blue. The importance of buffers can be connected to this exercise. Because most antacids are buffered, they will not raise the pH beyond 7 or 8, and the solution will remain green. If you use a nonbuffered antacid such as milk of magnesia, the pH will continue to rise as the tablet dissolves, and the color of the solution will change from green to blue and finally to violet (pH 10). Since the Universal Indicator can be reused, you can add more vinegar to the bottle and repeat the color change.

2. There is a great deal of misinformation/misrepresentation concerning nutrition in the popular press. Have students bring in articles to critique.

3. Have students analyze and critique the diets they consume (there are many good diet analysis programs available) and discuss ways to improve their diet.

RESOURCES AND REFERENCES

Blaser, M.J. "The Bacteria Behind Ulcers." *Scientific American*, February 1996. Acidophile bacteria that cause ulcers and may also cause stomach cancer.

Booth, D.A. "The Physiology of Appetite." *British Medical Bulletin*, May 1981. Addresses some implications for appetite physiology of the fact that the appetites and satieties are partly learned and are not congenital.

Brown, J.L., and E. Pollitt. "Malnutrition, Poverty and Intellectual Development." *Scientific American*, February 1996. The importance of nutrition during early development.

Christian, J.L., and J.L. Greger. *Nutrition for Living*, 4th ed. Redwood City, CA: Benjamin/Cummings, 1994. A popular nutrition text.

Diamond, J. "The Athlete's Dilemma." *Discover*, August 1991. Have endurance athletes reached the limits of what metabolism can support?

In the Company of Whales. New York: The Discovery Channel, 1989. A 90-minute videotape that describes the work of Dr. Roger Payne on many whale species: their natural history, ecology, and interactions with humans.

McKinney, Shortie. *MacDiet Academic.* Santa Barbara, CA: Intellimation, 1993. Computer software; a Macintosh HyperCard database that enables a student to study the nutritional details of his or her own food intake as they relate to the student's general dietary needs and level of physical activity. Includes a database of over 2500 foods and 24 nutrients.

National Geographic Society. *Digestive System.* Washington, DC: National Geographic Society, 1988. A 17-minute videotape that traces the events in digestion.

Pennington, Jean A.T., and Helen Nichols Church. *Bowes and Church's Food Values of Portions Commonly Used*, 14th ed. New York: Harper & Row, 1985.

Sanderson, S., and R. Wasserug. "Suspension-Feeding Vertebrates." *Scientific American*, March 1990. An emphasis on whales.

Tufts University Diet & Nutrition Letter. An excellent source of up-to-date nutrition information.

Uvnäs-Moberg, Kerstin. "The Gastrointestinal Tract in Growth and Reproduction." *Scientific American*, July 1989. Describes why this set of organs can be considered to be the largest endocrine gland in the body, and how it has a significant role in the readjustment of metabolism that accompanies pregnancy as well as fetal and infant growth.

Whitney, E.N., and S.A. Rolfes. *Understanding Nutrition*, 7th ed. New York: West, 1996. One of the best basic human nutrition texts.

Willett, W.C. "Diet and Health: What Should We Eat?" *Science*, April 22, 1994. A critical analysis of international diets.

22

RESPIRATION: THE EXCHANGE OF GASES

APPROACH

In light of the great interest in health and exercise, nonscience majors will be interested in the material in this chapter. Despite this, the majority of Americans get insufficient exercise.

In addition to tracing anatomical structure and function, which most students will have been previously introduced to, two basic concepts of body function are covered. The countercurrent principle, introduced in Module 22.4, is an important concept that functions widely. This principle is well illustrated. The general principle of diffusion has been previously introduced, and its role in gas exchange is mentioned often in this chapter. Be sure to emphasize one aspect that is not detailed; in spite of its simplicity, students may not recognize it: As molecules are used up, the concentration of a substance decreases, and as a waste product accumulates, its concentration increases. These processes drive both ends of the respiratory process where diffusion occurs: in the lungs and in the tissues.

CHAPTER OBJECTIVES

Determine the relative advantages and disadvantages of gas exchange in aquatic and terrestrial environments.

Identify four types of respiratory surfaces found in different animal groups.

Describe the countercurrent principle of substance exchange.

Name the organs involved in the vertebrate respiratory system, and explain the function of each.

Show how the circulatory system of an adult functions in gas transport, emphasizing the circulatory pattern and the role of hemoglobin in holding gases and buffering blood.

Describe the aspects of the bird's respiratory system that make it particularly efficient.

Explain the nature of fetal gas exchange and the drastic changes in gas exchange that occur with the birth of a baby.

LECTURE OUTLINE

I. Introduction.

    A. *Review:* Cellular respiration: Animals need to obtain oxygen and glucose and rid themselves of waste carbon dioxide (Chapter 6; Master 6.2A; Acetate 6.8).

    B. Life at high altitude imposes many changes on the organs and tissues that function in respiration (*Opening Essay*).

        1. People born in and adapted to high altitudes have relatively large hearts, more red blood cells, and greater hemoglobin levels.

        2. A short period of conditioning will help those living in lower altitudes acclimate to higher altitudes. Faster heart rate and larger capillary diameter are replaced over time with deeper and more rapid rates of

breathing, more capillaries, and higher numbers of red blood cells and levels of hemoglobin.

    3. Many animals are capable of exchanging gases from environments humans would find inhospitable. Some birds can stand the cold and low oxygen concentrations of altitudes of 20,000–30,000 feet. They have more efficient lungs, hemoglobin with a very high affinity for oxygen, a larger number of capillaries, and muscle proteins that hold oxygen.

C. Gas exchange involves breathing, the transport of gases, and the servicing of tissue cells: an overview (*Module 22.1; Acetate 22.1*).

    1. Breathing involves inhaling O_2 and exhaling CO_2.

    2. The transport of gases involves diffusion into and transport by hemoglobin in the red blood cells of the circulatory system.

    3. The blood supplies every cell with O_2 and picks up waste CO_2.

II. Animals exchange O_2 and CO_2 through moist body surfaces (*Module 22.2*).

A. Respiratory surfaces vary among animal groups. However, what all respiratory surfaces have in common is that they must be moist, they must be thin, and they must be extensive. Gases must be dissolved in water before they can diffuse in or out. In each part of Acetate 22.2A–D, the circle represents a cross section of the animal's body in the region of the respiratory surface, and the green color represents the respiratory surfaces.

    1. Earthworms and other "skin-breathers" (Module 19.6) must live in moist environments to keep their skins moist. Small size or flatness provides the high ratio of respiratory surface to body volume required for efficient gas exchange between environment and cells (Acetate 22.2A).

    2. Gills have evolved in most aquatic animals to increase the respiratory surface. They generally project from the body surface (Acetate 22.2B).

    3. Tracheae are specialized breathing tubes found in insects. These branched tubes bring external gases directly to the inner cells without the aid of the circulatory system (Acetate 22.2C).

    4. Lungs are found in the majority of terrestrial vertebrates (Acetate 22.2D). They are composed of branched tubes ending in tiny internal sacs lined with a moist epithelium. Gases are carried between the lungs and body cells by the circulatory system.

B. Gills are adapted for gas exchange in aquatic environments (*Module 22.3*).

    1. The chief advantage of exchanging gases with water is that energy does not have to be expended to keep the transfer surface wet.

    2. However, the concentration of O_2 is only 3–5% of its concentration in air, and the warmer and saltier the water, the less O_2 it can carry. Consequently, gills must be very efficient to extract O_2 from water.

    3. A fish ventilates its gill surfaces with water by "inhaling" water with its opercula closed and mouth opened, and "exhaling" the water across the gills with its mouth closed and opercula opened.

    4. Oxygen-poor blood enters each gill filament, crosses the lamellae (red blood cells travel single file here), picking up O_2 and leaving CO_2 (Acetate 22.3).

C. Countercurrent flow in the gills enhances O_2 transfer (*Module 22.4*).

1. Countercurrent exchange is a general principle of transfer found in many animal systems.

   NOTE: For example, a countercurrent system is used in thermoregulation (Module 25.2) and to enhance water reuptake in the kidneys (Module 25.11).

2. It is the transfer of a substance from a fluid flowing in one direction to another fluid moving in the opposite direction.

3. The opposite flows maintain a diffusion gradient that enhances the transfer of the substance, O_2 in the case shown (Acetate 22.4).

   NOTE: To impress students with the efficiency that results from this arrangement, diagram a transfer system where both fluids flow in the same direction.

4. The mechanism is so efficient in fish that their gills remove more than 80% of the oxygen dissolved in the water flowing through them.

D. The tracheal system of insects provides direct exchange between the air and the body cells (*Module 22.5*).

   1. Air contains much more O_2 than an equal volume of water, and air is easier to move than water. Thus, terrestrial animals expend less energy in ventilating their respiratory surfaces.

   2. Tracheae in an insect branch throughout the body, conveying air directly to body cells (Acetate 22.5A, B).

   3. Included in the system are tracheal air sacs that work like bellows when muscles around them alternately contract and relax, moving air out and in.

   4. Water is conserved, and respiratory surfaces remain moist because only the ultimate narrowest tubes, the tracheoles, contain fluid. It is across the tracheoles that gas exchange occurs.

III. Terrestrial vertebrates have lungs (*Module 22.6*).

A. Since lungs are restricted to one part of the body, unlike tracheaa, the circulatory system must be involved in transporting the gases to and from body cells.

B. Amphibians supplement their lungs with skin breathing, but all other terrestrial vertebrates (and aquatic reptiles and mammals) have efficient lungs only.

C. The organs of the human respiratory system (Acetate 22.6A).

   1. The nasal cavity filters, warms, humidifies, and samples odors of incoming air.

   2. The pharynx controls the passage of air through the mouth region and into the larynx (Module 21.6).

   3. Exhaling through the vocal cords of the larynx produces sounds.

   4. A branched system of tubes (trachea and bronchi) lead into the lungs.

      NOTE: This branched system is another example of the hierarchical organization of life (Module 1.1).

   5. Lungs include the ultimate branches of the bronchioles and the grapelike clusters of alveoli. Gas exchange occurs across the alveolar surfaces, a total surface equal to the size of a racketball court (Acetate 22.6B, Figure 22.6C).

6. All surfaces of the respiratory system are lined by moist epithelium. In all but the alveoli and smallest bronchioles, this tissue is covered by cilia and a thin film of mucus that helps eliminate dust, pollen, etc.

7. A muscular diaphragm helps move air in and out of the lungs.

D. O_2 in inhaled air dissolves in a film of moisture lining the alveoli, then diffuses across the epithelial cells and into a web of capillaries that surrounds the alveolus. CO_2 diffuses the other way.

E. Smoking is one of the deadliest assaults on our respiratory system (*Module 22.7*).

1. A breath of air in a polluted city may contain thousands of chemicals, many potentially harmful. Air pollutants such as sulfur dioxide, carbon monoxide, and ozone are associated with serious respiratory diseases, and asbestos fibers and radioactive radon gas have been linked with lung cancer.

2. Cigarette smoke is one of the worst sources of toxic air pollutants. Components are known to irritate epithelial cells and destroy the cilia along the bronchi.

3. Emphysema is a disease of cigarette smokers characterized by the alveoli becoming brittle and eventually rupturing.

4. Lung cancer is ten times more likely to occur in smokers than nonsmokers (Figure 22.7).

   NOTE: The leading cause of death among smokers is cardiovascular disease. The list of the adverse effects of smoking is a long one; some of the more important ones to emphasize are the effects of both active and passive smoking on prenatal development, infants, and children. These effects include (but are not limited to) an increased probability of cleft lip and cleft palate, decreased transfer of vitamin C to the fetus, increased risk of low birth weight babies, increased risk of allergies, delayed lung development, and increased risk of pneumonia. Smoking has been linked to SIDS (sudden infant death syndrome). Smokers are more likely to experience an ectopic pregnancy than nonsmokers, and smokers taking birth control pills are more likely to experience adverse side effects (Module 27.8).

F. Breathing ventilates the lungs (*Module 22.8*).

1. During inhalation, the rib cage expands, the rib muscles and diaphragm contract, and the chest expands. The lungs also increase in size. These changes reduce the air pressure within the alveoli, and air flows in as a result of the higher pressure outside (Acetate 22.8A).

2. During exhalation, the rib muscles and diaphragm both relax, decreasing the volume of the rib cage and forcing air out.

   NOTE: The elastic cartilage holding the rib cage together helps increase and decrease the rib cage's volume.

3. The normal volume of each breath is about 500 mL. The maximum volume that one can inhale and exhale, the vital capacity, is about 3500 mL and 4800 mL for college-age females and males, respectively. The air that remains in the lungs after complete exhalation is the residual volume. This is proportionally greater (relative to vital capacity) in older and more diseased people.

4. Gas-exchange systems of birds are more efficient than those of mammals. Birds maintain a one-way flow of air between two air sacs in addition to lungs. The air tubes within the lungs have no residual volume of air because all the air travels through the lungs in one direction (Acetate 22.8B).

G. Breathing is automatically controlled (*Module 22.9*).

1. Although breathing can be consciously controlled, most of the time automatic control centers in our brain regulate our breathing movements (Acetate 22.9).

2. These control centers are located in the lower parts of the brainstem, the pons and the medulla oblongata. About 10–14 times a minute, nerves from those areas signal the diaphragm and rib muscles to contract.

3. Increased cellular respiration causes increased concentrations of CO_2 in the blood. CO_2 reacts with water to form carbonic acid, lowering the pH. The medulla (Module 28.16) senses the pH drop and increases the rate and depth of breathing, thus eliminating more CO_2 from the blood in the lungs.

   NOTE: This is one of the mechanisms that results in panting during and after strenuous exercise. Also contributing to the acidity of the blood during strenuous exercise is the buildup of lactic acid (Module 6.15).

4. During severe depression of O_2 levels in the blood, sensors on arteries near the heart signal the breathing control center. This response may occur at high altitudes, where required levels of O_2 cannot be obtained by normal breathing.

IV. Gas transport.

A. Blood transports the respiratory gases, with hemoglobin carrying the oxygen (*Module 22.10*).

1. The human circulatory system functions in gas transport. One side of the heart pumps O_2-poor, CO_2-rich blood from the body to the lungs, and the other side of the heart pumps O_2-rich, CO_2-poor blood from the lungs to the rest of the body (Acetate 22.10A).

2. Every gas in a mixture accounts for a portion (that gas's partial pressure) of the mixture's total pressure. At each location (lungs and tissues), gases are exchanged as they diffuse along their own partial pressure gradient.

3. O_2 is not very soluble in water. Hemoglobin in red blood cells has a much higher affinity for O_2. Hemoglobin consists of four polypeptide chains, each chain attached to a heme chemical group with an iron atom in its center. Each iron atom can carry one O_2 molecule (Master 22.10B).

B. Hemoglobin helps transport CO_2 and buffer the blood (*Module 22.11*).

1. Chemical moderation of pH level is known as buffering. By helping transport CO_2, hemoglobin helps buffer the blood.

2. Within red blood cells, CO_2 reacts with water to form H_2CO_3. This breaks into acidic H^+ and basic HCO_3^- ions more quickly in red blood cells under the control of an enzyme there. Hemoglobin picks up most H^+ ions and allows most HCO_3^- ions to diffuse back into the plasma. This

provides a buffer in the blood that will react with any H^+ ions that are picked up elsewhere (Acetate 22.11A).

    3. When blood flows through the lungs, the process is reversed. H^+ ions are given up by the hemoglobin, reacting with the HCO_3^- ions to form H_2CO_3. This is then converted back into CO_2, and the CO_2 diffuses from the blood to the air (Acetate 22.11B).

  C. The human fetus exchanges gases with the mother's bloodstream (*Module 22.12*).

    1. The fetus lies within a watery bath of amniotic fluid. Its lungs are filled with this fluid.

    2. Capillaries from the fetal blood supply (through the umbilical cord) mix with capillaries of the uterus in the placenta (Acetate 22.12).

    3. A countercurrent arrangement of these capillaries facilitates the transfer of gases between fetus and mother.

    4. When the baby is born, when placental transfer stops, CO_2 concentration in the blood increases, lowering the pH and stimulating the breathing center, causing the baby to take its first breath.

Preview: Refer to this module when discussing fetal changes that take place during the third trimester of pregnancy (Module 27.17).

CLASS ACTIVITIES

1. A bell-jar model of lungs with a diaphragm can be built or purchased from biological supply companies. To build a model, take a Y-shaped tube and place balloons on the two ends (these will be the lungs), secure the stem of the Y-shaped tube in a stopper, and place this in the neck of the bell jar. Secure a piece of rubber across the wide, open end of the bell jar (this will be the diaphragm); the bell jar itself plays the role of the thoracic cavity. When the rubber diaphragm is pulled on, the volume of the bell jar increases and air will enter the balloons. When the rubber diaphragm is allowed to return to its original position, the volume of the bell jar will decrease and air will be forced out of the balloons. This is an elegantly simple way of demonstrating the forces that fill and empty mammalian lungs.

2. Handheld spirometers can be used to determine lung capacities. Compare males with females, smokers with nonsmokers, and those who exercise on a regular basis with those who do not.

RESOURCES AND REFERENCES

Bartecchi, C.E., T.D. MacKenzie, and R.W. Schrier. "The Global Tobacco Epidemic." *Scientific American*, May 1995. Tobacco use is on the rise.

Feder, Martin E., and Warren W. Burggren. "Skin Breathing in Vertebrates." *Scientific American*, November 1985. Skin breathing can supplement or replace breathing through lungs or gills in a wide variety of vertebrates from fish to mammals. Special adaptations of the skin and circulatory system help regulate gas exchange.

Houston, C.S. "Mountain Sickness." *Scientific American*, October 1992. How high altitude interferes with homeostasis.

Stroh, Michael. "Breathing Lessons." *Science News*, May 9, 1992. The use of computers to develop models for how breathing works.

U.S. Department of Health and Human Services. *The Health Benefits of Smoking Cessation.* U.S. Department of Health and Human Services, Public Health Service, Centers for

Disease Control, Center for Chronic Disease Prevention and Health Promotion, Office on Smoking and Health. DHHS Publication No. (CDC) 90-8416, 1990.

U.S. Department of Health and Human Services. *Reducing the Health Consequences of Smoking: 25 Years of Progress. A Report of the Surgeon General*. U.S. Department of Health and Human Services, Public Health Service, Centers for Disease Control, Center for Chronic Disease Prevention and Health Promotion, Office on Smoking and Health. DHHS Publication No. (CDC) 89-8411, 1989.

U.S. Department of Health and Human Services. *The Health Consequences of Involuntary Smoking*. U.S. Department of Health and Human Services, Public Health Service, Centers for Disease Control. DHHS Publication No. (CDC) 87-8398, 1987.

U.S. Department of Health and Human Services. *The Health Consequences of Using Smokeless Tobacco*. U.S. Department of Health and Human Services, Public Health Service. DHHS Publication No. (CDC) 86-2874, 1986.

23

CIRCULATION

APPROACH

Circulation is central to all other body systems. If you want to cover one system well, this should be the one. Be sure to indicate to students the various ways in which the circulatory system interacts with other systems (particularly the respiratory, immune, and endocrine systems). Emphasizing this chapter is also important because of the prevalence of circulatory diseases and the need for students to understand them.

CHAPTER OBJECTIVES

Contrast the internal transport systems found in different animals.

Compare the cardiovascular systems of fish and mammals.

Describe the location and function of the organs in the human cardiovascular system, tracing the flow of blood through the entire system.

Explain the features of the heartbeat, particularly as they are perceived when listening to the beat and measuring blood pressure.

Outline what happens during a heart attack and the common symptoms of a heart attack.

Describe how the blood supply is regulated to tissues.

Name the various components of blood, and explain their roles.

LECTURE OUTLINE

I. Introduction.
   A. Most animals have a circulatory system, for the internal transport of gases, nutrients, and waste.
   B. Gravity has had major effects in shaping the evolution of circulatory systems in terrestrial organisms as different as corn snakes and giraffes (*Opening Essay*).
      1. Strong hearts are able to pump against the force of gravity, even in tall animals.
      2. Muscles used in normal activities contract around veins and force blood back to the heart through one-way valves.
      3. Tight skin and abundant connective tissue keep blood vessels from enlarging.
      4. In the corn snake, veins have no valves, but tail vessels constrict during a climb, and a snake will wriggle after a climb to increase circulation.
   C. The circulatory system associates intimately with all body tissues (*Module 23.1*).

   Review: Chemical exchange between an animal and its environment (Module 20.11).

1. Diffusion is inadequate for transporting chemicals over distances greater than a few cell widths.
2. Capillaries are the smallest vessels and form an intricate net among the cells of every tissue (Figure 23.1A, Master 23.12A).
3. The various components of blood, particularly red blood cells, come in close enough contact with associated cells that materials can diffuse between them, via the interstitial fluid (Acetate 23.1B).
4. In most tissues, O_2 and nutrients diffuse from blood to tissue, and CO_2 and metabolic wastes diffuse from tissue to blood.
5. The circulatory system also functions in homeostasis by exchanging molecules with the interstitial fluid and by moving the blood through organs such as the liver and kidneys, where the blood's contents are regulated.

II. Several types of internal transport have evolved in animals (*Module 23.2*).

A. Not all animals have a circulatory system like a mammal.

1. The gastrovascular cavity of the cnidarians provides a path for external water to bathe all the cells (Module 21.3). Digestion also occurs in this cavity, and the contents are moved about by flagella on the cells lining it. In jellyfish, this path can be intricately branched, but it is not closed (Acetate 23.2A).
2. Many invertebrates (particularly arthropods and mollusks) have open circulatory systems. Blood is pumped by one or more hearts through open-ended vessels and flows out among the cells. There is no separate interstitial fluid. Pores in the hearts function as valves, opening when the hearts relax to pull in blood from the tissues (Acetate 23.2B).
3. Other invertebrates and all vertebrates have closed circulatory systems (also called cardiovascular systems). Blood is confined to vessels, which keep it distinct from the interstitial fluid (Acetate 23.2C).

   NOTE: Point out that in the overhead transparencies, O_2-rich blood is indicated by red and O_2-poor blood by blue.

4. In closed systems, arteries carry blood away from the heart, veins return blood to the heart, and capillaries convey blood between these two vessel types within each organ.
5. A fish system includes four gills, each with thousands of gill capillaries, on each side of the head, and a two-chambered heart (atrium receives and ventricle pumps out).

B. Vertebrate cardiovascular systems reflect evolution (*Module 23.3*).

1. The switch from gill breathing in aquatic vertebrates to lung breathing in terrestrial vertebrates was accompanied by drastic changes in the circulatory systems.

   NOTE: Although just fish and mammal systems are compared in this module, a progression of evolutionary adaptations in the circulatory systems of amphibians, reptiles, and birds link these two extremes.

2. Fish have a single circuit of blood flow, with the heart receiving and pumping only O_2-poor blood (Acetate 23.3A).

NOTE: Point out that the overhead transparencies show the circulatory system as though the animal were facing you, with its right side on your left. Historically such illustrations were oriented to illustrate animal dissections, with the open body cavity facing up.

    3. Mammals have two circuits, with a four-chambered heart having two atria and two ventricles. The pulmonary circuit carries blood from the right side of the heart to the lungs, and the systemic circuit carries blood from the left side of the heart to the rest of the body. This double system provides rapid delivery of O_2-rich blood to body tissues of highly active mammals (Acetate 23.3B).

III. A trip through the mammalian cardiovascular system (*Module 23.4*).

  A. The heart (Acetate 23.4A).

    1. This organ is composed mostly of cardiac muscle tissue.

    Review: Cardiac muscle fibers are connected to one another by specialized cell junctions (Module 4.19). These specialized cell junctions are called intercalated discs. Intercalated discs are a combination of anchoring junctions and communicating junctions.

    2. Thin-walled atria receive blood, which then flows to the ventricles.

    3. Thick-walled ventricles pump blood to other organs.

    4. Valves between chambers and between ventricles and main arteries maintain the flow in one direction.

  B. Body blood flow (Acetate 23.4B).

    1. The flow of blood follows this path: (a) right ventricle to lungs via pulmonary arteries, (b) lungs to left atrium via pulmonary veins, (c) left atrium to left ventricle, (d) left ventricle to all body organs via the aorta, (e) body organs to right atrium via superior and inferior vena cavae, (f) right atrium to left atrium.

    NOTE: An artery conveys blood away from the heart. A vein conveys blood to the heart. The pulmonary arteries and veins are exceptions to the general statement that arteries carry oxygen-rich blood and veins carry oxygen-poor blood. Other exceptions are the umbilical arteries and vein.

    2. The left ventricle is correspondingly larger because it pumps blood to a larger, more distant volume of tissue than the right ventricle.

    3. The aorta is the largest vessel in the human body, and branches from it supply blood to various body regions.

    NOTE: The first arteries that branch off the aorta supply blood to critically important organs, first the heart itself and then the head, including the brain and many sense organs.

  C. The structure of blood vessels matches their functions (*Module 23.5*; Master 23.5).

    1. Capillaries, which supply cells, have thin-walls composed of a single layer of epithelial cells wrapped in a thin basement membrane. Such a thin surface facilitates the diffusion of molecules to and from the interstitial fluid.

    2. Arteries contain blood under the pressure produced by the heart; they are thick-walled, with a smooth epithelial layer and layers of connective and

smooth muscle tissue for reinforcement and to regulate blood flow by constriction.

3. Veins contain blood under pressure and are similar in structure to arteries but many also contain valves to prevent backflow.

   Preview: Veins are under less direct pressure from the heart. Thus, it is not direct pressure from the heart that is responsible for the venous return of blood but several other factors. Factors that play a role include negative pressure created by the enlargement of the thoracic cavity during inhalation, skeletal muscle movement pushing blood along, and valves assuring unidirectional blood flow. See Module 23.9 and Master 23.9B.

D. The heart contracts and relaxes rhythmically (*Module 23.6*; Acetate 23.6).

   1. It passively fills with returning blood and actively contracts, pumping out blood.
   2. During diastole (lasts about 0.4 sec), the heart is relaxed, and blood flows into all four chambers, with all valves open.
   3. Systole begins as the atria contract (about 0.1 sec), forcing blood into the ventricles, and continues as the ventricles contract (about 0.3 sec), forcing the atrioventricular (AV) valves closed and the semilunar valves open.
   4. Cardiac output is about 75 mL per beat, or 5.25 liters per minute in the average person.
   5. The "lub-dupp" sound of a beating heart is from the closing of the AV valves ("lub") and the forcing of blood out the semilunar valves ("dupp").
   6. A heart murmur, sounding like a quiet hiss to the trained ear, occurs when a valve malfunctions, allowing blood to squirt back into a preceding chamber.

E. The pacemaker sets the tempo of the heartbeat (*Module 23.7*; Master 23.7).

   1. The pacemaker is a specialized region of cardiac muscle in the wall of the right atrium, also known as the sinoatrial (SA) node.
   2. When the SA node contracts, it sends out electrical signals, first to the atria, making them contract, and then to the AV node, which acts as a relay.

      NOTE: The relay function of the AV node is needed because the atria and ventricles are separated by nonconductive connective tissue. Also note that the conduction system of the ventricles is more extensive than that of the atria.

   3. The signals are delayed 0.1 sec in the AV node and then travel along special muscle fibers to the cardiac muscles of the ventricles, causing them to contract.
   4. The SA node sets the normal rate of contractions. The brain also can send signals to modify the basic rate, depending on body activity.
   5. If the pacemaker does not function correctly, an artificial one can be implanted next to the heart. This provides a regular electrical signal to trigger the beat.

      Preview: The medulla oblongata is also involved in the regulation of heart rate (Module 28.16).

IV. **What is a heart attack?** (*Module 23.8*)

   A. A heart attack is the death of cardiac muscle cells and the resulting failure of the heart to deliver enough blood to the rest of the body. Heart attacks follow clogging of the coronary arteries, blocking blood flow to regions of cardiac muscle (Master 23.8A).

   Review: Dietary influence on cardiovascular fitness (Module 21.20).

   B. Such clogs occur if blood clots back up behind constrictions (due to lipid buildup) in these arteries.

   C. Symptoms of heart attack: a squeezing ache in the center of the chest; radiating pain to the shoulder, arm, neck, or jaw; and sweating, nausea, shortness of breath, dizziness, or fainting.

   NOTE: Females and males typically exhibit different warning symptoms. For females, typical warning signs are the following: shortness of breath or fatigue when engaging in activities previously found easy (e.g., climbing stairs), angina that comes and goes (with or without exertion), and bouts of nausea and heartburn. For males, typical warning signs are the following: angina and pain that may radiate to the neck, upper abdomen, or left arm. However, males may exhibit symptoms that are typical of females and females may exhibit symptoms that are typical of males.

   D. Cardiac muscle cells do not regenerate but leave noncontracting scar tissue.

   E. Relief to heart disease patients includes coronary artery bypass surgery, and angioplasty and laser surgery to open up constricted coronary arteries.

   F. Cardiovascular disease has decreased somewhat as a result of increased awareness of the roles of diet and exercise in health, early diagnosis of problems, and the availability of trained personnel to help in emergencies.

V. **Blood pressure.**

   Preview: The hypothalamus plays a role in the regulation of blood pressure (Module 28.16).

   A. Blood exerts pressure on vessel walls (*Module 23.9*).

      1. Pulse is the rhythmic stretching of the arteries caused by the pressure of blood from the heart during systole.

      2. Blood pressure is caused by the pumping of the heart against the resistance offered by smaller vessels in the tissues supplied with blood.

      3. Blood pressure is greatest in the aorta and decreases along the path back to the vena cavae (Acetate 23.9A).

      4. Velocity decreases sequentially into the capillaries, and then increases in the pattern shown because of frictional resistance and because the cross-sectional area of the capillary beds is greater than that of larger vessels.

      5. Blood pressure in veins is near zero, but blood returns to the heart with the aid of muscular contraction, valves, and the lifting of the chest cavity during breathing (Master 23.9B).

   B. Measuring blood pressure can reveal cardiovascular problems (*Module 23.10*).

      1. A blood pressure of 120/70 indicates that the force of the heart's beat during systole is 120 mm of mercury (mm Hg) and the general background pressure of the blood in arteries during diastole is 70 mm Hg.

Circulation

NOTE: Normal blood pressure is usually considered to be 120/80. Blood pressure slightly lower than this is an indicator of good cardiovascular fitness. Healthy young females tend to have blood pressures ≈8–10 mm Hg less than this.

2. Blood pressure is measured with a sphygmomanometer. Pressure of the cuff cuts off the blood flow in outer arteries (no pulse is heard). Pressure is reduced in the cuff until the force of systole first pushes blood through (the turbulent sounds of blood flow are heard). Further reduction in the cuff's pressure reaches a point where the sounds of turbulent blood flow are no longer heard; this marks diastole (Master 23.10).

NOTE: The sounds that are heard when measuring blood pressure are referred to as Korotkoff sounds.

3. Low blood pressure is a persistent systolic blood pressure below 100. This is usually not dangerous, but may result from poor nutrition or glandular disorders.

4. Hypertension is a persistent blood pressure of 140/90. It makes the heart work harder against greater resistance due to blockages and reduced flexibility.

VI. Supply of blood in tissues.

A. Smooth muscle controls the distribution of blood (*Module 23.11*).

Review: Types of muscle, including smooth muscle, are discussed in Module 20.6.

1. In all tissues except the brain and heart, blood supply varies.

2. Arteriole constriction can reduce the flow to capillaries. This flow is under the control of nerves and hormones (Master 23.11A).

3. In another mechanism, some blood flows through the center of a capillary bed, but precapillary sphincter muscles control the passage of most blood into the bed. For example, after a meal, precapillary sphincters let more blood pass into the capillaries that supply the villi of the small intestine (Master 23.11B).

NOTE: At the same time, blood supply may be diverted from the outer extremities. Thus, on a cold day, you will feel extra chilled after a meal.

B. Capillaries allow the transfer of substances through their walls (*Module 23.12*).

1. *Review:* Movement of materials across membranes by diffusion, endocytosis, and osmosis (Modules 5.14, 5.15, 5.17, and 5.19).

2. Capillaries are the only vessels with walls thin enough (Master 23.12A).

3. Some substances simply diffuse across the capillary wall to and from blood and interstitial fluid; others are moved across by exocytosis (Module 5.19).

4. Water and some small molecules (salts, sugars, and O_2) "leak" through small cracks between the epithelial cells surrounding capillaries.

Preview: This fluid is returned to the cardiovascular system via the lymphatic system (Module 24.3).

5. Blood pressure tends to actively force fluid out of capillaries. Osmosis (Module 5.15) tends to cause fluids to move in. At the arterial ends of capillary beds, blood pressure is relatively higher, and at the venous ends, osmotic pressure is higher (Master 23.12B).

VII. Blood consists of cells suspended in plasma (*Module 23.13*).

A. An average adult human contains 4–6 liters of blood (Master 23.13).

1. About 45% of this is cellular (red and white blood cells and platelets).

2. About 55% of this is plasma, of which 90% is water and 10% dissolved molecules. Ions of salts maintain osmotic balance and pH and regulate the permeability of membranes. Proteins help in blood clotting and are important in body defense, among other things.

3. *Preview:* The function of these important immune system proteins is covered in Chapter 24.

B. Red blood cells transport oxygen (*Module 23.14*).

1. Erythrocytes (red blood cells) are the most numerous blood cell type; there are about 25 trillion present at one time in the average person. They are formed in the bone marrow and lose their nuclei as they develop. Each cell circulates about 3–4 months before being removed in the liver.

2. The biconcave-disk structure of red blood cells provides for a maximum ratio of surface area to volume, allowing maximum gas exchange.

3. Each red blood cell contains about 250 million molecules of hemoglobin.

4. *Review:* The function of red blood cells in exchanging and carrying gases (Modules 22.10, 22.11).

5. Red blood cell production is under the control of a negative-feedback mechanism that is sensitive to the amount of oxygen reaching tissues. This mechanism is mediated by production of the hormone erythropoietin in the kidneys.

6. Low levels of hemoglobin or number of red blood cells is known as anemia. Iron deficiency is the most common cause, and iron supplements usually correct the problem.

C. White blood cells help defend the body (*Module 23.15*).

1. Five types of leukocytes (white blood cells) are distinguished by nuclear shape and staining properties. They are also produced in the bone marrow. As a group, they spend most of their time outside the circulatory system fighting infections and preventing cancer cells from growing (Master 23.13).

2. Basophils help fight infections by releasing chemicals.

3. Neutrophils and monocytes are phagocytic cells that move into tissues, searching out and "eating" foreign bacteria and other debris.

4. Eosinophils are phagocytic cells that search out parasitic protozoans and worms and may help reduce allergy attacks.

Preview: These, and other white blood cells, are discussed in Module 24.1.

5. Lymphocytes, some of which produce antibodies, are key cells in the immune process (see Chapter 24).

D. Blood clots plug leaks when blood vessels are injured (*Module 23.16*).

1. The blood-clotting mechanism involves materials carried in the blood: platelets and the plasma protein fibrinogen.

NOTE: Blood clotting also requires Ca^{2+}.

2. Minor damage to a blood vessel exposes connective tissue to blood. Platelets adhere to this tissue and release a substance that makes nearby platelets sticky. If major damage occurs, a chain of enzymatically regulated reactions forms a more complex plug, a fibrin clot (Acetate 23.16A).

3. The platelet clot activates a protein, prothrombin, converting it to the enzyme thrombin, which in turn converts fibrinogen into the threadlike protein fibrin. These threads trap additional blood cells.

   NOTE: Blood clotting is one of the few examples of a positive feedback mechanism, another being labor (Module 27.18).

4. Hemophilia is an inherited disease in which individuals lack this mechanism (Modules 9.19, 13.9).

VIII. Stem cells offer a potential cure for leukemia and other blood cell diseases (*Module 23.17*).

A. White blood cells, red blood cells, and platelets all arise in the bone marrow from stem cells (Master 23.17).

B. Leukemia is cancer of the bone marrow cells that produce white blood cells. The leukocytes are in abnormally high numbers, and these in turn may interfere with red blood cell production, causing the person to be anemic.

C. Standard treatment for leukemia involves radiation and chemotherapy (Module 8.11) or bone marrow replacement (following radiation, removal, and the introduction of donor marrow). Bone marrow replacement requires the patient to be on lifelong treatment with drugs that suppress the rejection of transplanted cells. Such drugs are not selective and suppress all immune function, making individuals who take these drugs more susceptible to infections.

D. A potential variation on the latter treatment involves a technique to remove and purify bone marrow from a patient with leukemia, isolating the stem cells. These are then reintroduced in bone marrow that has been radiated to kill off all cancerous leukocytes. Since these are the patient's own cells, there is no risk of rejection.

CLASS ACTIVITIES

1. To measure venous blood pressure, hold the arm straight out at heart level. Raise the arm until the veins collapse. Measure the distance moved; the units will be mm H_2O. Multiply this distance moved by 0.07 mm Hg/mm H_2O to convert to venous blood pressure as mm Hg. Venous blood pressure usually ranges from 0 mm Hg to 20 mm Hg. Note that venous blood pressure is much lower than arterial blood pressure.

2. To see the effect of the unidirectional valves in veins, let an arm hang down the side of the body. After a short period of time, the veins will be visibly filled with blood. Raise the arm above the head, and the veins will empty of blood.

RESOURCES AND REFERENCES

The Anatomy and Physiology of the Heart Videodisc. Burnaby, British Columbia, Canada: British Columbia Institute of Technology, 1990. Provides access to hundreds of specimens, video clips of surgery, autopsy, medical images, animations, and explanations of functions.

Biology Explorer: Cardiovascular System. E. Arlington, MA: Logal Software, 1991. Part of a series of interactive, experimental simulations, this Macintosh computer program allows the user to make hypotheses, plan experiments, observe results, and draw conclusions about the various facets of the cardiovascular system.

Cantin, Marc, and Jacques Genest. "The Heart as an Endocrine Gland." *Scientific American*, February 1986. The atria secrete a recently discovered hormone, atrial natriuretic factor, that interacts with other hormones to fine-tune control of blood pressure and volume.

Erickson, Deborah. "A Better Red." *Scientific American*, February 1992. Describes another candidate in the search for a blood substitute.

Gillis, Anna Maria. "As Good as Blood?" *BioScience*, September 1993. The creation of blood substitutes has been harder than scientists predicted, but recent improvements are encouraging.

Golde, D. "The Stem Cell." *Scientific American*, December 1991. The master cell of blood cell production.

Harken, A.H. "Surgical Treatment of Cardiac Arrhythmias." *Scientific American*, July 1993. A new way to correct lethally fast heartbeats.

Lillywhite, H.B. "Snakes, Blood Circulation, and Gravity." *Scientific American*, December 1988.

Radetsky, P. "The Mother of All Blood Cells." *Discover*, March 1995. Potential benefits of research on stem cells.

Robinson, Thomas F., Stephen M. Factor, and Edmund H. Sonnenblick. "The Heart as a Suction Pump." *Scientific American*, June 1986. A new model suggests that some energy from each contraction is stored within the muscle and provides the power for a suction that aids filling.

Woodworth, Robert H. *Circulatory System of a Frog*. Concord, NH: Pssayo Productions, 1988. Available from Carolina Biological Supply Co. A 16-minute videotape that provides photographic sequences of heart action, pulsating arteries, and red and white blood cell flow. View a thrombus occurring within a vein, and the effect on blood flow when blockage occurs.

Zivin, J., and D. Choi. "Stroke Therapy." *Scientific American*, July 1991. The causes, effects, and experimental treatment of strokes.

24

THE IMMUNE SYSTEM

APPROACH

The topic of this chapter is constantly in the news, and it is likely that the subject of immunity will be covered in every introductory course, at least to some extent. The structure of the chapter allows you to cover the material at various levels of detail. If you plan on spending just one lecture on this material, first cover Modules 24.1–24.5, skip the molecular and cellular details in modules 24.6–24.14, and continue with Module 24.15.

Distinguishing between the two types of immune system responses (humoral and cell-mediated) can confuse students, particularly if you choose to go into the details of both systems. Be sure you differentiate between each system's target molecules and cells each time you change systems. Another point of confusion is the term *antibody*. Strictly speaking, only the proteins released by B cells are antibodies. The surface proteins of T cells function like antibodies.

A molecular picture of the antibody–antigen complex (Master 24.6) may help students understand how antibodies function, when the concept is first introduced in Module 24.4.

The material covered in this chapter is directly tied to the material covered in the previous chapters of this unit. Individuals who are in good cardiovascular condition and have good dietary habits have a more effective immune system and as a result are less likely to become ill than individuals with poor dietary and exercise habits.

CHAPTER OBJECTIVES

Describe the nonspecific defenses against infection, and discuss how each works.

Explain the immune response, describing the organs and tissues involved and the overall characteristics of the response.

Distinguish between the terms *antigen*, *antibody*, and *allergen*; and *immunity* and *vaccination*.

Differentiate between the humoral and cell-mediated immune responses, as to cells involved and cells and molecules targeted.

Discuss the diseases and problems attributable to immune system dysfunction: transplanted-organ rejection, allergies, and immunodeficiency diseases, including AIDS.

LECTURE OUTLINE

I. Introduction.

    A. Humans and other animals depend on several elaborate systems of defense.

        1. Nonspecific defenses do not distinguish individual infectious agents.

        2. The immune system recognizes specific invaders, and it attacks and eliminates them.

    B. The AIDS epidemic (*Opening Essay*).

1. *Review:* The HIV life cycle, noting that the transfer of HIV occurs in body fluids, including blood and semen (Module 10.21; Acetates 10.21A, B).

2. Acquired immune deficiency syndrome results from HIV infection, eventually undermining the entire immune system, after a latent period of several (up to 10) years.

3. Most AIDS patients die from other infectious diseases that their ravaged immune systems cannot combat.

4. For a sexually active person, using condoms and having only one, non-promiscuous partner are the best defenses against AIDS.

II. Nonspecific defenses against infection include the skin, phagocytes, and antimicrobial proteins (*Module 24.1*).

A. The skin and the internal linings of some organs.

1. Provide tough, physical barriers.

2. Provide general chemical defenses in the form of glandular secretions (stomach acids, sweat, and other secretions) that inhibit or kill microbes. Sweat, saliva, and tears contain lysozyme.

B. Phagocytes.

1. *Review:* White blood cells (Module 23.15).

2. Neutrophils and monocytes phagocytize bacteria and viruses in infected tissue.

3. Macrophages develop from monocytes and phagocytize bacteria and virus-infected cells (Figure 24.1A).

4. Natural killer cells (NK cells) attack cancer cells and virus-infected cells.

5. All of these types of white blood cells leave the blood and scavenge invading cells in the interstitial fluid and body tissues.

C. Antimicrobial proteins.

1. Interferons are proteins produced by virus-infected cells that help other cells resist viruses. Interferons produced by recombinant DNA technology may provide an approach to combating viral infections (Figure 24.1B).

2. Inactive complement proteins circulate in the blood and are activated by the immune system or by microbes. Some coat the microbes, making the microbes more susceptible to attack by macrophages; others lethally damage microbial membranes. Complement also amplifies the inflammatory response.

D. The inflammatory response mobilizes nonspecific defense forces (*Module 24.2*).

1. Response is triggered by any infectious agent or break in the barriers.

2. The damaged cells release chemical alarms such as histamine. Histamine induces blood vessels to dilate and become leakier, facilitating the flow of blood and fluid to the affected region (Acetate 24.2).

3. Other chemicals attract phagocytes.

4. Local clotting reactions seal off the infected region and allow repairs to begin.

Review: Clotting (Module 23.16).

5. Local action of this response is the disinfection and cleaning of injured areas, which become inflamed from the increased blood supply.

6. Systemic action of the response, due to microbes or their toxins circulating in the blood, results in increased white blood cells and fever. Moderate fevers may stimulate phagocytosis and inhibit the growth of microbes.

III. The lymphatic system becomes a crucial battleground during infection (*Module 24.3*).

1. The lymphatic system consists of an open, branching network of vessels, lymph nodes, and associated glands. The system has two main functions: to return excess fluid from the interstitial fluid to the circulatory system and to fight infection (Acetate 24.3A).

2. Lymph nodes are concentrated areas of branched ducts containing large numbers of lymphocytes (B cells and T cells) and macrophages. During an infection, these areas become activated and swell, causing the tenderness and aches and pains associated with a systemic infection (Acetate 24.3A-C, Figure 24.3D).

3. In addition to lymph nodes, other lymph organs include the tonsils, appendix, spleen, and bone marrow.

4. Lymph (the fluid of the lymphatic system) enters the system through open, lymphatic capillaries. The largest lymph ducts empty into circulatory system veins in the shoulders (Acetate 24.3A).

5. Lymph is similar to interstitial fluid, except that it is lower in oxygen and contains fewer nutrients. As it circulates through the lymphatic organs, microbes from infected sites and cancer cells may be phagocytized by macrophages. Also, within these lymphoid organs, lymphocytes may be activated to mount a specific immune response.

IV. The immune response counters specific invaders (*Module 24.4*).

A. General description and definitions.

1. The immune system recognizes specific invaders more efficiently than the nonspecific defenses, and it amplifies the inflammatory and complement responses. The immune system is characterized by extreme specificity, memory, and prompt response on second exposure to an antigen.

2. An antigen is any molecule that elicits an immune response. Such molecules include those found on the surfaces of viruses, bacteria, mold, etc.

    Preview: An autoimmune response occurs when the antigen(s) that elicit(s) an immune response is (are) that body's own molecule(s). Autoimmune diseases include Type I diabetes and rheumatoid arthritis (Module 24.16).

3. The system responds to antigens by producing antibodies that attach to the antigen and help counter its effects.

4. In the future, the primed system remembers the antigen and reacts to it.

5. Immunity refers to resistance to specific invaders. Active immunity is achieved by exposure to the invader or to parts of the invader incorporated in vaccinations. Passive immunity is achieved by a person's getting the antibodies from someone else. For instance, a fetus may achieve passive immunity to antigens from its mother through the placenta.

V. Lymphocytes mount a dual defense (*Module 24.5;* Acetate 24.5).

A. Lymphocytes arise from stem cells in the bone marrow (Modules 23.17 and 30.5).

B. Humoral immunity is defense against bacteria and viruses free in the blood or interstitial fluid. It is mounted by B (mature in the bone) lymphocytes, or B cells. B cells release antibodies that function dissolved in the blood.

C. Humoral immunity can be transferred passively by transferring antibody-containing plasma from an immune individual to a nonimmune individual (or by antibodies moving across the placenta: see Module 24.4).

D. Cell-mediated immunity is defense against bacteria and viruses inside body cells, against fungi and protozoans, and against cancer cells. It is mounted by T (mature in the thymus gland) lymphocytes, or T cells. T cells circulate in the blood and mount a cellular attack on repeated foreign invaders, and also promote the functioning of other aspects of the immune response.

Preview: The functioning of the thymus gland is also discussed in Module 26.3.

E. Both B cells and T cells must mature before they are able to function in defense of the body. This involves a process by which these cells become capable of recognizing and responding to a specific antigen. Such mature cells are said to be competent.

F. A human has 100 million to 100 billion different B cells and T cells.

VI. General course of the immune response.

NOTE: This describes humoral immunity. The general course is the same for cell-mediated immunity. Details differ and are covered in Modules 24.9–24.12 for the humoral system, and Modules 24.13 and 24.14 for the cell-mediated system.

A. Antigens have specific regions where antibodies bind to them (*Module 24.6*).

1. Antigens are usually proteins or large polysaccharides.

2. Antibodies usually identify localized regions, the antigenic determinants, on part of the antigen molecule, by means of a "lock-and-key" fit (Master 24.6).

3. An antigen may have several antigenic determinants. Each antibody has two identical antigen-binding sites.

B. Clonal selection musters defensive forces against specific antigens (*Module 24.7*).

1. Upon exposure to an antigen, a tiny fraction of the lymphocytes are able to bind to it and are activated (Acetate 24.7).

2. These cells proliferate, forming a clone of genetically identical effector cells.

3. These effector cells secrete antibodies specific to the antigen.

C. A type of "memory" results in immunity (*Module 24.8*).

1. The effect of the proliferation of the effector cells is the primary immune response. This may take several days, and it results in the release of modest levels of antibodies (Acetate 24.8A).

2. The secondary immune response occurs when the body is later exposed to the same antigen. This response is a bit faster than the primary response, lasts longer, and produces much higher levels of antibodies.

3. During the primary response, some of the cloned cells function as effector cells, and some become memory cells. The latter remain in the lymph

nodes ready to be activated by a second exposure to the antigen (Acetate 24.8B).

VII. Details of the B-cell system.

    A. Overview: B cells are the warriors of humoral immunity (*Module 24.9*).

        1. *Review:* The role of the B-cell system in attacking free antigenic molecules and those on the surfaces of bacteria and viruses free in body fluids (Module 24.5).

        2. Overall, the system works by combining clonal selection and immunological memory. A clone is composed of some effector plasma cells that immediately produce antibodies to the antigen, and a smaller number of memory cells (Acetate 24.9).

        3. Using a military metaphor, plasma cells are "frontline" warriors, the antibodies are their weapons, memory cells are "reservists," and the lymphatic system (indeed, the whole body) is the battleground.

        4. Plasma cells may secrete up to 2000 antibody molecules per second during their 4-to-5-day lifetime.

    B. Antibodies are the weapons of humoral immunity (*Module 24.10*).

        1. *Review:* Tertiary and quaternary structures of proteins (Acetates 3.18C,D).

        2. Each antibody is made of four polypeptide chains, two "heavy" chains and two "light" chains. The quaternary structure of these chains results in a Y shape (Master 24.10B).

        3. Each of the four chains of the antibody has a C (constant) region and a V (variable) region. A pair of V regions, at the tip of each arm of the Y, forms an antigen-binding site.

        NOTE: Genetically, these variable regions are assembled following transcription and translation of combinations of a few each of several dozen genes. Each B-cell or T-cell line activates one set of such genes and continues to activate the same set over its lifetime.

        4. The constant region helps destroy and eliminate the antigens.

        5. Based on the nature of the C region, human antibodies are divided into five major classes, each with a particular role.

    C. Antibodies mark antigens for destruction (*Module 24.11*).

        1. Antibody–antigen complexes are eliminated by several mechanisms. All mechanisms involve a specific recognition phase followed by a nonspecific destruction phase (Acetate 24.11).

        2. Neutralization physically blocks harmful antigens, making them harmless.

        3. Agglutination clumps groups of cells to ease their capture by phagocytes.

        4. Precipitation clumps free molecular antigens together so they precipitate out of solution and can be captured by phagocytes.

        5. Activated complement proteins attach to foreign cells and lyse them.

    D. Monoclonal antibodies are powerful tools in the lab and clinic (*Module 24.12*).

        1. Antibodies used in diagnosis and research were first produced in animals by injecting the antigen and removing the mixture of B cells, some of which produced the right antibodies.

2. Techniques were developed to separate the one B-cell type specific to an antigen. This cell is fused with a culturable tumor cell to form a hybrid cell that both is culturable and produces the desired antibody (Acetate 24.12A).

3. Monoclonal antibodies are useful in medical diagnoses, such as pregnancy tests (which test for the presence of a specific hormone) and may provide a way to target drugs to certain cells that cause disease (including cancer).

   NOTE: They are also used to attach radioactive labels to molecules, thus allowing the molecular positions to be determined by electron microscopy.

VIII. T cells mount the cell-mediated defense and aid humoral immunity (*Module 24.13*).

A. *Review:* The role of T cells in attacking antigens from bacteria and viruses inside body cells and those of protozoans and fungi (Module 24.5).

B. The mechanism of the T-cell system results from the close cooperation of a number of cell types.

C. Overall, the results are similar to the B-cell system by combining clonal selection and immunological memory.

D. Three types of cells are involved.

   1. Cytotoxic T cells attack pathogen-infected cells.

   2. Helper T cells activate other T cells, stimulate the release of antibodies by B cells, and interact with antigen-presenting cells (APCs).

E. Helper T cells are of central importance to the cell-mediated immune response.

   1. APCs are macrophages that combine and display on their cell surface "self proteins" in combination with a portion of the foreign antigen.

   2. Helper T cells recognize only one combination of a self protein and a foreign antigen as presented by an APC (Acetate 24.13A). The binding to a self-nonself complex is one of the two signals required to activate a helper T cell.

   3. The other signal required to activate helper T cells can be either the secretion of a protein (such as interleukin-1) by the APC, or the binding of a separate receptor protein on the helper T cell's plasma membrane with a portion of the self protein on the APC. Either way, the result is activation of the helper T cell via signal transduction pathways (Module 11.13).

   4. Activated helper T cells bind with B cells and stimulate antibody production (humoral immunity). The activated helper T cells also secrete proteins (such as interleukin-2) that activate cytotoxic T cells (cell-mediated immunity), stimulate B cells, and stimulate helper T cells to grow and divide, resulting in the production of both memory cells and active helper T cells (Acetate 24.13B).

   5. Cytotoxic T cells recognize and bind to infected body cells in much the same way that helper T cells bind to APCs: They recognize only a combination of a self protein (different from the APC self protein) and the foreign antigen as presented by the infected cell. The cytotoxic T cells then secrete perforin, a protein that makes holes in the target cell, causing lysis (Acetate 24.13C).

The Immune System

IX. Cytotoxic T cells may prevent cancer (*Module 24.14*).

   A. *Review:* The molecular and cellular bases of cancer (Module 8.11).

   B. Some of the changes that occur in cancer cells involve the outer membrane.

   C. If these changes result in the cancer cell's not appearing as "self" to the T-cell system, they may be eliminated by the cytotoxic T-cell fraction (Figure 24.14).

   D. How often this built-in system functions and why it sometimes fails are the subjects of considerable research on possible cancer cures.

X. The immune system depends on our molecular fingerprints (*Module 24.15*).

   A. The ability of our immune system to distinguish self from nonself enables it to battle foreign invaders without harming healthy body cells.

   1. There are two types of self proteins. Class I proteins occur on all nucleated body cells. Class II proteins are found only on B cells, activated T cells, and macrophages.

   2. Both are unique to each individual and are determined by the action of about 20 genes, each with 50 or more alleles.

   B. Organ transplant procedures trick the system into recognizing foreign tissue as self.

   1. By finding a donor whose self proteins match the recipient's.

   2. By using drugs that suppress the immune response against the transplant. Most such drugs interfere with the beneficial effects of the system. Cyclosporine suppresses only the cell-mediated response.

   NOTE: Cyclosporines were first isolated from molds in the kingdom Fungi.

   3. Approaches that have yet to be perfected would (a) use monoclonal antibodies to target and eliminate the T cells that attack transplants, and (b) find a way to use isolated stem cells to establish a new, second immune system that would recognize the new tissue as self (Module 23.17).

   C. Immunological malfunction or failure causes diseases (*Module 24.16*).

   1. In autoimmune diseases, the immune system turns against its own body cells. Such diseases include insulin-dependent diabetes (Module 26.9), rheumatoid arthritis (damage to joints, bones, and cartilage: Module 30.4), lupus (against nucleic acids, resulting in a buildup of toxic substances in kidneys and joints), rheumatic fever (response to self proteins that are similar to those on the outer envelope of certain streptococci bacteria, to which a person has been previously exposed), and possibly multiple sclerosis (Module 28.2).

   NOTE: In addition to rheumatic fever, many other autoimmune diseases, such as Type I diabetes, appear to be triggered by a viral infection.

   2. In immunodeficiency diseases, part or all of the immune system is lacking. Such diseases include Hodgkin's disease (a lymphocyte cancer) and severe combined immunodeficiency (SCID, in which both T cells and B cells are inactive or absent).

   NOTE: A version of SCID, X-SCID is an X-linked disorder (Module 9.19).

3. There is some indication that physical and emotional stress can also weaken the immune system.

D. Allergies are overreactions to certain environmental antigens (*Module 24.17*).

1. These antigens are allergens (pollen, dust, insect toxins, cat saliva, proteins).

    NOTE: Children subjected to cigarette smoke as well as the children who were prenatally exposed to cigarette smoke are more likely to develop allergies.

2. Allergic reactions follow two stages: (a) A person is first exposed to the allergen, eliciting B cells to form an immunological clone against the allergen, and the antibodies produced to attach to histamine-producing mast cells; (b) the person is secondarily exposed to the same allergen. The antibody–mast cell unions join with the allergen and produce histamine in greater amounts than in a normal inflammatory response, causing the symptoms of allergies: nasal irritation, itchiness, and tears (Master 24.17).

3. Antihistamines interfere with histamine action and give temporary relief.

4. The precipitous release of histamine can cause anaphylactic shock in some people.

E. AIDS leaves the body defenseless (*Module 24.18*).

Review: AIDS is an RNA virus (Module 10.21).

1. HIV has a preference for helper T cells. Once HIV has reproduced and attacked all of these, the body has neither humoral nor cell-mediated immunity.

2. HIV infection and AIDS are incurable. But the use of drugs may postpone the development of AIDS, and the use of several drugs in combination shows particular promise.

    NOTE: HIV seems to be particularly good at mutating the genes responsible for its surface proteins, thereby avoiding the effects of drugs. Simultaneous mutation to configurations that thwart different drugs seems to inactivate the effects of any one mutation.

3. Vaccines have not been successful, perhaps because of the complexity and changeability of HIV's surface proteins.

4. The best current weapon is education.

RESOURCES AND REFERENCES

AIDS: Everything You and Your Family Need to Know. New York: HBO/Ambrose Video, 1993. Former Surgeon General Dr. C. Everett Koop provides level-headed specifics in this up-to-date videotape. Available from Carolina Biological Supply Company.

"AIDS: The Unanswered Questions." *Science*, May 28, 1993; and June 28, 1996. Special sections are devoted to AIDS and HIV.

Anderson, R., and R. May. "Understanding the AIDS Pandemic." *Scientific American*, May 1992.

Beardsley, T. "Better Than a Cure." *Scientific American*, January 1995. The importance of vaccinating children.

Diamond, Jared. "The Mysterious Origin of AIDS." *Natural History*, September 1992. Why did a virus of African wild monkeys take so long to infect us?

Edelson, Richard L., and Joseph M. Fink. "The Immunologic Function of Skin." *Scientific American*, June 1985. Describes why the body's largest organ is more than a passive protective coating.

Gloub, E.S., and D.R. Green. *Immunology, A Synthesis*, 2nd ed. Sunderland, MA: Sinauer, 1991. A strong, process-oriented text.

The Human Immune System: The Fighting Edge. Princeton, NJ: Films for the Humanities and Sciences, 1993. A 52-minute videotape that details the stories of four people whose immune systems do not function correctly: a toddler lacking an immune system, a man with B-cell lymphoma, an AIDS patient, and a woman who nearly died from a wasp sting.

"Life, Death and the Immune System." *Scientific American*, September 1993. A special, single-topic issue covering all the aspects of the immune system discussed in this chapter, and more.

Malkin, Leonard. *Biochemistry of the Immune System*. Troy, MI: Helix Corporation, 1985. Interactive tutorial software emphasizing the biochemistry and genetics of the immune system. Apple II and IBM versions. Available from Carolina Biological Supply Company.

Marchalonis, John, and Samuel F. Schluter. "Origins of Immunoglobulins and Immune Recognition Molecules." *BioScience*, November 1990. Some recognition systems emerged early in evolution; others are restricted to particular phyla.

National Geographic Society. *Our Immune System*. Washington, DC: National Geographic Society, 1988. A 25-minute videotape that examines the anatomy and physiology of the immune system, focusing on allergies and the problems of AIDS.

"The New Face of AIDS." *Science*, June 28, 1996. A special section devoted to AIDS and HIV.

Penniski, E. "Teetering on the Brink of Danger." *Science*, March 22, 1996. Experimental evidence for the new danger hypothesis for immunity.

Rennie, John. "Triple Whammy." *Scientific American*, May 1993. Describes a new AIDS therapy involving three different drugs that shows promise.

Robbins, A., and P. Freeman. "Obstacles to Developing Vaccines for the Third World." *Scientific American*, November 1988.

Roitt, I., J. Brostoff, and D. Male. *Immunology*, 2nd ed. St. Louis, MO: Mosby, 1989. A well-illustrated text.

Schwartz, Ronald H. "T Cell Energy." *Scientific American*, August 1993. When cells of the immune system "see" antigens in the absence of the right cosignals, they shut themselves down instead of attacking. Future therapies might capitalize on that reaction.

Stanfield, Robyn L., Terry M. Fieser, Richard A. Lerner, and Ian A. Wilson. "Crystal Structures of an Antibody to a Peptide and Its Complex with Peptide Antigen at 2.8 Å." *Science*, May 11, 1990. Includes stereo pairs showing the space-filling and molecular skeleton conformations of an antibody to its antigen.

"What Science Knows About AIDS." *Scientific American*, October 1988. A single-topic issue.

25

CONTROL OF THE INTERNAL ENVIRONMENT

APPROACH

The processes of thermoregulation and osmoregulation are straightforward and easy to understand. The example of the countercurrent heat-exchange mechanism reiterates the importance of this general principle. Before covering osmoregulation, be sure to check students' understanding of osmosis and diffusion and what is meant by active and passive.

If you choose to cover the anatomy and physiology of the human kidney in detail, let students know how much terminology they are responsible for. If you prefer to cover kidney structure and function more briefly, try working only through Module 25.9 (perhaps including the overview of function provided in Module 25.10) and skipping the details covered in Module 25.11.

CHAPTER OBJECTIVES

Define thermoregulation, and discuss the ways animals either avoid the problem or carry out the process by adjusting either heat production or heat loss and gain.

Define osmoregulation, and discuss the ways animals carry out this process by regulating the water and solute content of their bodies.

Describe the organs of the human excretory system and their roles in osmoregulation and waste removal.

Identify the structural and functional details of the human kidney.

Explain the role of the liver in homeostasis.

LECTURE OUTLINE

I. Introduction.

　A. Animals survive fluctuations in the external environment because they have internal, homeostatic controls.

　　1. Thermoregulation is the control of temperature.

　　2. Osmoregulation is control of the concentration of water and dissolved solutes.

　　3. Excretion is the disposal of nitrogen-containing wastes.

　B. Animals vary in the ways they control body temperature (*Opening Essay*).

　　1. The concept of cold-blooded and warm-blooded animals is better replaced by the terms *ectothermic* and *endothermic*, because these terms focus on the processes animals use to control body heat.

　　2. Ectotherms warm themselves by absorbing heat from the surrounding environment.

　　3. Endotherms derive most of their body heat from their own metabolism.

4. The blood temperature of an endotherm might be cold, if it were using a mechanism to shut down heat production. The blood temperature of an ectotherm might be warm, if it were to inhabit the warm ocean.

5. An organism we might think of as an ectotherm is, in fact, an endotherm. The great white shark has ways to reclaim heat produced by metabolism, enabling its muscles to provide more power when swimming in cold water.

6. A polar bear is one of the most well-adapted endotherms. Water caught in dense fur insulates against heat loss in the sea. Above water, the white fur transmits heat and the black skin absorbs it.

II. Thermoregulation.

A. Heat is gained or lost in four ways (*Module 25.1*; Master 25.1).

1. Conduction is the direct transfer of heat between surfaces in contact.

2. Convection is the transfer of heat from air or liquid moving past a surface.

3. Radiation, the emission of electromagnetic energy, can transfer heat between two bodies not in contact.

4. Evaporative cooling is the loss of heat from a surface of liquid as the liquid is transformed into gas.

5. In each mechanism, heat is conducted from an area of higher temperature to one of lower temperature.

B. Thermoregulation depends on both heat production and heat gain or loss (*Module 25.2*).

1. Endotherms generally alter their rate of heat production. For example, hormonal changes in birds or mammals will change metabolic rate. Muscle movement (moving around or shivering) generates additional heat (Figure 25.2A).

Review: Thermoregulation and homeostasis (Module 20.13).

Preview: Thermoreceptors (Module 29.3).

2. Endotherms and ectotherms may change their rate of heat gain or loss. Adjustments in coat thickness by muscles that raise and lower hairs change the insulating power of fur. Changes in blood flow to the skin can increase or decrease the temperature of the skin and subsequent heat loss by convection. Cooling also occurs by evaporation during sweating or panting (see Module 20.13).

3. The great white shark has a countercurrent (see Module 22.4) heat-exchange mechanism that controls the amount of heat in its skin. Heat carried by blood flowing outward from the animal's core is picked up by cooler blood flowing inward (Acetate 25.2B, C).

C. Behavior often affects body temperature (*Module 25.3*).

1. Relocation to more suitable positions and migration to more suitable climates are used to increase or decrease body heat.

2. Bathing uses convection and, later, evaporative cooling to remove excess heat. An elephant has special, highly vascularized ears to help cool its blood.

D. Reducing the metabolic rate saves energy (*Module 25.4*).

1. Ectotherms, such as the gray tree frog, can spend much of the winter frozen. The extremely low metabolic rate means that the frog uses almost no energy all winter. A frog version of antifreeze prevents ice crystals from rupturing its cells.

2. In contrast to ectotherms, endotherms can remain active during severe weather. However, a large expenditure of energy is required to keep the body warm. Endotherms have several different adaptations that reduce energy.

3. Torpor is the temporary reduction in body activity to bypass times of cooler temperatures. Bats and hummingbirds use torpor to escape the requirements of keeping their bodies warm during cold days (bats) and nights (hummingbirds).

4. Hibernation is a type of long-term torpor practiced by squirrels and other mammals during cold winter months.

5. Aestivation is a similar type of long-term torpor practiced during times of reduced food and water during summer months.

III. **Osmoregulation: All animals balance the uptake and loss of water and dissolved solutes** (*Module 25.5*).

   A. *Review:* Mechanisms of transport: diffusion, osmosis, active transport (Modules 5.14–5.19).

   B. Background.

   1. The metabolic reactions of life depend on certain solute concentrations in cells.

   2. Cells of animals would lyse (burst) if there were a net gain of water, and crenate (shrivel) and die if there were a net loss of water.

   Review: Osmosis (Module 5.15).

   3. Animals acquire water in food and drink. Aquatic animals also gain and lose water through osmosis across body surfaces.

   4. Animals lose water in urine, feces, perspiration, and breath. Terrestrial animals also lose water through evaporation.

   C. Mechanisms of regulating water content.

   1. Osmoconformers are all aquatic animals that maintain their cells at solute concentrations essentially the same as the surrounding water. Examples include invertebrates such as jellyfish, flatworms, mollusks, and arthropods.

   2. Osmoregulators maintain their body fluids with solute concentrations different from that of their surroundings. Examples include all freshwater animals, all land animals, and most marine vertebrates.

   3. Freshwater fish constantly take in fresh water by osmosis. They control internal water and solute balance by taking up ions from food in the digestive system and gills, and by the production of large amounts of dilute urine via the excretory system (Acetate 25.5A).

   4. Saltwater fish have the opposite problem: They lose water by osmosis to their surroundings. They control internal solute concentration by drinking water, pumping out ions through the gills, and producing concentrated urine (Acetate 25.5B).

5. Some fish, such as salmon, inhabit both fresh water and salt water at different phases in their lives. When they change habitats, they change strategies, using the freshwater mechanisms or saltwater mechanisms as necessary.

6. Land animals have osmoregulatory problems like marine fish. Only two groups of animals are successful land dwellers: the arthropods and the vertebrates. Both have solved osmoregulatory problems by having thick skins that inhibit water loss, complex excretory systems, and adaptations that protect fertilized eggs and developing embryos from drying out.

D. Sweating can produce serious water loss (*Module 25.6*).

1. Water loss is a more serious problem than salt loss to an athlete.

2. Drinking water is the best way to rehydrate. If salts are needed, they should be very dilute or taken after water.

E. Many freshwater animals face seasonal dehydration (*Module 25.7*).

1. Some animals are able to escape the problem of severe desiccation in their environments. One way is to lay drought-resistant eggs.

2. A few animals, such as water bears (tardigrades), escape the problem by drying up along with their environment. To protect their proteins against denaturation and otherwise keep internal structures intact, they replace water molecules with sugar molecules.

IV. Excretion.

A. Animals must dispose of nitrogenous wastes (*Module 25.8*).

1. Nitrogen-containing wastes come mostly from the breakdown of proteins and nucleic acids.

2. Aquatic animals can dispose of nitrogenous wastes in the form of toxic ammonia, which diffuses readily in water (Acetate 25.8).

3. Terrestrial animals must use energy to convert amino groups of proteins into less toxic compounds.

4. Urea is highly soluble in water, 100,000 times less toxic than ammonia and excreted by mammals, amphibians, and a few fishes.

5. Uric acid is also less toxic than ammonia and is excreted by birds, insects, many reptiles, and snails. Because it is not soluble in water, uric acid is excreted in crystalline form, and thus has the benefit of conserving water.

B. The excretory system plays several major roles in homeostasis (*Module 25.9*).

1. The major roles include forming and excreting wastes in urine, and regulating the concentrations of water and solutes in body fluids.

2. The human excretory system includes two kidneys, a bladder, interconnecting ducts, and the associated vessels of the circulatory system. The system extracts about 180 liters of fluid ("filtrate") from the 1000–2000 liters of blood passing through it per day, concentrating and storing 1.5 liters of urine for disposal (Acetates 25.9A, B, C).

3. The kidney is the processing center of the excretory system. Blood enters and leaves each kidney through the renal artery and renal vein. Urine passes from the kidney to the bladder in the ureter and from the bladder outside through the urethra.

4. Within the kidney are thousands of blood-filtering units, nephrons, composed of tubules and associated blood vessels. Each extracts and refines a small amount of filtrate and releases a small quantity of urine (Acetate 25.9D).

5. The nephron consists of a blood-filtering region (Bowman's capsule) and a filtrate refinery (the proximal tubule, loop of Henle, and distal tubule).

6. The blood vessel parts of the nephron include a ball of capillaries (glomerulus) where blood pressure forces water and solutes out of the blood and into the tubule, and a second, looser, capillary network that surrounds the region of the loop of Henle and helps refine the filtrate.

C. Overview: The key functions of the excretory system are filtration, reabsorption, secretion, and excretion (*Module 25.10*; Master 25.10).

1. During filtration, water and other small molecules are forced by blood pressure through capillary walls into the nephron tubule.

2. During reabsorption, water and solutes still valuable to the body are reclaimed from the filtrate.

3. During secretion, excess ions, drugs, and toxins are secreted from the blood into the nephron tubule.

4. Upon excretion, the urine passes from the kidneys to the outside, by the ureters, urinary bladder, and urethra.

D. From blood filtrate to urine: a closer look (*Module 25.11*).

1. In Figure 25.11, capillaries are not shown; red arrows represent active reabsorption, blue arrows represent active secretion, pink arrows represent passive reabsorption (osmosis), and the intensity of the gray represents solute concentration in the interstitial fluid (Acetate 25.11).

2. Small molecules travel between blood and nephron filtrate through the interstitial fluid.

3. Blood pressure forces water and most solutes into the Bowman's capsule of the nephron tubule from the glomerulus.

4. NaCl and H_2O are reabsorbed from—and excess H^+ ions secreted to—the filtrate in the proximal and distal tubules. The proximal and distal tubules are also the sites of secretion of toxins such as ammonia. The proximal tubule reabsorbs water and nutrients such as glucose and amino acids. The distal tubule may also secrete drugs such as penicillin.

5. The loop of Henle is the principal site of water reabsorption. NaCl and urea reabsorption along the descending limb of the loop of Henle only heightens the rate at which water is reabsorbed from the filtrate. Excess water in the interstitial fluid is carried away by the blood. The ascending limb of the loop of Henle is impermeable to water, and NaCl is first passively and then actively reabsorbed, maintaining the concentration gradient encountered by the descending limb of the loop of Henle (another example of a countercurrent mechanism).

NOTE: The need of an animal to conserve water, and thus the type of environment in which it lives, can be approximated by looking at its loops of Henle. The longer the loops of Henle, the greater the need for water conservation, and the more arid the environment in which the animal lives.

6. This results in a high concentration of NaCl in the interstitial fluid surrounding the distal tubule and the first part of the collecting tubule. Water can then be reabsorbed from the filtrate in the upper collecting duct.

7. Until the collecting duct enters the medulla, urea tends to remain in the filtrate because the nephron is relatively impermeable to it.

8. In the medulla, the collecting duct is permeable to urea and some urea moves into the interstitial fluid. This increases the concentration of the interstitial fluid of the medulla even more, enhancing water reabsorption.

9. Much of the reabsorption of water is under the control of ADH (antidiuretic hormone). ADH is released from a control center in the brain.

   Preview: ADH is produced by the hypothalamus and released from the posterior pituitary (Table 26.3 and Module 26.5).

   NOTE: The reason alcohol acts as a diuretic is that alcohol consumption inhibits the release of ADH.

E. Kidney dialysis can be a lifesaver (*Module 25.12*; Master 25.12).

   1. Kidney failure or impaired function can lead to death from a buildup of solutes and toxins in the blood.

   2. Kidney dialysis machines function like kidneys, but receive and return blood from a person's vein.

   3. The machine removes small molecules from the blood, using selectively permeable membranes to allow only water and desired solutes to pass out with the dialyzing solution.

V. **The liver is vital in homeostasis (*Module 25.13*; Master 25.13).**

   A. This organ has a number of functions because of its central location between intestine and heart, and because its cells are capable of a wide range of metabolic activities.

   B. The liver supports the vital activities of the kidney.

      1. Prepares nitrogenous wastes (synthesizes urea from ammonia) for disposal by the kidneys.

      2. Converts certain toxic compounds (alcohol, drugs) into inactive compounds that the kidneys will remove.

   C. The liver also regulates glucose levels (maintained at about 0.1%) in the blood by interconverting glycogen and glucose.

CLASS ACTIVITIES

1. You should be able to find recipes for artificial urine in physiology laboratory manuals. These recipes can be used to compare the chemistry of the urine of a healthy person with the urine of (for example) a diabetic or individuals with kidney damage resulting in (for example) proteinuria or hematuria.

RESOURCES AND REFERENCES

Carey, F.G. "Fishes with Warm Bodies." *Scientific American*, February 1973. How some fish species conserve metabolic heat.

Ferris, Judith. *The Kidney: Structure and Function*. Jericho, NY: BioLearning, 1981. Interactive software that presents the relationship between kidney structure and function through simulation of several lab investigations. Apple II and IBM PC versions. Available from Carolina Biological Supply Company.

Heinrich, B. "Some Like It Cold." *Natural History*, February 1994. How winter moths keep warm enough to stay active.

Homeostasis. Princeton, NJ: Films for the Humanities and Sciences, 1993. A six-part series (advanced-placement high school level) of 10-minute videotapes that explore the internal systems of balance and regulation.

Line, L. "Staying the Winter." *National Wildlife*, February-March 1995. How chickadees and other songbirds survive northern winters.

McClanahan, L.L., R. Ruibal, and V.H. Shoemaker. "Frogs and Toads in Deserts." *Scientific American*, March 1994. Amphibian adaptations to dry environments.

McKenzie, A. "Seeking the Mechanisms of Hibernation." *BioScience*, June 1990.

Smith, H.W. *From Fish to Philosopher*. Boston, MA: Little, Brown, 1953. A classic book on vertebrate evolution as revealed by kidney structure and function.

Vickers-Rich, P., and T.H. Rich. "Australia's Polar Dinosaurs." *Scientific American*, July 1993. How did dinosaurs survive cold dark environments?

26

CHEMICAL REGULATION

APPROACH

Along with the nervous system, the endocrine system plays a central role in coordinating the activities of all other systems. If you cover the nervous system, seriously consider covering the material in this chapter as well, even in short courses. Most students will have a strong interest in this topic (perhaps more so than the nervous system) because hormones play such an immediate role in the social and academic lives of college students.

Instead of simply naming the hormones and stating their effect, investigate the elaborate cascade effects and feedback controls of a few particular hormone systems. Rather than exhaustively covering all the material in Table 26.3, use the table to highlight the general features of all hormones.

In a short treatment of this subject, it is better to cover a few glands well than to try and cover them all superficially. Of primary importance in this chapter is the overall control of the endocrine system by the central nervous system, and the concept of antagonistic hormonal control. Be sure to cover these points.

CHAPTER OBJECTIVES

Define the term *hormone*, and distinguish hormones from other chemical messengers.

Differentiate between the general chemical structure and mode of action of the two major classes of hormones.

Identify the locations of the endocrine glands, noting the interconnections with other systems and organs that are particularly important (circulatory and nervous systems, brain, spinal cord, kidneys, and pancreas).

Describe the nature of the overall control of the endocrine system by the hypothalamus and pituitary gland.

Outline the general roles played by each of the major endocrine glands (thyroid, pancreatic cells, adrenal glands, and sex organs).

Explain what is meant by antagonistic hormones, giving an example of the action of at least one pair of antagonists.

LECTURE OUTLINE

I. Introduction (*Opening Essay*).

    A. One particular chemical governs the mating behavior of bullfrogs.

        1. Pineal gland activity is stimulated by darkness.

        2. Melatonin is released by the pineal gland into the blood, and levels increase at night to induce the mating call activity.

    B. Melatonin is produced by most vertebrates, including humans.

1. Humans do not have a pineal "eye" like frogs, but their pineal glands do secrete melatonin on a daily, rhythmical basis, peaking about 2:00 in the morning. One hypothesis is that higher melatonin levels normally help induce sleepiness.

2. Melatonin levels also cycle seasonally in humans and other mammals.

3. Mounting evidence suggests that abnormal cycles with higher-than-normal levels of melatonin cause the rare disorder known as seasonal affective disorder (SAD). SAD is characterized by irritability, depression, and weight gain and is brought about by the short days of winter. Exposure to bright lights for an hour or so will reduce the symptoms.

C. Chemical signals coordinate body functions (*Module 26.1*; Acetates 26.1A, B, C).

1. Hormones are chemicals produced by endocrine glands or neurosecretory cells. They have a controlling effect on another part of the body (at specific target cells).

2. Collectively, all hormone-secreting tissues constitute the endocrine system. This system is particularly important in controlling whole-body activities such as metabolic rate, growth, maturation, and reproduction.

3. The nervous system transmits information along nerve cells. At the end of a nerve cell, the message travels in the form of diffusing neurotransmitters, chemical messengers produced by one nerve cell that have an effect on the next nerve cell or on a cell it activates. Neurotransmitters do not usually travel in the bloodstream but are local regulators.

   Preview: The functioning of a neurotransmitter is discussed in Modules 28.7–28.10.

4. Local regulators are secreted into the interstitial fluid and cause changes in cells near the point of secretion. Prostaglandins are a type of local regulator with a wide range of functions.

   NOTE: Prostaglandins play a role in the inflammatory response, and it is the inhibition of prostaglandins that is a major cause of ulcers (Module 21.9).

5. *Preview:* The nervous system is the subject of Chapters 28 and 29.

6. The two systems, endocrine and nervous, coordinate most of their activities. The nervous system provides split-second control, and the endocrine system provides control over longer duration, from minutes to days.

D. Hormones affect target cells by two main signaling mechanisms (*Module 26.2*).

1. Steroid hormones are lipids made from cholesterol. Most interact with the DNA in the nuclei of their target cells, stimulating the synthesis of new proteins by those cells (Acetate 26.2A).

2. Nonsteroid hormones are made from amino acids. Nonsteroid hormones (the first messenger) never enter their target cell, but interact with a surface protein of that cell, inducing the production of a second, internal messenger (usually cAMP) that produces the desired effects (Acetate 26.2B). Hormones such as these act by a signal transduction mechanism (see Module 11.13).

3. There are three classes of nonsteroid hormones. Amine hormones are made of single modified amino acids. Peptide hormones are short amino acid chains. Protein hormones are long amino acid chains.

4. There are over 50 known hormones.

II. Overview: The vertebrate endocrine system (*Module 26.3*).

A. Endocrine glands secrete into the blood.

Review: Contrast this with the nonendocrine (exocrine) glands of the GI tract (Chapter 21), which secrete into body cavities.

B. The endocrine system is composed of more than 12 glands, spread throughout the body (Acetate 26.3A).

1. Some (e.g., pituitary, thyroid) are endocrine specialists.

2. Some (e.g., pancreas, testes) have both endocrine and nonendocrine (exocrine) functions.

C. General features shared by the endocrine systems of all vertebrates (Table 26.3).

1. Chemical class. Only the sex organs and the adrenal cortex produce steroid hormones. Most hormone action is by means of second messengers.

2. Target cells ("Representative Actions" column in Table 26.3). Some (e.g., sex hormones) affect body tissues generally. Some have specific targets in or out of the endocrine system.

3. The endocrine and nervous systems are closely associated ("Regulated by" column in Table 26.3).

D. Two glands are not discussed later in the chapter.

1. The pineal gland secretes melatonin, which links environmental light conditions with activities that show daily or seasonal rhythms (see the Opening Essay). A major role is to cue reproductive activity. Rising levels cue reproduction in sheep and deer that breed in the fall. Lowering levels cue reproduction in mammals that breed in the spring.

Preview: Biological clocks in animals are discussed in Module 37.9.

2. The thymus gland is important in the immune system, stimulating the development and differentiation of T cells in early childhood. This gland virtually disappears but still remains functional in adults.

Review: The role of the thymus and T cells (Module 24.5).

III. The hypothalamus, closely tied to the pituitary, connects the nervous and endocrine systems (*Module 26.4*).

A. The hypothalamus is the endocrine system's master control center (Master 26.4A).

1. It receives information from nerves about the internal condition of the body and about the external environment.

2. It signals the pituitary gland, which in turn secretes hormones that influence many body functions, including those of other endocrine glands.

B. The posterior lobe of the pituitary gland.

1. Consists of an extension of the hypothalamus composed of nervous tissue.

2. Stores and secretes hormones made in the hypothalamus.

C. The anterior lobe of the pituitary gland.

1. Composed mostly of glandular tissue.

2. Synthesizes its own hormones, several of which control other endocrine glands.

D. Hypothalamus control over the pituitary.

1. Releasing hormones make the anterior pituitary secrete its hormones.

2. Inhibiting hormones make the anterior pituitary stop secreting hormones.

3. For example, secretion by the hypothalamus of TRH (thyroid-releasing hormone) induces the anterior pituitary to secrete TSH (thyroid-stimulating hormone). TSH causes the thyroid to release thyroxine into the blood. Thyroxine increases most cells' metabolic rates. Increased thyroxine levels have negative-feedback control on the release of TSH (Acetate 26.4B).

NOTE: The release of the hormones stored in the posterior pituitary is under the control of nerve impulses from the hypothalamus.

E. The hypothalamus and pituitary have multiple endocrine functions (*Module 26.5*).

1. Neurosecretory cells extend from the hypothalamus into the posterior pituitary and synthesize the hormones oxytocin and antidiuretic hormone (ADH: Module 25.11). Oxytocin induces contraction of the uterine muscles during childbirth and causes the mammary glands to release milk during nursing. ADH helps the kidneys retain water (Acetate 26.5A).

NOTE: Oxytocin has a role in males as well as females. This hormone is sometimes referred to as the "cuddling hormone" since it plays a role in affectionate behavior.

2. A second set of neurosecretory cells of the hypothalamus secretes releasing and inhibiting hormones that are carried by small vessels to the anterior pituitary. Under the control of releasing hormones, the anterior pituitary can release TSH, adrenocorticotropic hormone (ACTH), follicle-stimulating hormone (FSH), or luteinizing hormone (LH), all of which activate other endocrine glands. These glands' hormonal secretions all exhibit negative-feedback control on the anterior pituitary (Acetate 26.5B).

3. Other hormones produced by the anterior pituitary include growth hormone (GH, which promotes the development and enlargement of all body parts in young mammals), prolactin (PRL, which stimulates mammals to produce milk, regulates larval development of amphibians, and regulates salt and water balance in fishes), and endorphins (the body's natural painkillers).

IV. The thyroid regulates development and metabolism (*Module 26.6*).

A. Thyroid hormones affect virtually all vertebrate tissues. Two very similar iodine-containing amine hormones are produced: Thyroxine and triiodothyronine have four and three iodine atoms per molecule, respectively.

B. Roles of thyroid hormones.

1. In amphibians, they trigger tissue reorganization during metamorphosis.

2. In mammals, they control the early development of bone and nerve cells.

Chemical Regulation

3. In adult mammals, they maintain normal blood pressure, heart rate, muscle tone, and digestive and reproductive functions.

C. Disorders.

1. Hyperthyroidism causes overheating, profuse sweating, irritability, high blood pressure, and weight loss.

2. Hypothyroidism causes lethargy, intolerance to cold, and weight gain.

3. Hypothyroidism often accompanies goiter, the enlargement of the thyroid that occurs when too little iodine is consumed in the diet. This condition results from an interruption of normal negative-feedback control on TSH release by the pituitary (Master 26.6).

D. Hormones from the thyroid and parathyroids maintain calcium homeostasis (*Module 26.7*).

1. Appropriate levels of blood calcium are essential for nerve and muscle cell functions, blood clotting, and active transport across cell membranes.

2. Together, secretions from these two types of glands keep Ca^{2+} ions at a concentration of 9–11 mg per 100 mL of blood.

3. Calcitonin from the thyroid and parathyroid hormone (PTH) are antagonistic; that is, they have opposite effects (Acetate 26.7).

    NOTE: In adults, calcitonin plays a relatively minor role in the regulation of Ca^{2+} levels.

4. Calcitonin lowers the Ca^{2+} level whenever that level rises above about 10 mg/100 mL. It causes Ca^{2+} to be deposited in bone and absorbed less by the intestine, and it causes the kidneys to reabsorb less Ca^{2+} as they form urine.

5. PTH raises the Ca^{2+} level whenever that level falls below about 10 mg/100 mL. It causes Ca^{2+} to be released from bone, absorbed more by the intestine, and reabsorbed more by the kidneys.

    NOTE: In the kidneys, PTH promotes the conversion of vitamin D to its active form. In turn, vitamin D promotes the absorption of calcium and phosphate from the alimentary canal, the retention of these minerals by the kidneys, and their release from bone into blood.

V. Pancreatic hormones manage cellular fuel (*Module 26.8*).

A. Insulin and glucagon are antagonistic hormones produced by islet cells in the pancreas (Acetate 26.8).

1. Insulin is a protein hormone produced by beta islet cells. Glucagon is a peptide hormone produced by alpha islet cells.

2. The set point that controls hormone balance is about 90 mg glucose/100 mL.

3. Rising blood glucose level (after a meal) stimulates the beta islet cells to secrete insulin. The blood glucose level falls because insulin stimulates all body cells to take more glucose from the blood. Most is converted by the liver into stored glycogen. Other cells metabolize glucose into energy, stored fats, or proteins.

4. Falling blood glucose level (during a fast) stimulates the alpha islet cells to secrete glucagon. The blood glucose level rises because glucagon makes liver cells convert glycogen to glucose, as well as convert fatty acids and amino acids to glucose.

B. Diabetes is a common endocrine disorder (*Module 26.9*).

1. Much of the function of insulin has been discovered in people with diabetes mellitus, which occurs in about five out of 100 people.
2. This disease occurs when there is not enough insulin produced to maintain proper absorption of glucose from the blood. The glucose concentration of blood becomes so high (hyperglycemia) that glucose is excreted by the kidneys.
3. Type I diabetes develops before age 15 and involves the destruction of beta islet cells, by disease or by hereditary immune system dysfunction. Type I diabetes is controlled by the injection of recombinant human insulin. Type II diabetes is usually associated with older people (40+ and obese) and occurs when body cells do not respond correctly to insulin. Type II diabetes is usually controlled by diet.
4. A glucose-tolerance test is used to detect diabetes (Master 26.9).
5. Hypoglycemia occurs in some people who secrete too much insulin. Symptoms appear 2–4 hours after a meal and include hunger, weakness, sweating, and nervousness. In severe cases, convulsions can lead to death in people whose brains do not receive enough glucose.

VI. **The adrenal glands mobilize responses to stress (*Module 26.10*).**

A. The adrenal glands are associated with the kidneys and are composed of two functionally different parts.

1. The adrenal medulla in the center produces the "fight-or-flight" hormones.
2. The adrenal cortex at the outside produces hormones that provide slower, longer-term responses to stress.

B. Stress produces a cascade effect. Stressful stimuli (negative or positive) activate certain hypothalamus cells. These cells send signals along nerve cells through the spinal cord to stimulate the adrenal medulla (Acetate 26.10).

C. The adrenal medulla ensures a rapid, short-term response to stress.

1. When stimulated, the adrenal medulla releases epinephrine (adrenaline) and norepinephrine (noradrenaline) into the bloodstream.
2. Both hormones stimulate liver and muscle cells to release glucose, making more energy available for cellular fuel. They increase blood pressure, breathing rate, and metabolic rate and change blood-flow patterns.
3. Epinephrine dilates blood vessels in the brain and skeletal muscles but constricts vessels elsewhere, directing blood to critical areas.

D. The adrenal cortex causes slower responses.

1. The adrenal cortex responds to endocrine signals (ACTH) from the pituitary.
2. When stimulated, the adrenal cortex secretes a family of steroid hormones, the corticosteroids. These hormones help the body function normally, whether stressed or not.

3. Mineralocorticoids affect salt and water balance. Glucocorticoids promote the synthesis of glucose from noncarbohydrate sources. In addition, high levels of the glucocorticoids can suppress the body's defense system and can control excessive inflammation.

E. Glucocorticoids offer relief from pain, but not without serious risks (*Module 26.11*).

1. Physicians often prescribe glucocorticoids to relieve the pain of athletic injuries.

2. Unfortunately, they depress the activity of the adrenal glands and may have dangerous side effects, such as psychological changes.

VII. The gonads secrete sex hormones (*Module 26.12*).

A. Sex hormones are steroid hormones that affect growth and development and regulate reproductive cycles and sexual behavior.

B. The three categories of sex hormones—androgens, estrogens, and progestins—are all found in both females and males, but in different proportions.

C. Females.

1. Females have a high ratio of estrogens to androgens.

2. Estrogens stimulate the development and maintenance of the female reproductive system and secondary sex characteristics, such as smaller body size, higher voice, breasts, and wider hips.

3. Progestins are most active in human females, where they prepare the uterus to support the developing embryo.

Preview: The role of these hormones in the menstrual and ovarian cycles is discussed in Module 27.5.

D. Males.

1. *Review:* Anabolic steroids, artificial analogs of testosterone (Module 3.10).

2. Males have a high ratio of androgens (e.g., testosterone) to estrogens.

3. Androgens stimulate the development and maintenance of the male reproductive system and secondary sex characteristics, such as deeper voice, more body hair, and larger skeletal muscles.

Preview: The role of theses hormones in regulating sperm production is discussed in Module 27.3.

E. The release of sex hormones is controlled by the hypothalamus and anterior pituitary. The anterior pituitary synthesizes FSH and LH, which stimulate the ovaries and testes to synthesize and secrete sex hormones.

CLASS ACTIVITIES

1. Melatonin has been getting a great deal of attention in the media. Have students bring in articles on melatonin from the popular press and compare what these articles say with what articles from scientific journals say. Relate this activity to the scientific process (*Review:* Modules 1.2 and 1.3).

2. Engage the students in thought experiments (the process of science: Module 1.2) concerning the potential effects of hypersecretion or hyposecretion of the hormones discussed in this chapter.

RESOURCES AND REFERENCES

Atkinson, M., and N. MacLaren. "What Causes Diabetes?" *Scientific American*, July 1991.

BBC Television. *Sense of Timing*. Deerfield, IL: Coronet Film and Video, 1989. A 30-minute videotape that examines the rhythms in the cycles of life; birds, fish, crabs laying eggs, and honeybee behavior, as tied to internal rhythms controlled by hormones.

Leinhard, G.E., J.W. Slot, D.E. James, and M.M. Mueckler. "How Cells Absorb Glucose." *Scientific American*, January 1992.

Malkin, L. *Biochemistry of Hormones*. Troy, MI: Helix Corporation, 1987. Three-part tutorial software on the biochemical aspects of hormonal activity. Apple II and IBM PC versions. Available from Carolina Biological Supply Company.

Morell, V. "Zeroing In on How Hormones Affect the Immune System." *Science*, August 11, 1995.

"Signal Transduction." *Science*. April 14, 1995. New research highlights.

Snyder, S.H. "The Molecular Basis of Communication Between Cells." *Scientific American*, October 1985. Emphasizes the relationship between the endocrine and nervous systems.

Snyder, S., and D. Bredt. "Biological Roles of Nitric Oxide." *Scientific American*, May 1992. Recent research on a signal molecule with functions in both the endocrine system and the nervous system.

Welch, W.J. "How Cells Respond to Stress." *Scientific American*, May 1993. Proteins produced by stressed cells.

Wilson, Jean D., Fredrick W. George, and James E. Griffin. "The Hormonal Control of Sexual Development." *Science*, March 20, 1981. Describes how different hormone balance turns on the development of sex characteristics after embryos have developed similarly during the first phase of gestation.

27

REPRODUCTION AND EMBRYONIC DEVELOPMENT

APPROACH

High school biology programs or health courses on human reproduction vary considerably in the level and content covered, not necessarily from a biological perspective. The close integration of text and illustrations in this chapter is particularly valuable when presenting this subject from a biological perspective, while providing additional information about personal and social contexts, which will be of great interest to college students.

Despite prior introduction, students will have misconceptions about much of this material. Modules 27.2 and 27.3 are organized from the "gametes' perspective," which will help you emphasize the biology. Be sure students understand meiosis (Modules 8.14 and 8.15) before covering the material on gamete formation in Module 27.4. Module 27.5 contains a particularly clear exposition of the complex hormonal control of the female cycle, made much easier by constant reference to the figure in that module. The ovarian and menstrual cycles of females, as well as sperm production by males, are controlled by negative-feedback mechanisms (Module 5.8). An understanding of negative feedback will greatly facilitate an understanding of the regulation of the female and male reproductive systems.

The second half of the chapter on development is fascinating, particularly if you show some time-lapse videos of the actual processes to back up the excellent illustrations in the text. In the latter context, explain to students the relationship between the cross-sectional and surface views of developing embryos.

CHAPTER OBJECTIVES

Identify the advantages and disadvantages of sexual versus asexual reproduction.

Describe the anatomy of the human male and female reproductive systems, and the formation and maturation of the gametes. Trace the paths the gametes take prior to conception.

Outline the hormonal control of changes in the ovaries and uterus.

Explain the nature of human sexuality in contrast to that of most animals, including the use of contraception in preventing unwanted pregnancy.

Trace the course of early development in humans (and experimental animals), describing the future outcome of each of the three tissue layers.

Describe the role of the following in development, giving specific examples of where each functions in developing vertebrates: changes in cell number, size and shape, induction, cell migration, cell aggregation, and pattern formation.

Outline the general development that occurs during each of the three trimesters of pregnancy.

Name and explain the advances in technology that have bypassed problems in human fertility and conception.

LECTURE OUTLINE

I. Introduction.

A. *Review:* Human life cycle (Module 8.14, Acetate 8.14), the roles of meiosis (Module 8.17), fertilization (Module 8.19), and mitosis, and the influence of master control genes on animal development (Module 11.12).

B. Reproduction in terrestrial animals involves distinct processes and adaptations (*Opening Essay*).

1. Mating behavior normally involves both copulation and processes that ensure fertilization and development. In the desert-grassland whiptail lizard, these behaviors are still present despite the fact that only females exist in the population (chapter-opening photo).

2. Reptile eggs include a food supply that nourishes the embryo, and three other membranes that support the embryo and help keep it moist (Acetate 27.0).

C. Sexual and asexual reproduction are both common (*Module 27.1*).

1. Reproduction allows a species to transcend the finite life span of its individuals.

2. Asexual reproduction is the creation of offspring whose genes all come from one parent. Asexual reproduction can be by budding, fission (Figure 27.1A), or fragmentation, and it is often accompanied by the regeneration of missing parts.

Review: Budding by *Hydra* is discussed in Module 8.12; binary fission by bacteria is discussed in Module 8.3.

Preview: Reproduction by fragmentation by plants is discussed in Module 31.14.

3. Asexual reproduction allows animals to reproduce without mates, to produce offspring quickly without expending energy for gamete production or fertilization, and to produce large numbers of successful genotypes. But populations developing from asexual reproduction are genetically homogeneous.

NOTE: Relate this to Chapter 38. Forests are logged and then replanted. Are they replanted with a diversity of trees or with trees of a single species? If the replanted trees are of a single species, is the population genetically diverse or genetically homogeneous (clones)? How does this relate to resistance to environmental factors (e.g., extremes of weather, pests)?

4. Sexual reproduction is the production of offspring by the fusion of two gametes (ova and sperm). The zygote and the offspring that develop from it contain a unique combination of genes from the parents. Species may alternate between asexual and sexual reproduction (e.g., rotifers) or reproduce hermaphroditically (earthworms). Fertilization can be external (most fish and amphibians) or internal (Figures 27.1B, C, D).

Review: Modules 8.17 and 8.19 discuss how meiosis and random fertilization increase genetic variation.

5. Sexual reproduction increases genetic variation in populations. But animals must find mates and expend extra energy to produce gametes and ensure that fertilization takes place.

Reproduction and Embryonic Development

II. Human reproductive anatomy, gamete formation, and hormonal control.

A. Reproductive anatomy of the human female (*Module 27.2*; Acetates 27.2A, C).

1. Eggs develop in ovaries. Each ovum develops in a separate follicle that contains other cells to protect and nourish the egg, and to produce estrogen.

    Review: The hormones released by the follicles are discussed in Module 26.12.

    Preview: The activities involving egg production and release are under the control of hormones from the pituitary (FSH and LH). The female hormone cycle is discussed in Module 27.5.

2. Women are born with 40,000–400,000 follicles, but only several hundred will release eggs, one every 28 days, starting at puberty (Figure 27.2B).

    NOTE: Some students may be under the misconception that menopause occurs when a female runs out of potential ova. This is not true; menopause occurs as a result of changes in the pattern of release of hypothalamic and pituitary hormones and ovarian response to these hormones.

3. Following ovulation, the remaining follicular tissue grows temporarily into the corpus luteum, which secretes progesterone and estrogen to maintain the uterine lining if pregnancy occurs.

4. The released egg traverses a short space to the oviduct opening. Fertilization normally occurs in the oviduct. Abnormal, ectopic pregnancies often occur in the oviduct, where they must be surgically removed.

    NOTE: Smokers are more likely to experience an ectopic pregnancy than nonsmokers.

5. The fertilized egg begins to develop and moves into the uterus. The embryo implants in the endometrium, and development is completed there. (After the eighth week, when body structures begin to appear, the embryo is known as a fetus.)

6. The uterus opens into the vagina through the cervix. The vagina receives the sperm during sexual intercourse and serves as the birth canal through which the newborn baby leaves the uterus.

7. A number of external structures (hymen, labia, clitoris, and Bartholin's glands) are associated with the vaginal opening, and most function during sexual arousal and intercourse.

8. The clitoris is homologous to the penis, consisting of a glans with a large number of nerve endings, and a prepuce (foreskin). Like the penis, it becomes engorged with blood during sexual arousal.

    NOTE: In addition to the homology of the clitoris and the penis, the ducts of the male and female reproductive system and the gonads are homologous. Thus, true hermaphroditism (reproductively functional as both female and male) is not possible in humans.

B. Reproductive anatomy of the human male (*Module 27.3*; Acetates 27.3A, B).

1. Sperm cells are continuously produced in the testes, held outside the abdominal cavity in the scrotum, which keeps the sperm below body temperature.

NOTE: This is why males with low sperm counts are told to wear boxer shorts rather than briefs.

2. The path sperm travel is from the testes to the epididymis, where they are stored while they develop motility and fertilizing ability. During ejaculation, sperm leave the epididymis and are moved by muscular contractions (peristalsis) to the vas deferens. The vas deferens passes into the body cavity and loops around the urinary bladder, where it merges with a duct from the seminal vesicle (one of the male sex glands). The union of the vas deferens with the duct from the seminal vesicle forms the ejaculatory duct. The ejaculatory ducts from the right and left sides of the body empty into the urethra. The urethra conveys both urine and sperm to the outside, through the penis.

    NOTE: The site where the vas deferens passes into the body cavity is the inguinal canal. The inguinal region is a weak area in the abdominal wall and is often the site of a rupture (hernia).

3. In addition to the paired seminal vesicles (which secrete a thick, clear fluid that lubricates the spermatic ducts and nourishes sperm), the other male sex glands are the prostate gland (which secretes a milky alkaline fluid that neutralizes the acidity of both the male and female reproductive tracts), and the paired bulbourethral glands (which secrete a few drops of fluid into the urethra during sexual arousal.

    NOTE: The seminal vesicles lie at the base of the urinary bladder. The prostate gland surrounds the first part of the urethra and is where the ejaculatory ducts merge into the urethra. The bulbourethral glands are found beneath the prostate gland. Enlargement of the prostate gland, benign prostatic hypertrophy (BPH), is an almost inevitable consequence of male aging (starting at about age 45).

4. Semen consists of sperm and the secretions of these glands. There are about 500 million sperm in the one teaspoon of semen discharged during a typical ejaculation.

5. During sexual arousal, erectile tissue in the penis swells with blood. Erection is essential for the insertion of the penis into the vagina. Like the clitoris, the penis consists of a glans that is richly supplied with nerve endings and a prepuce (foreskin) that covers the glans.

6. Ejaculation occurs in two stages. First, at the peak of sexual arousal, muscles in the epididymis, seminal vesicles, prostate gland, and vas deferens contract, forcing sperm and secretions into the vas deferens behind a closed sphincter at the base of the penis and urethra. Second, that sphincter opens, and strong muscular contractions force the semen along the urethra and out of the penis (Master 27.3C).

7. Sperm and androgen production are under the control of hormones from the pituitary (FSH and LH) (Master 27.3D). FSH increases sperm production by the testes. LH promotes the secretion of androgens, mainly testosterone. Androgens stimulate sperm production.

    Review: These hormones are also discussed in Module 26.12.

8. Unlike in females, the hormonal regulation of sperm production permits the testes to produce hundreds of millions of sperm every day from puberty onward.

C. The formation of sperm and ova requires meiosis (*Module 27.4*).

1. *Review:* The cellular genetics of gamete formation and the mechanics of meiosis (Modules 8.13–8.15, Acetate 8.15).

2. Both sperm and ova are haploid and develop by meiosis from diploid cells in the gonads (testes and ovaries).

3. Spermatogenesis takes about 65–75 days. Diploid cells in the seminiferous tubules of the testes ultimately develop into sperm cells.

4. Diploid cells near the outer tubule wall multiply by mitosis. About 3 million per day develop into primary spermatocytes. Each primary spermatocyte divides by meiosis, passing through a secondary spermatocyte stage after the first meiotic division. Immature sperm develop full motility in the epididymis (Acetate 27.4A).

    NOTE: If mature sperm are not ejaculated, they are broken down and reabsorbed in the epididymis tissue.

5. Oogenesis of all ova begins before birth. Diploid cells in the ovaries develop by mitosis into primary oocytes, one per follicle. Meiosis begins but is arrested in prophase at the time of a female's birth. Meiosis continues once a woman is mature, producing a single egg each month, until menopause.

6. At birth, each ovary contains all the follicles a woman will have, each with an egg at the beginning of meiosis, at the primary oocyte stage. FSH from the pituitary stimulates one dormant follicle to develop. The first meiotic division of the primary oocyte results in one secondary oocyte and one polar body. The second meiotic division produces an ovum and three polar bodies (one from the second oocyte and two from the first polar body). The polar bodies receive virtually no cytoplasm and do not function further (Acetate 27.4B,C).

    NOTE: Instead of putting all their eggs in one basket, mammals put all their primary oocyte's cytoplasm in one egg.

7. One cell develops at a time and is induced (in mid-meiosis, as a secondary oocyte) to leave the ovary (ovulate) by secretion of LH from the pituitary. This hormone continues to cause the ruptured follicle to develop into a corpus luteum, which begins to secrete estrogen and progesterone to prepare and maintain the uterine lining for implantation of the fertilized egg. The final meiotic division of the secondary oocyte occurs immediately following fertilization.

8. Spermatogenesis and oogenesis differ in the number of gametes produced by each meiotic division (four versus one), how often meiosis occurs (continuously versus once per month), and how many primary cells there are (unlimited versus a set number, present at birth).

D. Hormones synchronize cyclic changes in the ovary and uterus (*Module 27.5*, Table 27.5, Acetate 27.5).

    Review: The hormones released by the follicles are discussed in Module 26.12.

    1. The menstrual cycle involves the monthly changes in the uterus. Day one is defined as the first day of menstrual bleeding when, due to the decline in estrogen and progesterone levels, the endometrium breaks down and is shed through the vagina. Menstrual discharge consists of blood, endometrial cells, and mucus. This usually lasts for 3–5 days and corresponds with the preovulatory phase of the ovarian cycle.

2. After menstruation the endometrium regrows (the proliferative phase) under the influence of estrogen. This regrowth reaches a maximum at about 20 to 25 days.

3. FSH initiates the preovulatory phase of the ovarian cycle. During the preovulatory phase, the secondary oocyte is developing and the follicle is stimulated to secrete low levels of estrogen. As the follicle grows, its secretion of estrogen increases, but estrogen levels are still relatively low. This exerts negative-feedback control on the pituitary and keeps FSH and LH levels low.

4. Estrogen levels peak just before ovulation; this results in a surge in LH and FSH levels. LH stimulates the completion of the development of the secondary oocyte. This surge in LH is required for ovulation. Ovulation occurs on day 14, after which the ruptured follicle develops into the corpus luteum.

5. The corpus luteum secretes both estrogen and progesterone. The resultant high levels of progesterone and estrogen maintain the uterus for the possibility of implantation of a fertilized egg. The high levels of estrogen and progesterone also inhibit the secretion of FSH and LH. Thus, FSH and LH levels fall and, as a result, new ovarian and menstrual cycles cannot be initiated.

6. Next, the corpus luteum gradually degenerates. This leads to a fall in estrogen and progesterone levels, which allows FSH and LH levels to increase and new ovarian and menstrual cycles to be initiated.

7. If an embryo is implanted in the uterus, the degeneration of the corpus luteum is delayed, and levels of estrogen and progesterone remain high. The corpus luteum is maintained by the secretion of human chorionic gonadotropin (HCG) by the developing embryo.

III. Human sexuality.

A. The human sexual response occurs in four phases (*Module 27.6*).

1. Most female mammals are receptive to males only on certain days and for only a brief period once or twice a year, when they are "in estrus."

2. Humans and several other primates are unusual in having no distinct breeding periods.

3. Human sexuality is emotional as well as physical. The behavior surrounding sexual activity may have evolved as a way to strengthen bonding between mates.

4. Sexual response during intercourse occurs in four phases: excitement, plateau, orgasm, and resolution. These phases involve changes in sensitivity of the nerves supplying, and activity of the muscle groups associated with, the sex organs and other tissues.

B. Sexual activity can transmit disease (*Module 27.7*).

1. AIDS is only one of many sexually transmitted diseases (STDs) (Table 27.2).

2. STDs are epidemic throughout the world. Many of them cause long-term problems if not treated. Most of these diseases infect both partners. Viral STDs are not curable, but bacterial, protozoan, and fungal STDs generally are.

3. Latex condoms can usually prevent the spread of STDs.

C. Contraception prevents unwanted pregnancy (*Module 27.8*).

1. Only complete abstinence is completely effective. Contraceptive methods vary in their reliability, ease of correct use, and the potential for undesirable side effects (Table 27.8).

2. Birth control pills work by preventing the release of gametes. This is a highly effective method of preventing unwanted pregnancy. Most birth control pills contain a synthetic estrogen and a synthetic progesterone (progestin). Possible short-term side effects include weight gain, headaches, or nausea. Long-term side effects may include an increased risk of breast cancer and cardiovascular problems, especially for woman at risk, such as smokers.

3. Another type of birth control pill, the minipill, only contains progestin. This pill is slightly less effective than the combination pill and works by altering the cervical mucus so that sperm cannot enter the uterus. Norplant® is an implanted minipill.

4. Sterilization in males is called a vasectomy. This involves cutting a section out of each vas deferens. In females, sterilization involves tubal ligation, cutting a section out of each oviduct.

    NOTE: A vasectomy is done under local anesthesia and does not noticeably alter the volume of the ejaculate. The procedure is reversible, but fertility may not be restored. Some studies indicate that following a vasectomy, an immune response to sperm may occur.

5. Methods that are known to have a high failure rate include the rhythm method, which involves abstaining from intercourse from a few days before to a few days after ovulation, and the withdrawal method.

    NOTE: One of the reasons withdrawal is ineffective is that the secretions from the bulbourethral gland can carry a sufficient number of sperm for conception to occur.

6. Barrier methods physically block sperm and egg from meeting. Examples of barrier methods include condoms, the diaphragm (a cap that covers the cervix), the cervical cap (a smaller, more closely fitting diaphragm), and the contraceptive sponge (placed in the vagina). In order to be effective, diaphragms, cervical caps, and sponges need to be used in conjunction with spermicides.

7. Methods that block implantation include IUDs (intrauterine devices) and MAPs (morning-after pills). IUDs are not commonly used since they may cause bleeding, infections, or uterine perforations. MAPs contain estrogen and progesterone and, if taken within 3 days of intercourse, may prevent fertilization. In the U.S., doctors can prescribe high doses of certain ordinary birth control pills for the same purpose.

IV. Early human development.

Review: Homeotic genes and signal transduction (Modules 11.12 and 11.13).

A. Fertilization results in a zygote and triggers embryonic development (*Module 27.9*).

1. Human sperm are well adapted to their function. They have a streamlined shape; a head that contains only the nucleus and the acrosome, the enzyme-containing body that will help the sperm penetrate the egg; and a long, motile flagellum with one associated mitochondrion wrapped

around it. Power is generated by respiration of the fructose from semen (Acetate 27.9B).

2. To reach the nucleus, the sperm must pass through three barriers around the egg: a jelly coat, a middle vitelline layer of glycoproteins, and the egg cell's plasma membrane. The acrosomal enzymes digest a hole in the jelly, species-specific proteins on the tip of the sperm bind with receptor proteins on the vitelline layer, and, finally, the sperm and egg plasma membranes fuse. This triggers changes in the egg that prevent the entry of more sperm; the egg's plasma membrane becomes impenetrable, and the vitelline layer hardens and separates from the plasma membrane, forming the fertilization membrane. (Acetate 27.9C, Figure 27.9A).

NOTE: Meiosis of the egg nucleus is completed at the time of membrane fusion.

3. The egg's metabolism begins to change to that needed for development at the time of membrane fusion.

4. Simultaneously, the egg and sperm nuclei fuse, producing the diploid nucleus of the zygote.

B. Cleavage produces a ball of cells from the zygote (*Module 27.10*).

1. With each division, the number of cells doubles. At first, the overall size of the collection of cells remains the same. In the sea urchin, doubling occurs every 20 minutes. In 3 hours, a small ball of cells is produced. Then the ball becomes hollow (the blastula), forming a fluid-filled cavity, the blastocoel (Acetate 27.10).

2. Decreasing the size of the cells increases the surface-to-volume ratio of each cell, thus enhancing the flow of materials in and out of the cells.

3. During cleavage, the embryo is partitioned into areas that differ in size and chemistry. These areas activate different genes to guide the development of different parts of the animal.

C. Gastrulation produces a three-layered embryo (*Module 27.11*).

1. Gastrulation adds more cells and sorts them into three distinct cell layers in the gastrula: ectoderm, endoderm, and mesoderm (Acetate 27.11).

2. Ectoderm forms the outer skin of the gastrula and ultimately the outer layer of the adult skin and the nervous system.

3. Endoderm lines the embryonic digestive tract and ultimately that of the adult.

4. Mesoderm partially fills in the space between ectoderm and endoderm and ultimately forms most other tissues and organs, including the skin's inner lining, the muscles, and the excretory system.

5. The mechanics of gastrulation differ from animal to animal. Gastrulation involves the migration of outer cells across the blastula surface and into a small pore or groove, the blastopore. Once the gastrula has formed, some of the inner cells break off of the primitive gut (archenteron) to form the mesoderm layer. Ultimately, the blastocoel is filled with mesoderm and endoderm cells.

V. The development of organs and organ systems.

A. Organs start to form after gastrulation (*Module 27.12; Table 27.12*).

1. Frog: A few hours after gastrulation, the notochord (cartilagelike material) forms from mesodermal cells. It supports the length of the embryo, and later the backbone forms around it (Master 27.12A, C).

2. A hollow nerve cord forms when two ridges of ectoderm cells above the notochord fuse and move below the remaining ectoderm (Master 27.12A, B, C).

3. Somites form as blocks of mesoderm that will give rise to segmented muscles and bone along the backbone and ribs. At this time, a hollow space develops in the mesoderm that will become the coelom (Master 27.12C).

    Review: Segmentation and a coelom are basic feature shared by all chordates (Module 19.15).

4. Additional development will result in actual muscle cells, a heartbeat, blood in an early circulatory network, and a tail fin.

5. A recognizable tadpole is visible after 5–8 days of development (Master 27.12D).

B. Changes in cell shape, cell migrations, and programmed cell death give form to the developing animal (*Module 27.13*).

1. Cells of the ectoderm elongate and become wedge-shaped. This change in size and shape forces the surface into a groove, and then into a neural tube, which is the start of the brain and spinal cord (Master 27.13A).

2. Many embryonic cells migrate to specific destinations by following chemical trails and moving by amoeboid movement.

3. Once migrating cells reach their destination, surface proteins on similar cells cause them to aggregate and become glued together.

4. The cells then differentiate and take on the characteristics of the tissue appropriate to their location.

    Review: Patterns of gene expression in differentiated cells (Module 11.6).

5. Programmed cell death, controlled by suicide genes, is also essential for development (Figure 27.13B). For example, the timely death of specific cells result in the formation of the spaces between fingers and toes.

C. Embryonic induction initiates organ formation (*Module 27.14*).

1. When cells differentiate, some genes are activated and others remain inactive.

2. Cells may be activated by induction, the influence of one group of cells on another, by either physical contact or chemical influence.

3. For example, a lobe (optic cup) of the early brain area induces the formation of the lens of the eye from outer ectodermal cells above. The developing lens, in turn, induces the cornea to form from other ectoderm (Master 27.14).

    NOTE: Another example is the induction of the neural tube by the notochord.

D. Pattern formation organizes the animal body (*Module 27.15*).

1. The overall goal of the research described here is to find out how the one-dimensional information encoded in DNA directs pattern formation, the development of the three-dimensional form of the animal.

2. Different positional signals arise from different patterns of cells surrounding a developing region in the embryo.

3. Vertebrate limbs develop from embryonic limb buds. These buds develop into limb patterns in the right place when each cell within the bud receives the correct chemical signal from the correct surrounding cells (Master 27.15A, B).

4. By experimentally removing wing pattern-forming zones from one animal and grafting them on another prior to complete wing formation, cells of the host's wing bud are induced to form two wing structures. The cells on either side of the wing bud are getting similar chemical signals from the host pattern-forming zone and the grafted donor pattern-forming zone.

VI. Human development and birth.

A. Human development: the first month of pregnancy (*Module 27.16*).

1. The human gestation period averages 38 weeks. Gestation is another word for pregnancy.

2. Gestation begins with fertilization in the oviduct. Cleavage begins 24 hours later. By the time the embryo has reached the uterus, it is a blastocyst of about 100 cells. These cells are partitioned into an inner mass that will develop into the baby, and an outer layer of cells, the trophoblast (Acetates 27.16A, B).

3. About a week after conception, the trophoblast secretes enzymes that help the embryo implant in the uterine endometrium lining. The trophoblast and endometrium tissues then begin to build the placenta, as the embryo develops further from the inner cells (Acetates 27.16C, D).

4. The embryo develops as a pad of cells with layers that will form ectoderm, mesoderm, and endoderm. Gastrulation occurs at 16 days.

5. A series of extraembryonic membranes protect the embryo and help it interact with its environment. Moving from outermost to innermost: the chorion (the embryo's part of the placenta), the amnion (protects the embryo in a bag of fluid), yolk sac (produces the first blood cells and cells that will give rise to gamete-producing cells of the gonads), and allantois (important in waste disposal). In other animals, these membranes may have other functions, such as a food source for the yolk sac in birds and reptiles (Acetates 27.16 E, F).

6. Cells of the chorion secrete human chorionic gonadotropin (HCG), which ensures that the corpus luteum continues to produce estrogen and progesterone during the first three months of pregnancy. The chorion also forms the embryonic part of the placenta, the chorionic villi that contain embryonic blood vessels that interweave with the maternal blood supply in the placenta's margin. Here nutrients, gases, and antibodies are absorbed from the mother's blood, and wastes diffuse into the mother's blood.

7. A number of viruses, including AIDS, can cross the placenta. Other harmful substances that can cross the placenta include blood alcohol, morphine, toxins from smoking, and antibiotics such as tetracycline.

NOTE: The consumption of alcohol during pregnancy can result in FAS (fetal alcohol syndrome). The symptoms of FAS can include narrow eye slits, sunken nasal ridge, heart defects, malformed arms and legs, central nervous system damage, slow pre- and postnatal growth, hyperactivity, poor attention span, and failure to understand the connection between a behavior and the result of that behavior.

B. Human development from conception to birth is divided into three trimesters (*Module 27.17*).

1. During the first trimester, all organs and appendages are built in essentially a human pattern. At the very beginning, a notochord, gill pouches, limb buds, and a tail appear, but the notochord is subsumed by the developing spine, the gill pouches grow into parts of the middle ear and throat, the limb buds develop into tiny arms and legs, and the tail is lost.

2. Main changes during the second trimester involve an increase in size and a general refinement of human features. The placenta stops producing HCG and takes over the job of secreting progesterone and estrogen from the corpus luteum. By the end of this time, the fetus has the face of an infant and a definite heartbeat.

3. During the third trimester, the fetus grows in size and strength, and the circulatory and respiratory systems go through changes that will enable the baby to switch to breathing air. The fetus gains the ability to maintain internal temperature; its bones harden, and its muscles thicken.

   Review: Module 22.12 discusses this switch from placental gas exchange to breathing air.

4. A typical baby is born 20 inches long and weighing from 6 to 10 pounds.

   NOTE: Human baby birth weight is an example of stabilizing selection (Module 14.20).

C. Childbirth is hormonally induced and occurs in three stages (*Module 27.18*).

1. Estrogen reaches very high levels during the final weeks of pregnancy. This induces the formation of receptors on the uterus that are sensitive to oxytocin, the hormone that stimulates the contraction of smooth muscles during labor. At the appropriate time, cells of the fetus, and later the pituitary, produce oxytocin, and cells of the placenta produce local regulators, the prostaglandins. Positive feedback results in the uterine contractions causing the release of ever larger amounts of hormones (Acetate 27.18A).

2. During the first stage of childbirth, the cervix dilates to about 10 cm. During the second stage, expulsion, strong uterine contractions of about 1-minute duration occur at 2- to 3-minute intervals, and the baby is born. The total time of labor varies from 20 minutes to over 1 hour. The third stage is the delivery of the placenta, usually no later than 15 minutes after the baby (Master 27.18B).

3. Following childbirth, decreasing levels of progesterone and estrogen cause the uterus to start returning to normal. Lower levels of progesterone induce the pituitary to produce prolactin, a hormone that, along with oxytocin, promotes milk production.

D. Reproductive technology increases reproductive options (*Module 27.19*).

1. Couples may not be able to conceive because of male or female infertility, or because of oviduct blockage, possibly due to scar tissue from an STD.

2. Hormone therapy or surgery can sometimes correct infertility.

3. In vitro fertilization of ova, removed from a woman, by sperm from a man results in an embryo that can implant in the woman's (or a surrogate's) uterus after about 2 days (at the eight-cell stage). This is an expensive process but has resulted in conception and birth of hundreds of babies.

CLASS ACTIVITIES

1. Examples of contraceptives can be brought to class. Describe the mode of functioning of these contraceptives (Module 27.8).
2. Show video *The Miracle of Life*.

RESOURCES AND REFERENCES

BBC Television. *Patterns of Development*. Deerfield, IL: Coronet/MTI Film and Video, 1988. A 25-minute videotape that describes early development in frogs and fruit flies; includes experimental studies of mutants, showing how body patterns develop.

BBC Television. *Shaping Up*. Deerfield, IL: Coronet/MTI Film and Video, 1990. A 24-minute videotape that describes the roles of cell migration, regeneration, cell division, and polarity in determining body pattern during development.

BBC Television. *What You Never Knew About Sex*. Deerfield, IL: Coronet/MTI Film and Video, 1990. A 24-minute videotape that explores the chromosomal, meiotic, behavioral, and developmental means of sex determination in a variety of animals.

Beaconsfield, Peter, George Birdwood, and Rebecca Beaconsfield. "The Placenta." *Scientific American*, August 1980. The development and functioning of this remarkable organ, composed of the tissues of two individuals.

Caldwell, M. "How Does a Single Cell Become a Whole Body?" *Discover*, November 1992.

Crews, D. "Courtship of Unisexual Lizards: A Model for Brain Evolution." *Scientific American*, December 1987. Research on all-female species.

Crooks, R., and K. Baur. *Our Sexuality*, 5th ed. Redwood City, CA: Benjamin/Cummings, 1993. A popular textbook on human sexuality.

Del Pino, Eugenia M. "Marsupial Frogs." *Scientific American*, May 1989. Some tropical frogs incubate eggs on the mother's back, often in a special pouch. Certain features of the adaptation recall pregnancy in mammals and the eggs and embryos of birds.

Development of the Amphibian Embryo. Chicago, IL: Clearvue, 1990. A 15-minute videotape that shows *Xenopus* larval development. Available from Carolina Biological Supply Company.

Fackelmann, K.A. "Cloning Human Embryos." *Science News*, February 5, 1994. Cloning experiments and the ethics of in vitro fertilization.

"Frontiers in Biology: Development." *Science*, October 28, 1994. Leading developmental biologists preview the next 5 years of research.

Gilbert, S.F. *Developmental Biology*, 3rd ed. Sunderland, MA: Sinauer Associates, 1991. An excellent textbook for upper-division courses.

Hart, S. "The Drama of Cell Death." *BioScience*, July-August 1994. The importance of programmed cell death in development and immunity.

Lagercrantz, Hugo, and Theodore A. Slotkin. "The 'Stress' of Being Born." *Scientific American*, April 1986. The stress of journeying through the birth canal is not usually harmful to infants, but the surge of "stress" hormones it triggers can be important to the newborn's survival outside the womb.

The Miracle of Life. Boston: WGBH Educational Foundation, 1986. A 60-minute videotape showing all internal activities involved with human fertilization, development, and birth. A videodisc version plus Macintosh software to access it are offered by Scholastic Software in its Interactive NOVA series. Available from Carolina Biological Supply Company.

Nüsslein-Volhard, C. "Gradients That Organize Embryonic Development." *Scientific American*, August 1996.

Reproductive Systems. Washington, DC: National Geographic Society, 1988. A 20-minute videotape in the "Human Body" series that details human fertilization and development.

Wasserman, Paul M. "The Biology and Chemistry of Fertilization." *Science*, January 30, 1987. Describes the precise sequence of the events that immediately precede and follow the fusion of egg and sperm, including species-specific recognition, membrane fusions, and enzyme-catalyzed modifications of cellular materials.

28

NERVOUS SYSTEMS

APPROACH

The mind-body connection can help focus attention on the subjects of this chapter and engage students in this challenging material. Encourage students to reinforce the facts about observed structures and functions by reflecting on what they know from experience about how their nervous system works.

Details of the physics and chemistry of nerve impulse transmission may be difficult for some students. To supplement the fine diagrams and descriptions of charge flow and imbalance, carefully explain why the inside of the neuron is more negative than the outside. Focus on the role of the sodium–potassium pump (Module 5.18) in expending energy to maintain this charge imbalance. Understanding the static view of the action potential (Module 28.5) and the progression of the action potential along a neuron (Module 28.6) depends on a full understanding of the graphic representations of time and space.

Although information about the effects of drugs and alcohol may not have an immediate influence on student behavior, it is essential information and will be of interest to students. The discussion of the interaction between drugs and neurotransmitters (Module 28.10 and elsewhere) could be easily expanded. For instance, the addictive properties of nicotine (which competes for some neurotransmitter receptors) are not mentioned in this chapter, although the physiological effects on lung tissue are covered in Chapter 22.

The parasympathetic and sympathetic subsystems are introduced as being antagonistic. Remind students that antagonism is a general feature providing balance to the activities of several systems (maintenance of blood pH, endocrine function, and musculature).

Keep in mind that the actual details of sensory input are discussed in Chapter 29 and that Chapter 28 simply describes the nervous system's architecture, the basic description of the nature of nerve impulses, and the overall function of the central nervous system.

CHAPTER OBJECTIVES

Describe the basic functions of the nervous system, and explain the close relationship of this system with the functions of the endocrine system.

Distinguish between the structure and functions of the vertebrate peripheral and central nervous systems.

Explain how neurons convey information in the form of propagating action potentials, including the molecular basis for the signal process at any one point on the neuron.

Name and discuss the mechanisms by which neurotransmitters work in the nervous system.

Describe the general effects of stimulants and depressants on nervous system functioning.

Trace the evolution of the nervous system, comparing the nerve net of cnidarians with the centralized systems of more advanced invertebrates and vertebrates.

Distinguish among the functions of the sympathetic and parasympathetic nervous systems.

Describe a simple reflex response as carried out by nerves of the spinal cord.

Trace the evolution of the three lobes of the vertebrate brain, emphasizing the changes in the cerebrum.

Identify the parts of the human brain and their functions, explaining the functional distinction between the right and left hemispheres of the cerebral cortex, and the formation of memory.

LECTURE OUTLINE

I. Introduction.

  A. The nervous system is basic to the functioning of any animal (*Opening Essay*).

  1. In order to survive and reproduce, an animal must respond appropriately to environmental stimuli, both internal and external.

  2. The nervous system coordinates immediate responses to stimuli with long-term responses from the endocrine system (Chapter 26).

  3. A squid's nervous system provides a precise sense of balance and eyesight, fine control of movement for feeding and retreat from predators, a defensive release of ink, skin color changes to provide camouflage, as well as all activities that maintain its internal environment.

  4. To effect such control over its body, a squid has a large brain and nerves that carry sensory input to the brain and commands away from the brain.

  5. Among other nerve cells, the squid has large cells with thick, long extensions called giant fibers. These fibers are easily managed and manipulated in the lab and have for a long time been used in research on nerve physiology (Master 28.0).

  B. Nervous systems receive sensory input, interpret it, and send out appropriate commands (*Module 28.1*; Master 28.1).

  1. A cubic centimeter of human brain tissue contains an elaborate network of several million nerve cells, each communicating with several others.

  2. Sensory input is triggered by the stimuli of receptors and involves the conduction of signals from the receptors to processing centers.

  3. Integration is the interpretation of these signals and the formulation of responses by the processing centers.

  4. Motor output is the conduction of signals from the processing center to effector (muscle or gland) cells that respond to the stimuli.

II. Neurons are the functional units of nervous systems (*Module 28.2*; Acetate 28.2).

  A. Common features of neurons.

  1. The cell body houses the nucleus and most of the organelles.

  2. Dendrites are short, numerous, and highly branched; they convey signals toward the cell body.

  3. Axons are long and usually unbranched (except at the very end); they convey signals away from the cell body toward other neurons or effector cells.

  4. Each axon branch ends in a synaptic knob that relays the signal.

  B. Neurons are found with supporting cells.

  1. There may be as many as 50 supporting cells for every neuron.

NOTE: These supporting cells are called neuroglia. There are six major types of neuroglia.

2. The supporting cells protect, insulate, or reinforce the neurons.

   NOTE: For example: Astrocytes are supportive cells within the CNS that connect neurons to blood vessels. Astrocytes have many functions, including neurotransmitter metabolism and K^+ balance. Recent studies have implicated astrocytes in learning and memory. Oligodendrocytes form the myelin sheath of the CNS. However, unlike Schwann cells, which form the myelin sheath of the PNS, oligodendrocytes do not guide the regrowth of damaged neurons. Microglia are phagocytic cells found within the CNS. There is evidence that microglia play a role in Alzheimer's disease. Ependymal cells are ciliated cells that aid in the circulation of cerebrospinal fluid (CSF). Satellite cells support clusters of nerve cell bodies (ganglia) in the PNS.

3. Those nerves that convey signals very quickly are enveloped by special supporting cells (Schwann cells) that form a myelin sheath. The Schwann cells are arranged like beads on a string, wrapped around the axon but leaving periodic, unmyelinated nodes of Ranvier. On axons of this type, the myelin sheath insulates the axon and the nodes of Ranvier are the only places on the axon where signals are transmitted (where the plasma membrane of the axon is depolarized).

   NOTE: Schwann cells are also called neurolemmocytes. Nodes of Ranvier are also called neurofibral nodes.

4. In people who have the debilitating autoimmune disease multiple sclerosis, the myelin sheaths are gradually degraded by the person's immune system.

   Review: Autoimmune diseases are discussed in Module 24.16.

C. Nerve signal speed.

   1. Nonmyelinated nerve cells: 5 m/sec.
   2. Myelinated nerve cells: 100 m/sec.

D. Nervous systems have central and peripheral divisions (*Module 28.3*; Acetate 28.3).

   1. The central nervous system (CNS) consists of the brain and spinal cord.
   2. The peripheral nervous system (PNS) consists of communications lines, nerves (ropelike bundles of neuron dendrites and axons), and ganglia (clusters of the cell bodies of these neurons). The nerves of the PNS relay sensory information from sensory receptors to the CNS, and information from the CNS to effectors.
   3. Neurons are divided among three functional types. Sensory neurons travel through the PNS and communicate information to the CNS. Interneurons are entirely within the CNS; they integrate data obtained from one or more sensory neurons and relay the integrated signal to other interneurons or to motor neurons. Motor neurons convey signals from the interneurons in the CNS, through the PNS, to effector cells.

III. The nervous signal along the neuron.

A. A neuron maintains a resting potential across its membrane (*Module 28.4*).

1. Like a battery, a neuron maintains potential energy (Module 5.1) as a difference in electrical charge across the plasma membrane. Cells in general have a negative resting potential, with more negative charges inside the cell than outside. A neuron has a resting potential of –70 millivolts (mV), about 5% of the voltage of an AA battery (Acetates 28.4A, B).

   NOTE: Remind the students that this is a localized charge and that the cell, as a whole, is not negatively charged; moreover, the interstitial fluid, as a whole, is not positively charged (ask the students if they stick to magnets).

2. The resting potential is maintained by negatively charged, large organic molecules remaining inside the cell and an excess of K^+ ions inside and Na^+ ions outside the cell. The K^+ ions are free to diffuse in both directions through K^+ channels across the membrane. Na^+ ions are actively transported out of the cell as K^+ ions are transported in by the sodium–potassium pump.

   Review: Active transport (Module 5.18).

B. A nerve signal begins as a change in the membrane potential (*Module 28.5*).

   1. In the 1940s, British biologists A. L. Hodgkin and A. E. Huxley worked out the details of nerve signal transmission using squid giant axons (fibers).

   2. A stimulus is any factor (electric shock, pressure, sudden temperature change, etc.) that results in the triggering of a nerve signal. A nerve signal involves the progressive formation of the action potential along a nerve.

   3. The graph traces the electrical changes over time at one point along an axon. These changes make up the action potential (Acetate 28.5).

   4. A typical action potential shows the following changes relative to the resting potential of –70 mV. Following a stimulus, the voltage rises in 2–3 milliseconds (msec) to the threshold potential, the minimum rise that will generate an action potential (in this case, to –50 mV). The threshold potential triggers the action potential, a rapid upswing to about 35 mV within 3–4 msec from the initial stimulus. The voltage then drops slightly below the resting potential, and returns to it about 7 msec after the stimulus.

      NOTE: It is only the axon that can achieve an action potential. The potentials that travel along dendrites and nerve cell bodies are graded potentials. Graded potentials travel only a short distance before dying out. However, graded potentials can be added together (summation) to result in an action potential.

   5. The specific ion movements that generate this action potential are controlled by the opening and closing of voltage-sensitive channels on ion channels. The stimulus triggers the opening of Na^+ channels. At first a few Na^+ ions move into the axon. If enough ions move in to reach the threshold potential, the increasingly positive charge within causes more and more Na^+ channels to open. The peak voltage triggers the closing of Na^+—and the opening of K^+—channels, allowing K^+ to diffuse out rapidly, thereby balancing the inward movement of Na^+. Next, there is a brief period during which the membrane potential is below –70 mV, followed by a return to the resting potential.

      NOTE: This hyperpolarization prior to a return to the resting potential is due to the slow closure of the K^+ channels.

C. The action potential regenerates itself along the neuron (*Module 28.6;* Acetate 28.6).

    1. The local spreading of the electrical changes is caused by inflowing Na^+.

    2. These changes trigger the opening of Na^+ channels just ahead of the action potential, generating a second action potential a little farther on.

    3. But the changes cannot be induced in the region behind the action potential where K^+ ions are moving out, so the action potential travels in just one direction.

    4. Action potentials are all-or-none phenomena. A signal with higher intensity reaches no higher peak voltage, but instead consists of an increase in the number of action potentials per millisecond.

IV. Transmission of nerve signals between cells.

  A. Neurons communicate at synapses (*Module 28.7*).

    1. A synapse is the junction between two neurons, or between a neuron and an effector cell.

    2. At electrical synapses, action potentials travel directly from one cell to another. In humans, electrical synapses are common in the heart and digestive tract, associated with cardiac and smooth muscle cells.

    3. At chemical synapses, action potentials are converted into a chemical signal. This chemical signal takes the form of neurotransmitter molecules (Module 26.1) that carry the message across a small gap (synaptic cleft) between the cells. The synaptic cleft prevents the spread of the action potential between cells.

    NOTE: The release of neurotransmitter into a synapse requires an influx of Ca^{2+}.

    4. In synaptic knobs at the ends of axons of transmitting neurons, neurotransmitters are stored in vesicles. The arrival of the action potential triggers the fusion of the vesicles with the plasma membrane, releasing the neurotransmitter into the cleft. The molecules diffuse across and bind to receptor molecules in the receiving cell's membrane. The neurotransmitters produce their effect by causing the opening of ion channels through which ions can diffuse and trigger a new action potential. The neurotransmitters are then broken down enzymatically, and as a result the ion channels close (Acetate 28.7).

    Preview: These events are similar to those that occur at a neuromuscular junction (Module 30.10).

  B. Chemical synapses make complex information processing possible (*Module 28.8*).

    1. Neurotransmitters can either open ion channels in the receiving cell's plasma membrane or trigger a signal transduction mechanism that will result in the opening of ion channels.

    Review: The mechanism by which signal transduction pathways function is discussed in Module 11.13.

    2. Excitatory neurotransmitters open Na^+ channels and trigger a new action potential in the receiving cell.

    3. Inhibitory neurotransmitters open Cl^- channels that decrease the tendency of the receiving cell to develop action potentials.

4. A single cell receives input from numerous synaptic knobs from hundreds of neurons. The cell receives various magnitudes and numbers of both inhibitory and excitatory signals. The receiving neuron or effector cell acts, depending on the summation of all incoming signals.

C. A variety of small molecules function as neurotransmitters (*Module 28.9*).

1. Most neurotransmitters are small, nitrogen-containing molecules. Acetylcholine slows heart rate and causes muscle cells to contract.

2. Several neurotransmitters are biogenic amines which also function as hormones: epinephrine, norepinephrine (increases heart rate), serotonin, and dopamine (affects sleep, mood, attention, and learning).

3. Biogenic amines are associated with various diseases. For example, Parkinson's disease is caused by a lack of dopamine; whereas schizophrenia has been linked to an excess of dopamine. Antidepressant drugs, such as Prozac, function by affecting neurotransmitter activity.

4. Aspartate, glutamate, glycine, and gamma aminobutyric acid (GABA) are amino acids with neurosecretory function in the brain. Aspartate and glutamate are excitatory. Glycine and GABA are inhibitory.

NOTE: Glutamate has been implicated in stroke-induced neuronal death.

5. Peptides such as endorphins (natural painkillers) and substance P (excitatory) are also neurotransmitters.

6. ATP, and the toxic gases NO (nitric oxide) and CO, have also been shown to serve as neurotransmitters. NO plays a role in penile erection and may play a role in learning.

NOTE: NO affects many other aspects of physiology, including playing a role in the regulation of blood pressure.

D. Many stimulants and depressants act on chemical synapses (*Module 28.10*).

1. Stimulants increase the activity of the central nervous system, often by altering or mimicking the effects of neurotransmitters. Cocaine, amphetamines (both of which increase the stimulatory effect of norepinephrine), caffeine, and theophylline are all stimulants (Table 28.9).

2. Depressants decrease the activity of the central nervous system, often by inhibiting signal transmission at chemical synapses. Barbiturates, tranquilizers (Valium and Librium), and alcohol intensify the inhibitory effect of GABA.

3. Heavy consumption of alcohol can physiologically damage the liver and kidneys, has been linked with cardiovascular disease and cancer, and can cause fetal alcohol syndrome in children of mothers who drink during pregnancy.

Review: The passage of substances, including alcohol, across the placenta is discussed in Module 27.16.

V. **Diverse nervous systems have evolved in the animal kingdom (*Module 28.11*).**

A. Neurons function in essentially the same way in all animals, but they are arranged in different patterns that provide different levels of integration and control.

B. Invertebrate nervous systems (Acetates 28.11A, B).

1. Cnidarians have a nerve net of individual neurons. The nerve net provides overall sensory function and control over limited muscular activity.

2. Flatworms are the first animal phylum to show cephalization (concentrating the nervous system in an anterior head) and centralization (the differentiation of CNS and PNS). Flatworms tend to move head-first through an environment, encountering new stimuli with light, touch, and chemical receptors. Integration of a response is coordinated by two small ganglia.

3. Insects have larger, more complex brains and integrating ganglia in each body segment, and they have many more complex sense organs (Acetate 28.11C).

C. The vertebrate nervous system (Acetate 28.11D).

1. This system is highly centralized into brain and spinal cord, all protected inside bony skeletal elements. The brain is the master control center, directing output through the spinal cord and including homeostatic centers, sensory centers, and centers of emotions and intellect.

2. The brain and spinal cord both include hollow regions that are filled with cerebrospinal fluid (CSF). These spaces in the brain, ventricles, are continuous with the central canal of the spinal cord (Master 28.11E).

NOTE: This stems from the developmental source of the nervous system as an infolding of ectoderm into a hollow nerve tube (Modules 27.12, 27.13).

3. The CNS is divided between white matter, with concentrations of myelinated axons and their synapses, and gray matter, with concentrations of neuron cell bodies.

VI. The vertebrate peripheral nervous system.

A. The peripheral nervous system of vertebrates is a functional hierarchy (*Module 28.12*).

1. The ganglia and nerves of this system form a vast, intercommunicating network. Cranial nerves carry signals directly to the brain. Spinal nerves carry signals only to the spinal cord.

2. Sensory neurons bring in information from either the external or the internal environment (Master 28.12).

Preview: Sensory input and sensory transduction (Modules 29.1 and 29.2).

3. Motor neurons are either under voluntary control (somatic system carrying messages to skeletal muscles) or out of direct, conscious control (autonomic system carrying messages mostly to glands and smooth muscles).

Preview: The somatic nervous system controls voluntary muscular movement (Module 30.10).

4. All spinal and most cranial nerves carry both sensory and motor neurons.

B. Opposing actions of sympathetic and parasympathetic neurons regulate the internal environment (*Module 28.13;* Acetate 28.13).

1. The parasympathetic division of the autonomic nervous system primes the body for digesting food and resting, activities that gain and conserve the body's energy supply. These include stimulation of all digestive processes, and slowing the heart and breathing rates.

NOTE: The parasympathetic division is associated with "rest and repose."

2. Neurons from this system leave the basal part of the brain and the lower part of the spinal cord. Most neurons release the neurotransmitter acetylcholine to affect their target organs.

3. The sympathetic division prepares the body for intense, energy-consuming activities, such as fighting a competitor or fleeing a predator. These include inhibiting of digestive activity, increasing the heart and breathing rates, and stimulating the liver to release glucose and the adrenal gland to release the fight-or-flight hormones epinephrine and norepinephrine.

   NOTE: The sympathetic division is associated with "fight or flight."

4. Neurons from this system leave the middle part of the spinal cord. Most neurons release the neurotransmitter norepinephrine to affect their target organs.

VII. The vertebrate central nervous system.

A. The spinal cord produces the simplest behaviors in vertebrates (*Module 28.14*).

  1. The spinal cord has two functions: to convey information to and from the brain and parts of the peripheral nervous system, and to produce simple responses (reflexes) to certain kinds of stimuli.

  2. The simplest reflex involves only two neurons, a sensory neuron that directly synapses with a motor neuron in the spinal cord. The knee-jerk reaction tested by doctors is such a response. This type of reflex cannot be overridden by conscious control (Acetates 28.14A, B).

  3. More complex reflexes involve an intermediary interneuron that integrates the response. The "pull-away" response to pain is one such reflex. This type of reflex can be modified by conscious control.

B. Complex behavior evolved with the increasing complexity of the vertebrate brain (*Module 28.15*).

  1. Many vertebrate activities (e.g., finding food and mates, avoiding predators, and raising offspring) require far more integration and conscious control than that provided by the autonomic nervous system or by spinal cord reflexes.

  2. The vertebrate brain evolved from three anterior bulges on the spinal cord: forebrain, midbrain, and hindbrain. These subdivisions can be distinguished in early stages of brain development in all vertebrates (Acetate 28.15A).

  3. Three trends appeared during the course of evolution of the brains of animals: an increase in relative brain size, subdivision of the basic regions into subregions with specific roles, and the increasing integrative power of the forebrain, particularly the cerebrum (Acetate 28.15B).

  4. In birds and mammals, the cerebral cortex is highly folded, increasing the surface area of gray matter. The human cerebral cortex occupies 80% of the total brain mass and has the largest surface area (relative to body size) of any vertebrate. The cerebral cortex of the porpoise is the second largest.

C. The structure of a living supercomputer: the human brain (*Module 28.16*).

  1. The human brain is composed of around 100 billion neurons, with a much larger number of supporting cells.

2. The three ancestral lobes of the brain are present, but they are highly evolved and overshadowed by the dominant cerebrum of the forebrain (Acetate 28.16A).

3. The hindbrain: The pons and medulla conduct information to and from the more forward portions through sensory and motor neurons. This region also controls such involuntary activities as breathing (Module 22.9), heart rates (Module 23.7), and digestion (Chapter 21), and helps coordinate whole-body movement (Chapter 30). The cerebellum is a planning area for whole-body coordination, integrating signals from sensory neurons (Chapter 29) with nervous activity in the rest of the brain.

4. The midbrain also relays sensory and motor signals, integrates some sensory input, such as sounds, and controls eye reflexes and sleep/wake cycles.

5. The forebrain is the site of the most sophisticated integration. The thalamus contains cell bodies of neurons that relay information to the cerebral cortex and filter signals that pass through it. The hypothalamus regulates homeostasis, particularly in controlling the hormonal output of the pituitary gland. The hypothalamus (Module 26.4) controls the pituitary gland, body temperature (Chapter 25), blood pressure (Module 23.9), hunger, thirst (Chapter 25), sexual urges, and responses to danger; is involved in the experiences of emotions; and contains a biological clock (regulating circadian rhythms). It is particularly sensitive to some addicting drugs such as cocaine.

NOTE: Much of the brain functions in relaying information from one part of the brain to another in the process of integration.

6. The cerebrum is composed of two hemispheres connected by the corpus callosum (Master 28.16B).

7. Basal ganglia, found beneath the corpus callosum, function in motor coordination. Degeneration of cells in the basal ganglia occurs in Parkinson's disease, a symptom of which is uncontrollable shaking.

D. The cerebral cortex is a mosaic of specialized, interactive regions (*Module 28.17*).

1. This region is a highly folded sheet of gray matter containing some 10 billion neurons and hundreds of billions of synapses. Its neural circuitry produces our most distinctive human traits: reasoning, language, imagination, artistic talent, and personality. It also creates our sensory perceptions by integrating sensory information with memory and analysis.

2. Localization of function within the cortex comes mostly from studying the effects of tumors, strokes, and accidental damage; from studying direct stimulation during surgery; and from studying brain activity using PET scans (Figure 20.10D). The cortex has no pain sensors.

3. Both hemispheres are divided into four discrete lobes, each of which has several functional areas. Regions often combine centers that receive signals with association areas that help integrate our sensory perceptions. These association areas are the sites of higher mental activities: evaluating consequences, making judgments, and planning for the future. Language also results from interactions among several areas, especially those areas associated with reading and speech (Master 28.17).

E. Roger Sperry discovered that the right and left cerebral hemispheres function differently (*Module 28.18*).

Nervous Systems 261

1. The two hemispheres look the same, their primary motor and sensory areas function the same, but left and right association areas function differently.
2. The "rational" left hemisphere houses language centers and has association areas for logic and mathematical abilities. The "intuitive" right hemisphere lacks these and has association areas that underlie our imagination, spatial perceptions, artistic and musical abilities, and emotions.
3. In the 1960s, neurologist Roger Sperry discovered these differences by studying "split-brain" patients. In these patients, the capabilities of each hemisphere remain separate. For example, the tactile sense of a key in the left hand, which goes to the right hemisphere, allows the key to be used correctly, but never makes it to the language center in the left hemisphere that identifies it as a "key" (Master 28.18B).

F. The brain is active during sleep (*Module 28.19*).
1. Humans require sleep, a brain state in which stimuli are received and, in part, acted on, but without awareness of the stimuli, as is the case during arousal.
2. Brain waves (electrical signals on the head's surface recorded by an electroencephalogram, or EEG) depend on mental activity. The less the mental activity, the more regular the EEG.
3. Alpha waves are characteristic of quiet, awake individuals. Beta waves are more agitated and are characteristic of awake individuals solving complex mental problems (Master 28.19B).
4. During sleep, activity cycles between two alternating types of sleep. Slow-wave (SW) sleep is characterized by delta waves and regular strong bursts of brain-wave activity. REM (rapid-eye movement) sleep is characterized by rapid and less regular brain-wave activity than SW sleep. It is during REM sleep that most dreams occur. Both REM and SW sleep seem to play a role in memory and learning, with SW sleep playing a role in memory storage and REM sleep playing a role in the learning of repetitive skills.

G. The limbic system underlies emotions, memory, and learning (*Module 28.20*).
1. The limbic system is a functional unit of several integrating centers and interconnecting neurons in the forebrain, including the thalamus, and parts of the hypothalamus and inner cerebrum (Master 28.20).

    NOTE: Feelings of pleasure and punishment are associated with the limbic system. Stimulation of these areas evokes intense reactions.

    Preview: Part of the hypothalamus functions as a biological clock (Module 37.9).

2. The hippocampus is involved in memory formation and learning. The hippocampus interacts closely with the prefrontal cortex, which functions in complex learning, reasoning, and personality.
3. The limbic system is closely associated with olfaction, as can be seen by the ability of odors to evoke both memories and emotions.
4. Short-term memory lasts only a short period of time.
5. Long-term memory requires the ability to store and retrieve information. The prefrontal cortex appears to be involved in the retrieval of stored

information. Overall, the process of memory formation and retrieval appears to be highly complex.

6. In memory formation, the hindbrain and midbrain serve as sensory filters, and the limbic system determines what emotions, if any, are to be associated with the stimulus.

7. Memory formation and learning are associated with repeated stimulation of neurons within the hippocampus. As a result, these neurons become increasingly sensitive to stimuli and more likely to reach an action potential. These changes are said to result in long-term potentiation (LTP) and may be involved in the formation of a permanent memory.

NOTE: This suggests that learning is greatly facilitated by repetition (i.e., studying, and not just the night before, or even just the week of, the exam).

Preview: Learning ranges from simple behavioral changes to complex problem solving (Module 37.4).

H. Studies of learning in invertebrates bridge the gap between neurons and behavior (*Module 28.21*).

1. *Aplysia*, a mollusk, has been used to study this link. *Aplysia* has a very simple nervous system of about 20,000 neurons, some arranged into ganglia.

2. The animal exhibits simple reflexive behavior, but it can modify this behavior over time. It will learn to ignore repeated mild prodding.

3. On the other hand, the normal withdrawal reflex can be strengthened by coupling the prodding with an electrical stimulus. Over time, the sensory neuron that detects the stimulus releases more neurotransmitter. Thus, it learns.

CLASS ACTIVITIES

1. Compare two types of nervous system by using the cnidarian *Hydra* and the planarian flatworm *Dugesia*. Place several individuals of each species in a watch glass on an overhead projector, and leave the preparation alone as you talk about the systems (Module 28.11). The animals will demonstrate their capabilities very well. Add a few *Daphnia* to introduce an even more sophisticated system, and to induce additional behavior of *Hydra*.

2. An impromptu demonstration of the "all-or-nothing" and cascading nature of the transmission of action potentials can be done with chalkboard erasers (make sure you have plenty on hand). Line them up as though they were dominoes (you can also use dominoes, but erasers work better), with each eraser upright on its narrow end and with a space of about two-thirds an eraser length between them. A slight push of the leading eraser will cause it to rock back and forth, but if it doesn't surpass the "threshold" push, the signal will not be passed further. A certain strength push will surpass the "threshold potential" and induce the transmission of an "action potential" (the sequential tipping over of all the erasers).

3. Reflexes (Module 28.14) such as the knee-jerk reflex can easily be demonstrated in class. Note the difference in response when the subject is not distracted as compared to when the subject is distracted (for example, reading a book). Another, similar, reflex can be demonstrated by having the subject stand next to a chair on one leg, with the shin of the other leg resting on the chair in such a way that the foot hangs off the back of the chair; tap the Achilles tendon, and the foot will move. Other easily demonstrable reflexes include the pupillary reflex and the gag reflex.

RESOURCES AND REFERENCES

The Addicted Brain. Princeton, NJ: Films for the Humanities and Sciences, 1987. A 26-minute videotape that describes the natural neurotransmitters and their drug analogs.

Alkon, Daniel L. "Memory Storage and Neural Systems." *Scientific American*, July 1989. How learning is accompanied by changes in electrical and chemical properties of nerve cells in the brain.

Aoki, Chiye, and Philip Siekevitz. "Plasticity in Brain Development." *Scientific American*, December 1988. The final wiring of the brain occurs after birth and is governed by early experience.

Barinaga, Marcia. "The Mind Revealed." *Science*, August 24, 1990. Some neuroscientists think that recently discovered oscillations of electrical potential hold the key to how the brain assembles sense impressions into a single object.

The Development of the Human Brain. Princeton, NJ: Films for the Humanities and Sciences, 1991. A 40-minute videotape that follows the physiological development of the human brain from conception to age 8.

Gillis, A.M. "Why Sleep?" *BioScience*, June 1996. Does sleep restock our brain with energy?

Goldstein, Gary W., and A. Lorris Betz. "The Blood-Brain Barrier." *Scientific American*, September 1986. Brain capillaries are unlike those of other organs. Their special properties enable them to serve as stringent gatekeepers between blood and brain. This article describes recent work to show how the feat is accomplished.

Horgan, J. "Can Science Explain Consciousness?" *Scientific American*, September 1992. Current hypothesis on how our brain makes us aware.

Human Brain Animations. Seattle, WA: University of Washington Health Sciences Center for Educational Resources, 1992. A videodisc with a collection of three-dimensional reconstructions of the human brain, including zooms, rotations, and variable coloration to emphasize different structures.

Inside Information: The Brain and How It Works. Princeton, NJ: Films for the Humanities and Sciences, 1992. A 58-minute videotape that explains the latest research on brain function; includes interviews with researchers in the fields of neuroscience, vision science, and computer neural-net technology.

Invertebrates: Conditioning or Learning? Washington, DC: National Geographic Society, 1975. A 15-minute videotape that describes experiments on learning in invertebrates.

Kalil, Ronald E. "Synapse Formation in the Developing Brain." *Scientific American*, December 1989. The refinement of existing synapses and formation of new connections depends on how young neurons generate impulses during development.

Kemp, M. "A Squid for All Seasons." *Discover*, June 1989. An important organism in neurobiology research.

Kimura, D. "Sex Differences in the Brain." *Scientific American*, September 1992. How hormones may induce brain differences in the sexes.

Koch, C. "What Is Consciousness?" *Discover*, September 1992.

LeDoux, J.E. "Emotion, Memory and the Brain." *Scientific American*, June 1994. Research on the chemical basis of memory.

Marvels of the Mind. Washington, DC: National Geographic Society, 1980. A 23-minute videotape that uses computer graphics and a unique synapse sculpture to examine the brain's structure and function.

"Mind and Brain." *Scientific American*, October 1992. The human brain is the most complex structure in the known universe. A special, single-topic issue devoted to many aspects of brain structure and function: development, visual image formation, biology of learning, language, memory, sexual differences, mental disorders, aging, consciousness, and artificial neural networks.

Nervous System. Washington, DC: National Geographic Society, 1988. A 20-minute videotape that describes how sensory input is integrated with memory and reasoning by the brain and results in the output of impulses to organs, muscles, and glands.

Sperry, R.W. "The Great Cerebral Commissure." *Scientific American*, January 1967. An early account of the different functions of the left and right hemispheres of the brain.

Travis, J. "Lighting Up Biological Clocks." *Science News*, August 12, 1995. Studying internal clocks using genes from light-producing organisms.

Williams, J. Michael. *Neuroanatomy Foundations Academic*. Santa Barbara, CA: Intellimation, 1992. Digitized, three-dimensional brain dissections, diagrams, and text cover the central nervous system in a Macintosh HyperCard format.

WNET. *The Brain*. Washington, DC: Corporation for Public Broadcasting, 1984. The Annenberg/CPB Project. An eight-part series of 1-hour videotapes on brain structure and function, including programs on stress and emotion, learning and memory, the two brains, and insanity.

29

THE SENSES

APPROACH

This chapter is an extension of Chapter 28, which dealt with the general architecture of the nervous system and the nature of nerve signals. The subject of this chapter is very accessible to students. The interpretation of several of the drawings and descriptions depends, more than usual, on students' having a good spatial sense. Ensure that they understand the arrangements of receptor cells relative to the surfaces (skin, eyeball, outer ear, tongue, and nostrils) with which they are familiar.

In preparation for your discussion of the senses of sight and sound, review the nature of electromagnetic radiation and sound waves.

Be sure to include some exercises on visual and auditory perception, like the optical illusion in Module 29.1, to spark interest and discussion. Check with your psychology department for sources of overhead transparencies or slides, if necessary.

CHAPTER OBJECTIVES

Describe how sensory stimuli elicit perceptions of the stimuli.

Name and differentiate the five types of sensory receptors.

Identify the receptors that are responsible for mechanical and thermal perception by the skin.

Define the three types of eyes that have evolved among animals, contrasting the qualities of perception by each.

Describe the vertebrate eye and its function as a single-lens eye, and explain how stimuli start to be integrated directly by it.

Describe the human ear, and trace a sound pattern through the auditory pathway.

Explain how smell and taste are perceived.

LECTURE OUTLINE

I. Introduction.

   A. Sensory information gathered by sensory receptors guides animals in their activities (*Opening Essay*; chapter-opening photo).

      1. Bears learn as cubs to feed at certain streams. They find the same streams year after year, using their well-tuned sense of smell. Their sense of smell and very fast reflexes make up for poor eyesight to aid in the capture of food.

         NOTE: The relatively large size of the nasal area of a bear's head.

      2. Salmon form a memory of the chemical "scent" of the river where they hatch. After migrating downstream and spending several years feeding and growing in the open ocean, they use a variety of senses to return to

their home spawning grounds. They use sight of the sun's angle (and perhaps a sense of Earth's changing magnetism) to find the river's ocean mouth. Once in the river, they use smell to follow the increasing concentration of the spawning ground's scent as they make the correct turns up the stream's branches.

B. Sensory inputs become sensations and perceptions in the brain (*Module 29.1*).

1. Receptor cells detect stimuli such as chemicals, light, muscle tension, sounds, electricity, cold, heat, and touch. They trigger action potentials that travel to the central nervous system.

2. A sensation is the awareness of a sensory stimulus.

3. A perception is the integration of a sensation with other information, including other sensations and memories. For instance, the perception of a fragrant rose results from the sum total of interconnected neurons in the visual and odor centers, and their associated memory areas in the brain.

4. The visual demonstration provided may first be sensed as just splotches and later integrated into the perception of a person riding a horse (Master 29.1).

II. Sensory receptor cells convert stimuli into electrical energy (*Module 29.2*).

A. This conversion is known as sensory transduction.

1. For example, in a taste bud sensing sugar, the sugar molecules enter the region of the sensory receptor cells, bind to specific proteins in the cell's membrane, cause ion channels to open, and induce a rise in membrane potential. The result is a receptor potential produced as a result of signal transduction (Acetate 29.2A).

2. Unlike action potentials, the stronger the stimulus, the larger the receptor potential.

3. Receptors synapse with sensory neurons and generate receptor potentials by increasing their release of neurotransmitter.

Review: Synapses and neurotransmitters (Modules 28.7–28.10).

B. The perception is transmitted to the brain, where it is integrated.

1. A receptor normally secretes neurotransmitter at a constant, low rate. The stimulus results in some higher rate. In the brain, the stimulus is sensed as a change in the frequency of action potentials arriving on the sensory neuron. The strength of the stimulus is also interpreted from how many sensory neurons send a signal of it (Acetate 29.2B).

2. Signals from different sensory receptors are perceived as different (sweet versus salty) depending on which interneurons in which region of the brain are stimulated.

NOTE: Distinguishing stimulus types depends both on genetically determined, "hard-wired" neuronal connections between association centers and on comparison with learned memories of other similar stimuli.

3. Sensory adaptation is the tendency of receptor cells to become less sensitive to constant stimulation, because stimuli are perceived as changes in rate. This keeps the body from becoming overloaded with background stimuli.

C. Specialized sensory receptors detect five categories of stimuli (*Module 29.3*).

    1. There are five general categories of sensory receptors: pain receptors, thermoreceptors, mechanoreceptors, chemoreceptors, and electromagnetic receptors.

    2. Skin contains receptors falling into three of these categories: pain receptors, thermoreceptors, and mechanoreceptors. In this case, each receptor is also the sensory neuron delivering the stimulus to the brain (Acetate 29.3A).

    3. Pain receptors indicate the presence of danger and often elicit withdrawal to safety. With the exception of the brain, all parts of the human body have pain receptors.

    4. Thermoreceptors detect either heat or cold and also monitor blood temperature deep in the body. The hypothalamus (Module 28.16) sets and monitors body temperature.

    Review: Thermoregulation as a homeostatic mechanism is discussed in Module 20.13; also see Module 25.2.

    5. Mechanoreceptors are diverse and respond to touch, pressure (including blood pressure), stretching of muscles (stretch receptors), motion, and sound. Hair cells detect movement of cilia or special projections of the cell membrane when exposed to stimuli from sound waves and other forms of movement. In one direction, more neurotransmitter molecules are released, and in the other, fewer (Master 29.3B).

    6. Chemoreceptors include sensory receptors in the nose and mouth, and internal receptors that monitor blood levels of chemicals. Chemoreceptors of insects can be extremely sensitive to just a few molecules (Figure 29.3C).

    7. Electromagnetic receptors are sensitive to energy of various wavelengths, including electricity, magnetism, and light (photoreceptors). Some fishes are sensitive to changes in electrical fields caused by environmental interaction with electrical currents the fish produce. The heads of a number of animals contain magnetite that they may use to sense changes in magnetic fields. Photoreceptors are sensitive to humanly visible wavelengths and to infrared (in invertebrates such as mosquitoes and in vertebrates such as snakes; Figure 29.3D) and ultraviolet (in many pollinating insects) wavelengths.

III. The eyes.

A. Three different types of eyes have evolved among invertebrates (*Module 29.4*).

    1. Eye cups (in planarian flatworms and other invertebrates) are composed of rounded shields of dark-colored cells that shade photoreceptor cells from one side. They do not form images but allow the animal to sense the intensity and direction of light so they can escape to shady hiding places (Master 29.4A).

    2. Compound eyes (found in insects and other arthropods) are composed of many tiny light detectors, each with its own covering (cornea) and lens, which focuses light onto a cluster of photoreceptor cells. The whole eye produces a mosaic image that is likely integrated in the brain into a coarse-grained whole. Compound eyes are extremely acute motion detectors (Figure 29.4B).

    3. Single-lens eyes (e.g., squid) function like cameras. Light rays from an object enter through a small pupil and are focused into an image on the

photoreceptor surface of the retina. This produces a fine-grained, integrated image in the brain (Figure 29.4C). This type of eye is very similar in structure to the human eye.

NOTE: This is an example of convergent evolution (Modules 16.11 and 19.9).

B. Vertebrates have single-lens eyes (*Module 29.5*).

1. The vertebrate eye evolved independently of the single-lens eye of invertebrates and differs in many details.

   NOTE: Humans have two eyes that both face forward to focus on the same object. This is known as convergence and allows for depth perception.

2. The outermost layer of the eyeball is the sclera. The sclera forms the white of the eyeball, and in front, the transparent cornea. The conjunctiva is a thin mucous membrane that lines the eyelids and the front of the eyeball, except the cornea. The conjunctiva helps keep the eye moist.

   NOTE: When the blood vessels of the conjunctiva are dilated, the eyes appear to be bloodshot.

3. The sclera surrounds a thin pigmented layer, the choroid. The iris, which gives the eye its color, is formed from the choroid. Muscles in the iris regulate the size of the pupil, the opening that lets light into the eye's interior.

   NOTE: The pigmented choroid absorbs light rays and prevents them from reflecting within the eyeball and blurring vision.

4. After going through the pupil, light passes through a transparent lens that focuses images on the retina, the layer of photoreceptors that lies on the inner surface of the choroid (Acetate 29.5).

   NOTE: Although transparent, the lens is composed of hundreds of cells arranged in layers like the scales of an onion.

5. The photoreceptor cells of the retina are most highly concentrated in a region called the fovea.

6. The photoreceptor cells of the retina transduce light energy into action potentials, the nerve impulses then travel along the optic nerve to the visual areas of the brain.

7. The chamber in front of the lens is filled with aqueous humor, a liquid similar to blood plasma that nurtures the lens. The chamber behind the lens is filled with vitreous humor, a jellylike fluid. Fluid in both chambers maintains the shape of the eyeball. Excess aqueous humor is caused by glaucoma and can lead to blindness by causing excess pressure on the retina.

   NOTE: The chamber that is filled with vitreous humor is a closed chamber. The silver specks, floaters, that occasionally appear are bits of embryonic tissue suspended in the vitreous humor.

C. To focus, a lens changes position or shape (*Module 29.6*).

1. In fish, the lens moves back and forth (relative to the retina), moving the fixed point of focus.

2. In mammals, the lens changes shape, thereby changing the distance at which images are focused. This process is called accommodation. Muscles attached to the choroid contract, reducing tension on the ligaments that

The Senses

support the lens, thus allowing it to take a more rounded shape and focus images of nearby objects on the retina. When the muscles relax, the lens is stretched into a more elongated shape to focus images of distant objects (Master 29.6).

NOTE: The pupil also accommodates to near and far vision. For near vision, the iris decreases the size of the pupil so as to eliminate peripheral light rays.

D. How do corrective lenses work? (*Module 29.7*)

1. Visual acuity is the ability to distinguish fine detail. So-called normal vision, or 20/20 acuity, indicates that at 20 feet, the eye can read letters on a chart normally readable at 20 feet. 20/10 is better than normal; at 20 feet, the eye can read letters normally readable at 10 feet. 20/50 is worse than normal; at 20 feet, the eye can read letters normally readable at 50 feet.

2. Nearsighted (myopic) people cannot focus on far objects because their eyeballs are too elongated and the lens cannot accommodate. Corrective lenses that are thinner in the middle correct this problem (Master 29.7A).

3. Farsighted (hyperopic) people cannot focus on near objects because their eyeballs are too short and the lens cannot accommodate. This problem is also caused as the lens becomes less elastic as it ages. Corrective lenses that are thicker in the middle correct this problem (Master 29.7B).

4. Astigmatism is blurred vision caused by lenses or corneas that are misshapen. Asymmetrical lenses are used to correct the problem.

E. Our photoreceptor cells are rods and cones (*Module 29.8*).

1. The human eye contains about 130 million photoreceptors (Acetate 29.8A).

2. There are twenty times more rods than cones. Rods are most sensitive to dim light and distinguish shades of gray, not color, using the light-absorbing pigment rhodopsin. They are most common in the outer margins of the retina and completely absent from the eye's center of focus (fovea). The best night vision is thus achieved by looking at things out of the "corner of your eye."

3. Cones are sensitive to bright light, and they distinguish color. Three types of cones can distinguish three predominant wavelengths using three kinds of the light-absorbing pigment photopsin. Groups of cones can distinguish thousands of different tints. Cones are less numerous in the retina's margins and are densest in the fovea. The best color vision and most acute vision are achieved by looking right at an object in bright light.

NOTE: Rods are more sensitive to light than are cones. This is why we do not see color when there is little light available.

4. Photoreceptors detect light when light is absorbed by a pigment that changes it chemically, triggering signal transduction pathways that alter membrane permeability and result in a receptor potential. Integration of the stimuli first occurs among interneurons that interconnect the output of several neighboring rods and cones. Many such integrated signals leave in a layer of neurons (on the surface of the retina), combining to leave the eyeball via the optic nerve (Acetate 29.8B).

NOTE: Does it make sense to have light pass through several layers of neurons before reaching the photoreceptors?

IV. The ears.

A. The ear converts air pressure waves into action potentials that are perceived as sound (*Module 29.9*).

   1. The outer ear collects sound waves and channels them to the eardrum. The middle ear amplifies the sound wave vibrations from the eardrum through three small bones—the hammer, anvil, and stirrup—to the oval window, a membrane that separates the middle and inner ears. Air pressure in the middle ear and outer ear is equalized by the Eustachian tube. The inner ear houses the hearing organ, which is composed of several channels of fluid wrapped in a spiral (the cochlea) and encased in bones of the skull. Vibrations of the oval window produce pressure waves in the fluid (Acetate 29.9A, Masters 29.9B, D).

      NOTE: Sound is composed of waves of air at different pressures.

   2. The pressure waves travel through the upper canal to the tip of the cochlea and then enter the lower canal and gradually fade away. Pressure waves in the upper canal push down on the middle canal, causing the membrane below this canal to vibrate. The vibrations stimulate the hair cells attached to the membrane by moving them against the overlying shelf of tissue (this whole structure is the organ of Corti). The hair cells develop receptor potentials and release neurotransmitter molecules, thereby inducing action potentials in auditory neurons (grouped together as the auditory nerve). Each region of the cochlea vibrates best at a given pitch (Acetate 29.9C).

   3. Young people are sensitive to pitches between 40 and 20,000 Hz. Dogs can hear to 40,000 Hz, and bats echolocate with sounds as high-pitched as 75,000 Hz.

      NOTE: Bats can use their ears much as we use our eyes. Bats can hear the patterns formed by sounds bounding off their environment and prey. A fishing bat can hear, in three dimensions, the ripples on the surface of water indicating the movement of a fish below the surface.

   4. The organ of Corti is sensitive to a considerable range of sound amplitudes without perceiving pain, from about 0 to 120 dB. Exposure to sounds of 90 dB (modestly amplified rock music) for long periods can cause hearing loss (Figure 29.9E).

B. The inner ear houses our organ of balance (*Module 29.10*).

   1. These organs lie next to the cochlea.

   2. Three semicircular canals detect changes in the head's rate of movement. Because they are arranged in three perpendicular planes, they detect movement in all directions. Receptor potentials are triggered when a jellylike mass of tissue (the cupula) suspended in the thick fluid in a canal moves against hair cells (Master 29.10).

   3. The utricle and saccule detect the position of the head relative to the force of gravity. Within these chambers, tiny limestone crystals are pulled by gravity in different directions (depending on the head's orientation) against hair cells.

C. What causes motion sickness? (*Module 29.11*)

   1. Motion sickness seems to stem from the brain's receiving mixed signals from equilibrium receptors of the inner ear and a different set of perceptions, usually visual.

2. People vary considerably in what it takes to produce motion sickness. As NASA has discovered, some people are able to consciously override the sick feelings stemming from the mixed messages.

V. Odor and taste receptors detect categories of chemicals (*Module 29.12*).

A. Underlying mechanisms of chemoreception are similar in all animal species. Receptor cells detect groups of chemicals, and the brain perceives the signals as odors or tastes by integrating the signals from a variety of receptors with stored data (memories, or "hard-wired" feelings) concerning these perceptions.

B. Odor (Acetate 29.12A).

1. Chemoreceptors in the nose detect airborne molecules, distinguishing about fifty general types of odor. Humans have a relatively poor olfactory sense, particularly compared to other animals like dogs, cats, and bears.

2. Olfactory chemoreceptors are in the upper portion of the nasal cavity and are covered with mucus secreted by neighboring cells.

3. Molecules enter the nose, dissolve in the mucus, and bind to specific receptor molecules on the chemoreceptor cilia. The binding triggers receptor potentials.

Review: Olfaction is tied to the limbic system, which is why it is particularly good at evoking emotions and memories (Module 28.20).

C. Taste (Masters 29.12B, C).

1. Chemoreceptors localized in taste buds detect molecules in food. There are four types of taste buds that detect sweet, sour, salty, and bitter. Each type is stimulated by a broad range of molecules in each category. They are arranged in precise areas of the tongue.

2. The perception of taste results from similar signal mechanisms as for smell.

NOTE: Our sense of taste is strongly influenced by associated smells. When a cold "clouds" our sense of smell, we lose much of our taste discrimination.

3. Many other animals have chemoreceptors in different locations. Flies taste with their feet. Moths smell with their antennae.

VI. *Review:* The central nervous system couples stimulus with response (*Module 29.13*).

A. Sensory receptors enable an animal to avoid danger, communicate with others, find food and mates, and maintain homeostasis.

B. Tracing the information in the diagram: A flash of light enters the bear's photoreceptors. These cells transduce the light stimulus into action potentials that travel along sensory neurons to the brain. Additional perceptions (smell, sound, touch) are integrated, with memories, by the brain. The brain then sends action potentials along motor neurons to effector cells, muscles in the paws, neck, and jaws (Master 29.13).

CLASS ACTIVITIES

1. As the optic nerve runs from the eye to the brain, the fibers from the nasal half of each optic nerve cross over to the other side of the brain. The optic nerve fibers from the nasal (left) half of the right eye cross to the left side of the brain, and the optic nerve fibers from

the nasal (right) half of the left eye cross to the right side of the brain. The place at which these fibers cross is the optic chiasma. This crossing of nerve fibers can be demonstrated by placing a tube (for example, a cardboard paper towel tube) in front of one eye and looking straight ahead out of both eyes. Put the palm of the hand on the same side as the eye that is not looking through the tube at arm's length and directly in front of that eye. Slowly move the palm closer to that eye, all the while continuing to look straight ahead out of both eyes. At a certain point, it will appear that there is a hole in the palm.

2. Having ears on either side of the head allows for the direction from which a sound arose to be localized. For example, if the sound reaches the right ear before reaching the left ear, the sound must be coming from the right. This can be demonstrated by having students, with their eyes closed, determine the location of a tuning fork that is vibrating on the right or left side of their head. If the tuning fork is placed directly above their head, they will be unable to determine where the sound is coming from.

3. The arrangement of the different types of taste buds can be demonstrated by, for example, placing sugar on the back of the tongue. The sugar will not be tasted unless it dissolves and diffuses to the front of the tongue.

4. The importance of olfaction to what is perceived to be taste can be demonstrated by asking a student who has a stuffy nose to identify foods that are put in his or her mouth. Try using a raw potato and an apple; they have similar textures, and without olfactory cues are difficult to tell apart.

RESOURCES AND REFERENCES

Barinaga, Marcia. "The Secret of Saltiness." *Science*, November 1, 1991. The physiology of a basic taste.

BBC Television. *Supersense Series*. Deerfield, IL: Coronet/MTI Film and Video, 1991. Six 30-minute videodiscs that cover a variety of aspects of sensory perception, including discs reviewing the senses of sight, hearing, and smell.

Freeman, D.H. "In the Realm of the Chemical." *Discover*, June 1993. On the evolutionary significance of the senses of taste and smell.

Freeman, W. "The Physiology of Perception." *Scientific American*, February 1991. How the brain processes sensory data.

Grady, D. "The Vision Thing: Mainly in the Brain." *Discover*, June 1993. How visual perceptions are produced.

Hammond, Ronald E. "Exploring Somatic Sensations." *Carolina Tips*, April 1990. A review of somatic sensation, with suggested laboratory activities to demonstrate many phenomena.

Hammond, Ronald E. "What's Special About Special Senses." *Carolina Tips*, May 1990. A review of the senses of sight, hearing, taste, and smell, with suggested laboratory activities.

Hirota, Theodore. *Perception, Version 2.0*. Bayport, NY: Life Science Associates. Computer software that provides demonstrations and experiments illustrating visual and auditory perception from an experimental psychology viewpoint.

Konishi, M. "Listening with Two Ears." *Scientific American*, April 1993. How the brain creates aural perceptions.

Koretz, Jane F., and George H. Handelman. "How the Human Eye Focuses." *Scientific American*, July 1988. Contains a nice illustration of the cellular construction of the lens and also discusses the several probable causes leading to the loss of focusing function with age.

Miller, Julie Ann. "A Matter of Taste." *BioScience*, February 1990. A review of the application of new laboratory techniques to clarify the functioning of this sense.

Nilsson, Dan-E. "Vision Optics and Evolution." *BioScience*, May 1989. Nature's engineering has produced astonishing diversity in eye design.

Olivo, Richard. *MacRetina*. Santa Barbara, CA: Intellimation, 1991. Macintosh computer software simulation of the firing of a retinal ganglion cell. The program allows experimental input.

Ramachandran, V. "Blind Spots." *Scientific American*, May 1992. Numerous optical illusions based on the blind spot.

Richardson, S. "The Small Flies." *Discover*, August 1995. How odors are sorted.

Santi, Peter, and Eric Webster. *Cochlear Anatomy*. Santa Barbara, CA: Intellimation, 1991. Macintosh computer software; an animated introduction to and dissection tool for the inner ear.

Schnapt, Julie, and Denis A. Baylor. "How Photoreceptor Cells Respond to Light." *Scientific American*, April 1987. A discussion of recent information on how an individual photoreceptor cell registers the absorption of a single photon of light.

Simmons, James A., M. Brock Fenton, and Michael J. O'Farrell. "Echolocation and the Pursuit of Prey by Bats." *Science*, January 5, 1979. Echolocating bats use different information-gathering strategies for hunting prey in environments that differ in the number of obstacles.

Suga, Nobuo. "Biosonar and Neural Computation in Bats." *Scientific American*, June 1990. Bats extract remarkably detailed information about their surroundings from biosonar signals. Neurons are highly specialized to perform this task.

30

How Animals Move

APPROACH

Whole-organism movement is a fitting topic to end this unit on animal systems because movement is the hallmark of the kingdom Animalia (the kingdom takes its name from the fact that animals move).

At the molecular level, the mechanism of muscle contraction is one of the best worked-out models in biology, and one of the easiest to grasp, as long as students understand the biochemistry and the three-dimensional arrangement of the protein filaments involved. Once again, reviewing the levels of protein structure (Modules 3.14–3.18) will help students see the relationship between molecular form and function, particularly if you have not reviewed this subject recently (e.g., with Chapter 24 on the immune system). Spend some time going over each hierarchical level of structure in Acetate 30.8, and be sure to point out that the last two (and subsequent diagrams showing movement) are simplified longitudinal sections.

Visual aids (skeletons, videos, and computer animations) are useful tools in presenting the organ and molecular bases of muscle movement and skeletal joint articulations. Emphasize the fact that muscles only pull, and that an antagonistic muscle (or, in some cases, elastic structures) must return each muscle to its relaxed position.

CHAPTER OBJECTIVES

Identify the various means of animal movement through water, on land, and through air.

Distinguish between the structure and function of the three types of skeletal systems: hydrostatic skeleton, exoskeleton, and endoskeleton.

Describe the overall arrangement of vertebrate skeletons into axial and appendicular subsystems.

Compare homologous structures in a human skeleton with those of a quadrupedal primate, explaining how the spine and knees of the human skeleton are less than optimally designed to carry out human activities.

Trace how human skeletons develop and grow.

Outline the hierarchical structure of a striated muscle.

Explain the molecular basis of the events involved when muscle cells are stimulated to contract.

LECTURE OUTLINE

I. Introduction.

    A. Most animal activities involve movement (*Opening Essay*).

        1. Leaf-cutter ants exhibit almost continuous movement as they carry out numerous activities (note all the verbs in the following sentences). Individual worker ants slice, pick up, and carry plant parts. They are

protected by guard ants that ride on their backs and gnash their jaws at potential attackers. Back in the nest, workers chew up the leaves, plant the fungus, cultivate the fungi, apply fungicides to inhibit competitors, harvest fungal growth, and carry away refuse. The fungal growth rather than the plant material is fed to the growing larvae.

    2. Insects are able to move and work efficiently because they have light, articulated exoskeletons with tiny, powerful internal muscles (Master 30.0).

    3. Three organ systems play key roles in movement. The nervous system issues commands to the muscular system, which exerts force against a firm skeletal system.

B. Diverse means of animal locomotion have evolved (*Module 30.1*).

    1. Although muscles and skeletons move in a variety of ways (eating, digestion, circulation), the focus in this chapter is on locomotion.

    2. Some animals remain in place and let the world come to them. Sponges use flagellated collar cells to move water through their bodies. Some cnidarians (such as *Hydra*) remain attached and move slowly during feeding activities.

    3. Locomotion in all its forms requires an animal to overcome two forces: friction and gravity. Water supports against gravity but offers considerable frictional resistance. Air offers little resistance but provides little support.

    4. Swimming involves legs as oars (many aquatic insects and mammals), jet propulsion (squids), whole body side to side (fishes), and up and down (whales) (Figure 30.1A).

    5. Land animals not only must overcome gravity but must maintain balance while they move forward. Terrestrial locomotion includes hopping on springlike back legs, and quadrupedal or bipedal walking and running (Figure 30.1B, Master 30.1C).

    6. Some land animals crawl (by undulating movements or by peristalsis; Module 21.7). During peristalsis, longitudinal muscles shorten and thicken regions, while circular muscles constrict and elongate other regions. In an earthworm, bristles anchor the short, thick regions, and regions anterior to them lengthen (Acetate 30.1D).

    7. True flight has evolved only in insects, birds, extinct reptiles, and mammals (bats). To fly, animals move wings in patterns that provide lift. Bird wings have cross-sectional shapes of airfoils. Air flowing past an airfoil has lower pressure above relative to below, providing lift (Figure 30.1E).

NOTE: Insects, bats (most of the time), and some birds (hummingbirds) produce lift in a different way ("fluttering," which is more like the lift a helicopter produces) by pushing their wings down against the air during a power stroke and slipping them up through the air during a return, nonpower stroke. This type of flight enables these animals to hover, a feat the rest of the birds cannot do without fluttering, and then inefficiently. Some other animal groups (fishes, amphibians, and other mammals) have evolved gliding, which moves the animal through the air or water without producing lift.

8. All types of movement are based on either the contraction of microtubules (see cilia and flagella in Module 4.18) or the contraction of microfilaments (amoeboid movement and muscle contraction).

II. The skeletal system.

A. Skeletons function in support, movement, and protection (*Module 30.2*).

1. Skeletons have many functions, including support, protection of soft parts, and movement. There are three main types of skeletons: hydrostatic skeletons, exoskeletons, and endoskeletons.

   NOTE: Skeletons can also play a role in mineral storage.

2. A hydrostatic skeleton consists of a volume of fluid held under pressure in a body compartment. Such skeletons work well for aquatic animals and animals that burrow by peristalsis. Earthworms have a body composed of fluid-filled sections. *Hydra* plays muscle cell contractions against a hydrostatic skeleton of its closed gastrovascular cavity (Figure 30.2A).

3. An exoskeleton consists of a rigid, external, armorlike covering. Muscles are attached to the inner surface of the exoskeleton. At joints, the exoskeleton is thin and flexible. Clams and snails have exoskeletons (shells) that are enlarged by secretions from the body margin (mantle). The hollow, tubular exoskeletons of arthropods (Module 19.12) are extremely light for their strength, but they do not grow with the animal. Periodically, during molting, the old skeleton is lost, and, following body growth, a new skeleton is hardened (Figures 30.2B, C). At this time, these animals are particularly vulnerable to predators, and remain so until the new exoskeleton hardens.

   NOTE: Although most shell-bearing mollusks move by manipulating a muscular foot, the scallop moves by rapid opening and closing of its shells, producing a jet-propulsive movement that is somewhat random.

4. An endoskeleton consists of rigid, internal supports, usually consisting of noncellular material secreted by surrounding cells. Sponges support their cells on spicules. Spicules are made of materials such as calcium salts or silica. Echinoderms have an endoskeleton of calcium plates under their skin (Figure 30.2D). Vertebrates have endoskeletons of bone, cartilage, or a combination of the two (Master 30.2E).

B. The human skeleton is a unique variation on an ancient theme (*Module 30.3*).

1. All vertebrate skeletons have consistent features, both the overall pattern described below and the number, shape, and articulation of the individual bones. The basic patterns are modified according to the needs of each animal.

   Review: Primitive and derived characters (Module 16.13).

2. In contrast to the frog skeleton, which supports a quadruped that moves by hopping, the human skeleton supports a biped that walks or runs.

3. The axial skeleton consists of a skull protecting the brain, the backbone (vertebral column) protecting the spinal cord and supporting the remaining skeletal elements, and the rib cage surrounding the lungs and heart.

4. The appendicular skeleton consists of the bones of the appendages (arms, legs, fins) and the bones that link the appendages to the axial skeleton (the shoulder [pectoral] and pelvic girdles).

NOTE: The shoulder girdle consists of the clavicle and scapula. Coming off the shoulder girdle are the humerus, radius and ulna, carpals, metacarpals, and phalanges. The pelvic girdle is formed by the coxal bone (os coxa), which consists of three fused bones: the ilium, the ischium, and the pubis. Coming off the pelvic girdle are the femur, patella (kneecap), tibia and fibula, tarsals, metatarsals, and phalanges.

A human skeleton can be determined to be that of a female or male by examining the pelvic girdle. There are several differences, but one of the easiest to use is the angle of the pubic arch. If the angle is greater than 90°, then it is the skeleton of a female; if the angle is less than 90°, then it is the skeleton of a male.

5. This is the basic pattern of bones found in most land vertebrates (tetrapods).

6. Comparing the bipedal human skeleton with that of the quadrupedal baboon underscores the evolutionarily distinctive features. The human skull is large and flat-faced. The backbone is S-shaped. The pelvic girdle is shorter, rounder, and oriented vertically. The bones of the hands and feet are different. The hands are adapted for grasping and manipulating, and the feet are adapted to support the entire body bipedally (Master 30.3B).

7. The versatility of the vertebrate skeleton comes in part from its movable joints. Ball-and-socket joints allow movement in all directions. Hinge joints are strong and restrict movement to one plane. Pivot joints allow bones to rotate, providing ease of manipulation (Master 30.3C).

C. Skeletal disorders afflict millions (*Module 30.4*).

1. Lower back problems stem from the uneven distribution of weight vertically on the backbone. The S-shape in this area cushions vertical loads but cannot bear the lateral forces during lifting.

NOTE: Knee joints are also notoriously poorly constructed to withstand the pounding some human activities require. These structures are examples of historical constraints on evolution.

2. One form of arthritis, inflammation of the joints, seems to be a normal part of aging, as joints become stiff and cartilage between bones wears down.

3. Crippling, rheumatoid arthritis is an autoimmune disease (Module 24.16) in which the immune system attacks and degrades the joints following stress or an infection.

4. Osteoporosis is due to hormonal changes (greatly reduced estrogen levels) during aging, particularly in women following menopause. It is characterized by the bones becoming thinner, more porous, and easily broken.

NOTE: For women, bone density begins to decline at about age 30 to 35. Ca^{2+} intake and exercise to offset this decline should be of concern to all woman, both pre- and postmenopausal.

D. Bones are complex living organs (*Module 30.5*; Acetate 30.5).

1. *Review:* Tissues, bone, and cartilage (Module 20.5) and the role of the thyroid and parathyroid glands in calcium homeostasis (Module 26.7).

2. Bones are composed of other tissues besides bone and cartilage. These tissues intermix with tissues of the circulatory system (vessels and blood) and nervous system (nerves).

3. Most of the outside surface is covered with fibrous connective tissue. When bones break or crack, this tissue is able to form new bone.

4. At either end of most bones, cartilage replaces connective tissue, forming a surface that cushions the joints (Acetate 20.5E).

5. Bone itself is mostly a noncellular matrix of calcium salts (which resist compression) and protein fibers (which resist cracking) surrounding the cells that secrete these materials (Acetate 20.5F).

6. The shafts of long bones are made of compact bone, with a dense matrix surrounding a hollow cavity containing stored fat (yellow bone marrow). The ends of long bones are made of an outer layer of compact bone and an inner area of spongy bone. Within cavities in the matrix of the spongy bone, specialized tissues produce blood cells (red bone marrow).

NOTE: The cavity of long bones reduces the weight of the body and makes movement easier.

E. Bone growth is a major feature of human development (*Module 30.6*).

1. Bones begin to form about one month after conception (Figure 30.6A).

2. The bones of the skull form from sheets of connective tissue.

3. The bones of the remaining skeleton form and grow by replacing cartilage.

4. In long bones, this starts with the laying down of a ring of bone (bone collar) around the shaft. Bone also replaces cartilage at the center of the shaft. The bone grows in length and thickness, and blood vessels penetrate the shaft. Within the shaft, the yellow marrow cavity begins to form. Blood vessels also penetrate the cartilage at the ends of the bone, and bone formation takes place there. The bone continues to grow as long as new cartilage is added to the region of cartilage between the shaft and the ends of the long bone (the dark blue area in Acetate 30.6B).

5. Skeletal growth stops at about age 18 in women and 21 in men.

III. Muscles and movement.

A. The skeleton and muscles interact in movement (*Module 30.7*).

1. Muscles are connected to bones by tendons.

NOTE: At joints, bones are held together by ligaments.

2. A muscle can only contract. To extend, it must be pulled by the contraction of an opposing muscle. Thus, movement of most parts of the body requires antagonistic pairs of muscles (Acetate 30.7).

NOTE: Nerves that enervate antagonistic muscle pairs have a built-in circuitry that prevents both muscles of a pair from contracting at the same time (this is referred to as reciprocal innervation). A strong electrical shock can bypass this circuitry and cause both nerves to induce their muscles to contract at the same time. This can break bones.

B. Each muscle cell has its own contractile apparatus (*Module 30.8*).

1. *Review:* Striated skeletal muscle tissue (Module 20.6, Acetate 20.6A) and Duchenne muscular dystrophy (Module 13.9).

2. Each muscle fiber is a single cell with many nuclei. Within each fiber are numerous, long myofibrils (Acetate 30.8).

3. A myofibril is composed of contracting units called sarcomeres, joined end-to-end at Z lines.

4. Each sarcomere is composed of thin filaments (coiled strands of two actin proteins and one regulatory protein) and thick filaments (parallel strands of myosin protein). This structure produces a pattern of light and dark bands in the muscle tissue. The dark bands (Acetate 30.8) consist of thick filaments and thin filaments (which do not extend to the center of the dark band). The light bands have only thin filaments and straddle the Z lines that connect adjacent thin filaments.

NOTE: These bands give the skeletal muscle its striated appearance.

C. A muscle contracts when thin filaments slide across thick filaments (*Module 30.9*).

1. In the 1950s, British biochemist A. F. Huxley proposed the sliding filament model of muscle contraction. The model has been supported by considerable later research, and many molecular details have been added to it.

NOTE: The model originally attempted to explain one set of observations seen in living muscle: When the muscle contracts, the dark bands stay the same length, while the light bands decrease in length. When the muscle is fully contracted, an even darker band appears in the middle of the dark bands.

2. Contraction shortens the sarcomere but does not shorten the thick and thin filaments, which slide between each other (Acetates 30.8 and 30.9A).

3. Energy-consuming interactions between the myosin molecules of the thick filaments and the actin molecules of the thin filaments cause them to slide along one another. The myosin molecules of the thick filament expose about 350 swollen "heads" per filament. These "walk" along the actin filaments with the expenditure of ATP. Each head can repeatedly move at about five movements per second. The process continues until the muscle stops contracting or is fully contracted (Acetate 30.9B).

4. Data suggest that ATP attaches to each head and, upon hydrolysis to ADP and phosphate, adds potential energy to the head ("cocks" it). Ca^{2+} opens a binding site on the adjacent actin molecule. When ADP and phosphate are released from the bound head, its energy is released, pulling the actin in a power stroke.

D. Motor neurons stimulate muscle contraction (*Module 30.10*).

1. *Review:* Neuron structure and function (Module 28.2).

2. Each muscle fiber is stimulated by just one neuron, but a single neuron can stimulate many fibers, up to several hundred in a large muscle moving the appendicular skeleton. Each such group of muscle fibers is known as a motor unit because each is stimulated to contract together (Acetate 30.10A).

NOTE: The fewer the number of muscle fibers per motor unit, the greater the degree of fine control over the muscle.

3. A weak contraction is produced by the stimulation of one motor unit. A strong contraction involves the simultaneous contractions of several motor units.

4. The synapses between neuron and muscle fiber are called neuromuscular junctions. The action potential is transmitted to the fiber through the release of the neurotransmitter acetylcholine.

Review: Neurotransmitters (Module 28.7).

5. At the cellular level, muscle fiber stimulation proceeds as follows. The released acetylcholine changes the permeability of the muscle fiber's plasma membrane. This induces an action potential along the muscle cell membrane and into tubular infoldings of the plasma membrane into the cell. Within the cell, the action potentials cause the endoplasmic reticulum (ER) to release Ca^{2+} into the cytoplasm, and this triggers the binding of myosin to actin. When action potentials stop, Ca^{2+} moves back into the ER (Acetate 30.10B).

E. Robots mimic animal movements (*Module 30.11*).

1. A robot capable of independent action must have a sensor apparatus, a mechanical locomotor system that mimics the movement of an animal, and a computer "brain" to integrate the responses and movements.

2. Even the most advanced robotic machines appear very simple compared to humans. A task such as buttering toast requires numerous subtle movements and readjustments, depending on the sense of having completed each prior movement. Performing such a task robotically both smoothly and independently is a major challenge (Figure 30.11A).

3. One approach in producing robots capable of moving independently through a complex environment has been to build machines patterned after the simple nervous and muscular systems of insects. Genghis has six legs, each pair first controlled by a ganglionlike circuit, not dissimilar to the ganglia of an insect's ventral nerve cord (Module 28.11), and all integrated with a central "brain" involving only 57 sensory and motor electronic circuits (Figure 30.11B).

F. The structure-function theme underlies all the parts and activities of an animal (*Module 30.12*).

1. A baseball game, with its requirement for split-second decisions and precise actions, demonstrates some of the remarkable evolutionary adaptations of the human body (Figure 30.21A).

2. Likewise, the cooperative work and movements of individuals in a leaf-cutter ant colony also demonstrate the remarkable adaptations of those animals.

CLASS ACTIVITIES

1. The demonstrations of skeletal structures, joints, and antagonistic muscles will be much easier, and more dramatic, if you refer to a human skeleton (and for the material in Figure 30.3B, a monkey or baboon skeleton). Some demonstration skeletons show antagonistic muscle insertions. If one is not available, use differently colored cords to demonstrate the locations and functions of antagonistic muscles on the upper arm or leg.

2. To demonstrate the function of hydrostatic skeletons and peristaltic movement, use earthworms on an overhead projector. Place them briefly in a shallow bowl of cool water. The light and temperature of the overhead will cause them to move rapidly away from the projector's heat. Other means of invertebrate movement, such as in shrimp or bivalve mollusks, can also be demonstrated in this way. Worms are available in bait

stores or backyard compost piles. The other animals may not be available in all locations.

3. An easy (and the classic) way of demonstrating the sliding filament theory of muscle contraction is to interlock the fingers of your hands and slide them back and forth.

4. Many students will be interested in the human muscle groups from an athletic and/or body-building perspective. Have a local orthopedic surgeon describe and demonstrate the areas of human musculoskeletal anatomy that are especially vulnerable to injury.

RESOURCES AND REFERENCES

Alexander, R. "Muscles Fit for the Job." *New Scientist*, April 15, 1989. How are particular muscles adapted for diverse functions in movement?

Amato, I. "Heeding the Call of the Wild." *Science*, August 30, 1991. Materials scientists study the structure and function of skeletons and other animal products.

Bretscher, Mark S. "How Animal Cells Move." *Scientific American*, December 1987. In contrast to whole-organism movement due to muscle cells described in this chapter, this paper describes how individual cells (such as leukocytes and cells migrating during development) move. Pieces of the outer membrane are recycled from the trailing to the leading edge.

Malkin, L. *Biochemistry of Muscle*. Troy, MI: Helix Corporation, 1987. Interactive computer software that illustrates the microscopic and molecular details of muscle contraction and relaxation. Available in IBM PC and Apple II versions from Carolina Biological Supply Company.

Muscular and Skeletal System. Washington, DC: National Geographic Society, 1988. A 20-minute videotape that describes the protective and supportive functions of the skeleton and skeletal muscle movement and compares the movements created by smooth and cardiac muscles.

Triantafyllou, M.S., and G.S. Triantafyllou. "An Efficient Swimming Machine." *Scientific American*, March 1995. Mechanics of swimming by fish.

31

PLANT STRUCTURE, REPRODUCTION, AND DEVELOPMENT

APPROACH

Like animals and animal tissues, the details of plant structure hinge on an understanding of the basic patterns of plant tissues. Students have direct experiences of their own animal tissues but lack a personal understanding of the tissues of plants. Because of this lack of familiarity, many students have misconceptions about plants.

Review plant cell structure and the alternation of generations in the life cycles of plants, recalling details from Chapters 4 and 18. If you have been covering animals in recent lectures, this is particularly important. More than usual, prior introduction to the structures of plants in the laboratory will help clarify details from your lectures.

There are two ways to proceed with the details of plant structure, which are not mutually exclusive. Stress the human connection to the parts of plants students are the most familiar with. Describe what plant parts humans actually use for food, clothing, and shelter (among other things) and how those plant parts became what they are. Alternatively, describe plant structures as the unique solutions to sedentary, photosynthetic life on land. The former approach is used periodically throughout the chapter, and you can add additional details. The latter approach begins with Module 31.5. Rather than listing all the terms for cells and tissue types, emphasize the cooperation among tissue systems in building organs (roots, stems, leaves, flowers) that function well.

During your discussion of the sexual reproductive events of angiosperms, reiterate what distinguishes—and what happens to—the various developing tissues. Otherwise, nonscience majors may become lost in the details and new terms and the confusing codevelopment of many structures (seed and fruit, for example).

CHAPTER OBJECTIVES

State the distinction between dicots and monocots.

Describe the basic structure of all plants, and discuss how this overall structure is an adaptation to the basic function of plants in terrestrial environments.

Name the six types of cells and three tissue systems found in plants.

Differentiate primary and secondary growth, describing where and how each kind of growth occurs.

Distinguish between the terms *pollination* and *fertilization*.

Trace the following events in the sexual life cycle of the flowering plant: pollination, double fertilization, seed and fruit development, and seed germination.

Explain the importance of asexual reproduction in natural populations of plants and in techniques for propagating plants used by humans.

LECTURE OUTLINE

I. Introduction.

    A. Plants are unique organisms (*Opening Essay*).

1. *Review:* Basic characteristics of the plant kingdom (Module 18.1).

2. Examining the giant sequoia, *Sequoiadendron gigantea*, helps underscore the unique capabilities and adaptations of plants. The tree known as General Sherman is the largest individual plant on Earth: 83 m tall, 10 m in diameter at the base, first branch at 40 m, weighing about 1400 tons, and alive for about 2500 years (chapter-opening photo, Figure 31.0).

3. Humans depend on plants for a variety of needs (lumber, fabric, paper, food, etc.), and many other organisms depend on them for nutrition and shelter.

   NOTE: Despite the importance of plants, on a worldwide basis, slightly more photosynthesis is carried out by photosynthetic protists (algae) and bacteria of aquatic habitats.

4. Giant sequoias are gymnosperms (naked seeds; Modules 18.8 and 18.9). Because angiosperms (covered seeds; Modules 18.11–18.15) make up 90% of the world's plant species, they are the focus here.

   Review: Module 18.3 discusses the basic differences between gymnosperms and angiosperms.

B. Talking About Science: Plant scientist Katherine Esau, preeminent student of plant structure and function (*Module 31.1*).

1. Dr. Esau was born in Ukraine in 1898, studied in Moscow, and moved to Germany in 1918 and the United States in 1922. Her early research focused on plant viruses in sugar beets. She continued working into her nineties.

2. She is best known for her studies of the structure and function of the food-conducting tissues (phloem) in plants, and of plant anatomy in general. Among other discoveries, she first described plasmodesmata (open connections between plant cells; Module 4.19) and the transmission of plant viruses from cell to cell through these structures.

C. There are two major groups of angiosperms: monocots and dicots (*Module 31.2*).

1. Monocots include orchids, bamboos, palms, lilies, and grasses (including most of the agricultural, grain-producing plants). They are distinguished by having one seed leaf (cotyledon) but also usually have parallel-veined leaves, scattered vascular bundles in stems, floral parts in multiples of three, and fibrous root systems (Acetate 31.2).

2. Most angiosperms are dicots. Dicots include most shrubs and trees (except conifers) and many herbaceous plants, including the many food plants and plants domesticated for their fibers. They are distinguished by having two cotyledons but also usually have net-veined leaves, vascular bundles in a ring in stems, floral parts in multiples of four or five, and taproot systems.

D. For all plants, as for animals, structure and function are related.

1. A close look at plant structure often reveals its function.

2. Conversely, function provides insight into the logic of a plant's structure.

II. Plant structure.

A. The plant body consists of roots and shoots (*Module 31.3*).

1. These structural adaptations allow plants to function in terrestrial habitats without drying out. Functionally, plants need to take up water and minerals from the soil, absorb light and take in carbon dioxide from the

air, and create their plant bodies with the molecules assembled from these raw materials and from those produced by photosynthesis.

2. The root system anchors the plant, absorbs and transports minerals and water, and stores food. The fibrous roots of monocots, and the taproot plus secondary roots of dicots, are effective in anchoring and absorption. The ultimate point of absorption is the root hair, an outgrowth of the epidermal cells, which increases the absorptive surface area (Acetate 31.3).

3. *Preview:* Most plants are also aided in nutrient uptake by mycorrhizal fungi (Module 32.11).

4. The shoot system consists of supporting stems, photosynthetic leaves, and reproductive structures (in angiosperms, flowers). Stems are composed of nodes, where leaves, flowers, or other stems are attached, and internodes. Leaves are composed of photosynthetic blades and short stalks that join the blade to the stem's nodes.

5. Buds are undeveloped shoots that contain potential nodes, internodes, and leaves. Two types occur. The terminal bud at the plant apex is the source of growth in height. The axillary buds, one in each angle formed by a leaf and the stem, are usually dormant but can produce new branches that add to a plant's width.

6. Apical dominance results from the release of hormones from terminal buds, which inhibits the growth of the axillary buds. One can cause a plant to be bushier by pinching off the terminal bud, thereby stimulating (removing the inhibition of) the development of the axillary buds.

B. Many plants have modified roots and shoots (*Module 31.4*).

1. In many dicots (e.g., carrots, turnips, beets, sweet potatoes), food is stored in modified taproots (Masters 31.4A, B). Plants store food in these structures in the form of carbohydrates such as starch.

    NOTE: Animals store food as the carbohydrate glycogen.

2. Stems can be modified for several purposes. Runners (e.g., strawberries) provide a means of asexual reproduction. Rhizomes (e.g., irises) and tubers (e.g., potatoes) are underground stems that store food in the form of starch.

3. Leaves may also be modified from their photosynthetic function. Some leaf bases (e.g., celery) store food. Tendrils (e.g., cucumber) are modified for grasping and climbing. Spines (e.g., cactus) are modified for protection.

C. Plant cells and tissues are diverse in structure and function (*Module 31.5*).

1. *Review:* The cellular picture of plasmodesmata and the molecular basis of turgor pressure (Modules 4.19 and 5.16).

2. Plant cells have a unique combination of features. Many are photosynthetic and contain chloroplasts. They often have a large, central vacuole that helps support the cell (and plant tissues) by maintaining turgor. All plant cells are bounded by a cell wall composed mainly of cellulose. Most plant cells that provide support have an additional, stronger secondary wall hardened with lignin that is laid down inside the primary wall. Pits with cytoplasmically continuous plasmodesmata often interconnect adjacent cells (Master 31.5A).

3. Plant cells can be grouped into six types, depending on wall morphology and chemistry, shape, and function. Tissues made of the cells may be simple (bearing the cell type name) or complex (composed of several cell types).

4. Parenchyma cells are abundant, unspecialized cells, with only primary walls, and generally with equal-sided shapes. They function in food storage, photosynthesis, and aerobic respiration (Master 31.5B).

5. Collenchyma cells resemble parenchyma cells but lack secondary walls and have thicker primary walls. They provide support for young parts of plants that are still growing (Master 31.5C).

6. Sclerenchyma cells have rigid secondary walls, hardened with lignin, and function in support and protection. Lignin is the main chemical component of wood. The elongated fibers that strengthen many plants are a type of sclerenchymal cell. The resistant cells in many seed coats and the stone cells of pears are made of sclerenchymal cells called sclerids (Master 31.5D).

    NOTE: Lignins are dietary fibers and phytoestrogens. As such they play a role in the human diet. As we have seen (Module 21.20), fibers are important in the diet. Being a phytoestrogen, lignin may reduce the risk of ovarian and breast cancer.

7. Water-conducting cells have elongated shapes and secondary walls, and function only when dead and connected end-to-end. Tracheids have tapered ends, covered with open pits. Vessel elements are wider and shorter and have completely open ends. Xylem tissue is largely composed of these water-conducting cells, with parenchyma and sclerenchyma cells for support and storage (Masters 31.5E, F).

8. Food-conducting cells (also known as sieve-tube members) are also arranged end-to-end but have relatively thin primary walls and no secondary walls, and are alive, but lack nuclei and ribosomes, when they function. The end of each cell contains numerous pits with plasmodesmata. Each food-conducting cell is found in association with at least one companion cell that makes certain proteins for it. Phloem tissue is largely composed of these food-conducting cells, with parenchyma and sclerenchyma cells for support and storage.

    Preview: Phloem function (Module 32.5).

D. Three tissue systems make up the plant body (*Module 31.6;* Acetate 31.6A).

1. The epidermis is the skinlike first defense against damage or infection. This tissue system is composed of a single, surrounding layer of cells.

2. The vascular tissue system is composed of xylem and phloem and conducts water and nutrients throughout the plant.

3. The ground tissue system fills the spaces between the epidermis and vascular tissue system, is mainly composed of parenchyma, and functions variously in photosynthesis, storage, and support.

4. Each tissue system is continuous from organ to organ throughout the plant.

5. Roots are surrounded by epidermal cells with root hairs and without a cuticle. The ground tissue (cortex cells) functions in conducting materials from the root surface into the central vascular tissue and in food storage.

The inner layer of the cortex (the endodermis) provides a selective barrier, regulating flow into the vascular tissue (Figure 31.6B).

Preview: The plasma membranes of root cells control solute uptake (Module 32.2).

6. On most stems and leaves, the epidermal cells have a waxy coating (cuticle). Stems of dicots and monocots differ in the relative distributions of ground tissue and vascular tissue systems. Dicots have bundles of vascular tissue in an outer ring supported in a ring of ground tissue cortex and surrounding a parenchyma pith. Monocots have bundles of vascular tissue scattered in a more uniform ground tissue (Figure 31.6C).

7. Leaves have a complex arrangement of the three tissue types that perfectly fits their function of photosynthesis. The lower epidermal tissue includes pore-forming (stoma-forming) guard cells (Module 32.4). The ground tissue is arranged in mesophyll layers, a lower, loose layer specialized for gas exchange and an upper, more compact layer specialized for photosynthesis. Tiny branches of the vascular tissue of stems enter each leaf and provide contact to and from photosynthetic cells and gas-exchange surfaces (Acetate 31.6D).

III. Plant growth.

A. Primary growth lengthens roots and shoots (*Module 31.7*).

1. Unlike animals, plants have indeterminate growth; that is, they continue to grow during their entire lives. To get to other places in their environments (up and out), individual plants must grow.

2. Three seasonal growth patterns occur in plants. Annuals complete their life cycle in one year. Biennials complete their life cycle in two years. Perennials continue to live and reproduce for many years.

3. Indeterminate growth is the result of plants having meristems, embryonic tissues composed of unspecialized cells that continue to give rise to new cells. Apical meristems are found at the root and shoot tips and in axillary buds. The lengthwise growth produced by these regions is primary growth. Differentiation within these regions has been recently shown to be regulated by master control (homeotic) genes, as in animals (Acetate 31.7A).

Review: Cell differentiation (Module 11.6) and homeotic genes (Module 11.12).

4. The apical meristem in the root divides cells downward, forming a root cap that is sloughed off during the growth through abrasive soil. Other cells from the meristem grow upward and form three concentric rings of embryonic tissue that will form the epidermis, cortex, and central vascular cylinder. Just above the meristem, these tissues elongate, pushing the root tip downward. Above that region, the three tissue systems differentiate; the cells of the vascular cylinder differentiate into primary xylem and primary phloem (Acetate 31.7B).

5. The apical meristem in shoots forms three downward-developing cylinders of embryonic tissues, also with zones of elongation and differentiation. Some of the apical meristem cells remain in lateral positions and develop the apical meristems in axillary buds (Acetate 31.7C).

Preview: The hormonal basis of shoot growth is discussed in Module 33.3.

B. Secondary growth increases the girth of woody plants (*Module 31.8*).

1. Secondary growth involves meristems that grow laterally in stems and is most evident in trees, shrubs, and vines.

2. The vascular cambium is a cylindrical meristem that develops from parenchyma cells between the xylem and phloem of shoots. As cells in this meristem divide inward, they form new layers of secondary xylem to the outside of the primary xylem. As cells of the vascular cambium divide outward, they form new layers of secondary phloem inside the primary phloem. The increase in thickness of the stem (forming the wood of a tree, shrub, or vine) is mostly layers of secondary xylem (Acetate 31.8A).

3. In regions that have distinct seasons, secondary xylem cells are larger in diameter during periods of favorable growth and smaller at other times. This results in distinct annual growth rings.

4. The new layers of phloem external to the vascular cambium do not accumulate but are sloughed off in bark at about the same rate they are produced. Within the secondary phloem, meristematic regions (cork cambium) divide off cork cells outward. Mature cork cells are dead and have waxy, thick walls that protect the stem surface from damage and infection, much like epidermal cells. However, they eventually slough off as new cork cambium develops in layers below (Acetate 31.8B).

NOTE: This also provides a way for a woody plant to increase the circumference of its protective layer as the diameter of the stem increases.

5. Wood itself can be divided into central heartwood, xylem that no longer functions in transport because it is plugged with resins, and sapwood, younger secondary xylem that actually conducts water. Wood rays are collections of parenchyma cells that extend laterally from heartwood into the sapwood, providing channels between these two regions. The heartwood of trees acts as an endoskeleton, providing a strong, rigid, yet flexible core upon which the living plant substance is supported.

6. Wood is the source of many products useful to humans. As a building material, it is unmatched for its combination of strength, hardness, lightness, insulating properties, durability, workability, and beauty.

IV. **Sexual reproduction.**

Review: Features of angiosperms (Modules 18.12–18.14).

A. Overview: The sexual life cycle of a flowering plant (*Module 31.9*; Acetate 31.9).

1. A flower is actually a compressed shoot with highly modified leaves (sepals, petals, stamens, and carpels) inserted at nodes separated by greatly reduced internodes.

2. Sepals are usually green and protect flower buds. Petals are usually large and showy and attract pollinators.

3. Stamens are male structures with pollen-bearing anthers at their tip. Pollen grains deliver sperm nuclei to females.

4. Carpels are female structures composed of stigma and ovule-bearing ovaries. Ovules carry the female egg.

5. Pollination occurs when pollen is delivered to the stigma of another flower. Fertilization occurs in the ovule. The fertilized egg develops into an embryo, and the ovule develops into a seed that holds the embryo. The

ovary develops into a fruit that aids in seed dispersal. The seed germinates in a favorable environment to complete the life cycle.

B. The development of pollen and ovules culminates in fertilization (*Module 31.10*).

1. *Review:* Alternation of generations. Like all plants, angiosperms alternate between diploid sporophytes that produce spores by meiosis and haploid gametophytes that produce gametes by mitosis. The gametes unite by fertilization to form a diploid zygote, which is the first cell of the next sporophyte generation (Modules 17.26 and 18.4).

2. The mature plant we see is the sporophyte. Angiosperm gametophytes are microscopic and are found inside the flower parts (Acetate 31.10).

3. The male gametophyte is the two-celled pollen grain. It develops following meiosis of cells in the anther. The resulting "spores" divide mitotically to produce two haploid cells, a tube cell and a generative cell. The outer wall of the pollen grain is thick and resistant.

4. The female gametophyte develops inside the ovule, a central cell surrounded by a coating of smaller cells. The central cell undergoes meiosis, but only one of the resulting haploid nuclei develops into a spore. The nucleus in the haploid spore enlarges and divides mitotically, forming the embryo sac. (The embryo sac, housed in and protected by the sporophyte, is the female gametophyte. The embryo sac contains a large central cell with two haploid nuclei. Another of the cells is the haploid egg.) All this happens in specialized ovary tissue at the base of the carpel.

NOTE: There is a similarity between the one-functional-cell result of meiosis producing female "spores" in flowering plants and the result during egg formation in mammals.

5. Pollination is the delivery of a male pollen grain to a receptive stigma of the female carpel. Pollen is usually wind- or animal-dispersed.

6. After pollination, the pollen grain germinates and grows into the stigma and ovary. The generative cell divides mitotically, forming two sperm nuclei. At the base of the ovule, the pollen tube releases both sperm nuclei.

7. The double fertilization that follows is a hallmark of the angiosperms. One sperm nucleus fertilizes the egg, forming the zygote that will develop into the embryo. The other sperm nucleus fuses with the two central nuclei, forming a triploid nucleus that will develop into tissue that nourishes the embryo.

C. The ovule develops into a seed (*Module 31.11*).

1. Within the ovule, the triploid cell develops into a nutrient-rich endosperm, and the diploid cell develops into the embryo (Acetate 31.11A).

NOTE: Be sure to point out that the ovule includes everything inside the original coating of small cells that surrounded the cell that underwent meiosis to form, ultimately, the eight haploid nuclei, including the egg. All nuclei other than the zygote and endosperm do not develop further.

2. The embryo first develops an anchor cell and very soon differentiates into a shoot with cotyledons. The old ovule coat develops into a resistant seed coat.

3. The seed develops up to a point and then becomes dormant. This allows time for dispersal and for the seasonal occurrence of conditions favorable for independent growth.

4. Seeds of dicots have two fleshy cotyledons that have absorbed the endosperm nutrients and taken over the role of nourishment. The seeds of monocots have one cotyledon, a protective sheath over the embryonic root and shoot, and contain a large endosperm (Master 31.11B).

D. The ovary develops into a fruit (*Module 31.12*).

1. A fruit houses and protects seeds and helps disperse them. During development, hormonal changes make the ovary grow and thicken (Figure 31.12A, Master 31.12B).

2. Fruits are highly varied in organization, depending on how many ovules, how many ovaries, how many carpels, or how many flowers are involved in the formation, and on the ultimate means of dispersal (wind, water, or animal).

3. A pea pod is an example of a simple fruit (Figure 31.12A,B).

4. A blackberry is an example of an aggregate fruit, fruit that develops from many united carpels.

5. A pineapple is an example of a multiple fruit, fruit that develops from many united flowers (Figure 31.12C).

NOTE: Plums and avocados are more normal fruits with single seeds. The winged maple "seed" and plumed dandelion "seed" are examples of fruits modified for wind dispersal. The seed of each is at the heavy end of the fruit.

E. Seed germination continues the life cycle (*Module 31.13*).

1. New plant life does not start with seed germination. Dormancy is broken, and a previously developing embryo starts developing again. The seed takes up water, endosperm or cotyledons begin to enzymatically digest stored nutrients, and the nutrients are transported to the growing parts.

2. In dicots, the cotyledons supply nutrients for the seedling. Young shoots leave the seed in a hooked shape that protects the terminal meristem from abrasion by soil particles. Once the shoot clears the soil surface, the first foliage leaves develop at the tip and begin photosynthesis (Masters 31.13A, B).

3. In monocots, the endosperm supplies nutrients for the seedling. Young shoots are protected from abrasion by a sheath that surrounds them until they break through the soil surface.

V. Asexual reproduction.

A. Asexual reproduction produces plant clones (*Module 31.14*).

1. *Review:* Asexual and sexual reproduction (Chapter 8 Opening Essay and Module 27.1).

2. Vegetative reproduction is an extension of the capacity of plants for indeterminate growth, and it is common among plants. It involves fragmentation into separate parts, each of which regenerates into a new whole plant.

Review: Animal reproduction by fragmentation is discussed in Module 27.1.

3. There are advantages to asexual reproduction. Offspring are suited to their immediate environments. Early life for vegetative offspring is less hazardous for the young because they are closer to being mature than seedlings, and they may be nurtured by the parent before separation.

4. A garlic clove is part of an underground stem (bulb) (Figure 31.14A). A root sprout of a coast redwood will grow to take the place of the parent, if the parent is lost (Figure 31.14B). The creosote bush of southwestern deserts reproduces vegetatively from its roots, forming very old clones (the one in Figure 31.14C is estimated at 12,000 years old). Dune grass propagates by underground runners (Figure 31.14D).

B. Vegetative reproduction is a mainstay of modern agriculture (*Module 31.15*).

1. Many ornamental trees and shrubs and houseplants are propagated by stem or leaf cuttings.

2. Plant tissue culture provides another way to grow offspring from a few meristematic cells, and this technique has been adapted to propagate genetically engineered plant cells.

3. Modern techniques have produced artificial seeds, tissue-cultured seedlings encapsulated with a food supply in tiny cases.

4. Vegetative propagation has one main disadvantage: Crop plants developed from cloning processes have inherently low levels of genetic diversity, which exposes them to potential devastation from disease, especially when planted in monoculture.

CLASS ACTIVITIES

1. Bring in a variety of fruits and vegetables, and discuss the developmental source of the tissues in each, and their specific functions. Strawberries are separate fruits (the "seeds") on a modified flower receptacle (the red part). Coconuts have liquid endosperm. Some fruits have been bred without seeds (bananas, navel oranges, seedless grapes) and are propagated vegetatively. There is no end of interesting trivia in the produce section of your local grocery. Consult a botany text for further details.

2. The three-dimensional arrangements of tissue layers in roots, stems and leaves are sometimes difficult to grasp from overhead transparencies alone. For smaller-sized classes, the use of large models of these organs (available in most departments) will help show the three-dimensional arrangement of all the basic tissues, and their potential continuity from structure to structure. Secondary growth from the vascular and cork cambiums is also easier to demonstrate with models that show these structures. If you have a color video camera attachment on a dissecting microscope, cross sections of twigs of different age and of wood can be examined in front of a smaller lecture section (depending on your monitor size). This is particularly valuable once you have discussed the function of the two types of cambium using transparencies.

3. Engage your students in the more esoteric aspects of wood anatomy by examining a number of different types of wood (local or exotic species such as oak, maple, pine, mahogany, or even plastic with simulated grain!) with which all students have some familiarity. Examine the grain without magnification, and ask the student how the particular board was cut (cross-section, tangential, or radial cuts are the most common) and what part of the wood the visible grain represents (it's all xylem, but is it heartwood or xylem sap wood, and is it spring wood or fall wood?). Finally, if you have a video camera, examine the surface under higher magnification. Plant anatomy textbooks have numerous illustrations that will help you interpret these sections.

4. An interesting demonstration of the continuity of root and stem (and perhaps even leaf) vascular systems can be made by allowing a squash plant to "ret" over winter. Pull up the old squash plant in your garden at the end of the summer and let it lie on the surface of the ground through the fall and into the following spring. Bacteria and fungi will decompose away the softer, cellulose walls first, leaving the pattern of the vascular bundles. At the junction of root and stem, quite a tangle occurs, but it is clear that the vascular systems of each part are completely connected. Be sure to discuss the biological process that resulted in this preparation.

RESOURCES AND REFERENCES

Bolz, D.M. "A World of Leaves: Familiar Forms and Surprising Twists." *Smithsonian*, April 1985. A delightful article on the adaptations of leaves, with exquisite photographs.

Dale, J. "How Do Leaves Grow?" *BioScience*, June 1992. Researchers are using the techniques of molecular and cell biology to answer questions about plant structure and growth.

Fosket, D.E. *Plant Growth and Development*. San Diego, CA: Academic Press, 1994. An undergraduate textbook.

Gillis, A.M. "Using a Mousy, Little Flower to Understand Flamboyant Ones." *BioScience*, May 1995. A plant species as a research organism.

Haring, V., J.E. Gray, B.A. McClure, M.A. Anderson, and A.E. Clarke. "Self-Incompatibility: A Self-Recognition System in Plants." *Science*, November 16, 1990. Models of genetically controlled mechanisms that prevent plant inbreeding.

Kaplan, D.R. "The Development of Palm Leaves." *Scientific American*, July 1983. Describes how both differential growth and selective cell death help form the elegant, compound leaves of palms.

Kaplan, D.R., and W. Hagemann. "The Relationship of Cell and Organism in Vascular Plants." *BioScience*, November 1991. Plants have a distinctly different mode of multicellular construction than animals. This paper reviews recent studies suggesting that plants should be studied from their own, unique cellular and organ perspective.

The Life Cycle of Plants. Princeton, NJ: Films for the Humanities and Sciences, 1987. A 28-minute videotape describing plant reproduction, growth, and development.

Life Cycles. Seattle, WA: Videodiscovery, 1985. A videodisc presenting a visual record of animal and plant reproduction, with considerable examples from the plant kingdom using film from the Oxford Scientific Film collection. Macintosh HyperCard stack software available for access.

Pettitt, J., S. Ducker, and B. Knox. "Submarine Pollination." *Scientific American*, March 1981. The unique differences in the pollinating system of sea grasses and other angiosperms that live in the ocean.

Pollination Biology. Seattle, WA: Videodiscovery, 1989. A videodisc version of "Sexual Encounters of the Floral Kind," a classic film describing all aspects of flower pollination and reproduction. The disc has the complete film in CLV format and indexed diagrams, stills, and film clips in CAV format. Videotape of film alone available from Carolina Biological Supply Company.

Pollination: The Insect Connection. Burlington, NC: Oxford Scientific Films/Carolina Biological Supply Company, 1986. A 14-minute videotape that demonstrates coevolutionary relationships between flowering plants and their insect pollinators.

Power, J.F., and R.F. Follett. "Monoculture." *Scientific American*, March 1987. A review of the advantages and disadvantages of this type of agriculture.

Raven, P.H., R.F. Evert, and S.E. Eichhorn. *Biology of Plants*, 5th ed. New York: Worth, 1992. The standard introductory botany textbook.

Schwarz-Sommer, Z., P. Huijser, W. Nacken, H. Saedler, and H. Sommer. "Genetic Control of Flower Development by Homeotic Genes in *Antirrhinum majus*." *Science*, November 16, 1990. Homeotic mutants help in the study of the sequential pattern of developing organs comprising flowers.

Taiz, L., and E. Zeiger. *Plant Physiology*. Redwood City, CA: Benjamin/Cummings, 1991.

"The Trees Told Him So." *Science News*, September 7, 1985. Growth rings may give clues to volcanic eruptions.

Vaughan, D.A., and L.A. Sitch. "Gene Flow from the Jungle to Farmers." *BioScience*, January 1991. Problems and methods of maintaining genetic diversity in crop plants.

Woodworth, R. *Floral Evolution in Dicots*. Concord, NH: Essayo, 1988. A 53-minute videotape that presents an overview of flowering plant evolution by comparing the details of floral anatomy, microscopic anatomy of stems and roots, and the fossil record.

INTERNET RESOURCES FOR UNIT VI

Images

gopher://muse.bio.cornell.edu:70/11/images

Many images of plants, animals, etc. Read "About This Biology Image Archive" before searching for images.

Internet Directory for Botany

http://herb.biol.uregina.ca/liu/idb.html

32

PLANT NUTRITION AND TRANSPORT

APPROACH

Plant nutrition is a topic that is usually accessible to nonscience majors, although understanding the physical forces involved in the movements of fluid in xylem and phloem may be a challenge. If necessary, review the mechanisms of diffusion and osmosis and the properties of water. Remind students that K^+ is a potassium ion.

Emphasize the structure-function link in the unique solutions to meeting plant needs on land. Make the material relevant by using practical examples of how plant nutrition directly affects student lives.

Focus on unusual cases to clarify the overall concepts. Expand on the roles of other organisms, particularly bacteria and fungi. The peculiar activities of carnivorous plants always interest students. The unusual activities of these plants will carry over to the subject matter of the next chapter, plant responses to environmental cues.

CHAPTER OBJECTIVES

Identify the needs of plants and the entry points of these needed materials.

Explain why nutrients entering roots must pass through at least one cell membrane before entering the vascular tissue.

Describe the mechanism proposed for the flow of water and inorganic nutrients through xylem, stressing the fact that these motive forces are almost entirely physical, requiring no energy expenditure by the plant.

State the role of stomatal guard cells in controlling the rate of water loss by transpiration.

Describe the mechanism proposed for the flow of water and sugars through phloem, stressing the fact that the plant does expend energy, pumping sugar across membranes, to accomplish this.

List the essential macronutrients and micronutrients for all plants, and name a role for each.

Explain how soil structure and factors relating to soil conservation directly influence the health of plants.

Describe the roles played by other organisms in plant nutrition, particularly those of fungi, bacteria, and consumed animals.

Discuss the science of agriculture.

LECTURE OUTLINE

I. Introduction.

    A. The peanut plant (*Opening Essay*).

        1. *Arachis hypogaea* is a legume (pea family) that bears seeds in pods that mature underground.

        2. Peanuts are rich in protein, which they obtain from nitrogen-supplying (fixing) bacteria that grow in root nodules.

3. Through the work of agriculturist George Washington Carver and others, the peanut has been championed for its nutritional properties, both to humans and livestock who eat peanuts, and to crops that follow the peanut in crop rotations and benefit from the peanut's contribution to soil fertility. In addition, particularly because of the research of Carver, peanut meal can be processed into a wide variety of marketable items, including milk substitute, oil, flour, dye, and cheese.

B. Plants acquire their nutrients from soil and air (*Module 32.1*).

1. *Review:* The early experiments of Jan Baptista van Helmont showed that the substance of mature plants is mostly composed of elements that were not obtained from soil. The hypothesis of English botanist Stephen Hales stated that this source was air. Also review the equation for photosynthesis (see Chapter 7, particularly the Opening Essay).

2. A terrestrial plant has an efficient evolutionary design to obtain resources from terrestrial sources. A plant gets CO_2 from air through its leaves, and water, minerals, and some O_2 from the soil through its roots. All other materials are produced from mixtures of these raw materials and particularly from the sugars produced by photosynthesis (Master 32.1A).

3. Plants, like all aerobic organisms, obtain energy from the respiration of sugars. Leaves are net producers of O_2 and do not need to absorb more. Roots take up atmospheric O_2 through the soil for their respiratory needs.

4. Mineral forms of nitrogen, magnesium, and phosphorus (among others) are needed to make proteins, nucleic acids, phospholipids, ATP, chlorophyll, enzyme cofactors, and hormones.

5. That some plants support bodies over 100 m in height is astounding but explainable (Figure 32.1B).

II. The paths nutrients take.

A. The plasma membranes of root cells control solute uptake (*Module 32.2*).

1. Because of its large root surface area (particularly root hairs), a plant can absorb enough water and inorganic ions to survive and grow (Figure 32.2A).

    Review: The tissues of roots (Module 31.6).

2. *Preview:* Virtually all plants in the wild obtain nutrients and water through mycorrhizal fungal interconnections that may bypass root hairs. Root hairs are more important in laboratory seed cultures and fertilized agricultural plants (Module 32.11).

3. Substances enter roots in solution and make their way toward vascular tissue by two routes, intracellularly and extracellularly, and combinations of these two. In the end, nutrients must pass through at least one membrane before arriving at the vascular tissue. This allows plants to control the entry of substances into their roots (Acetate 32.2B).

4. The intracellular pathway goes through the cell membrane of a root hair and then, by means of cytoplasmic continuity through plasmodesmata, throughout the cytoplasmic content of cortex cells and endodermal cells, and finally into the xylem vessels.

5. The extracellular pathway goes through the porous cell walls of all epidermal and cortex cells, but it is forced, by the impervious Casparian

strip in the walls of endodermal cells, through cell membranes and then into the xylem.

B. Transpiration pulls water up xylem vessels (*Module 32.3*).

1. *Review:* The properties of water, diffusion, and osmosis (Modules 2.9–2.11, 5.14, and 5.15).

2. Getting water from the soil, against the force of gravity, up to the leaves where it is needed, is a major adaptation of land plants. Water and dissolved nutrients, xylem sap, travel in the xylem. Several mechanisms combine to produce forces that move the water.

3. In some plants, root pressure pushes the column of water up the xylem. Root cell membranes actively pump inorganic ions into the xylem, and osmosis causes water molecules to follow.

4. The main motive force through xylem is by the transpiration-adhesion-cohesion mechanism. Transpiration is the evaporation of water from internal leaf cell surfaces and diffusion out stomata. Water molecules in xylem stick together by cohesion, and the column of water in xylem is pulled up. The adhesion of water molecules to the cellulose molecules of the cell walls helps counteract the downward pull of gravity on the water column (Acetate 32.3).

NOTE: Remember, xylem cells are dead and have no membranes.

C. Guard cells control transpiration (*Module 32.4*).

1. Transpiration works for and against plants. Transpiration can result in the loss of more than 200 L of water per day.

2. Stomata are changeable openings in the leaf surface. Their size is regulated by two surrounding guard cells that change shape in response to changing environmental conditions (Acetate 32.4).

3. Guard cells buckle outward, opening the stomata; they actively pump in K^+, and water follows by osmosis, increasing the turgor. Guard cells close the stomata when they lose K^+.

Review: Osmosis (Module 5.15).

4. Increased sunlight and decreased internal CO_2 cause guard cells to take up K^+, and if the plant loses water too fast, the guard cells close. In addition, an internal daily timing mechanism triggers K^+ uptake in the morning, stomatal opening, and K^+ release, stomatal closing, in the evening.

Preview: This internal timing mechanism is a type of biological clock. The biological clocks of plants are discussed in greater detail in Modules 33.10–33.12.

D. Phloem transports sugars (*Module 32.5*).

1. Phloem is composed of sieve-tube members arranged end-to-end, each bounded at the end with a pitted plate. Each sieve-tube member's plasma membrane is continuous with the next, so the phloem sap can flow easily from one to another (Figure 32.5A).

Review: The structure of phloem is also discussed in Module 31.5.

2. Phloem sap contains a watery solution of inorganic ions, amino acids, hormones, and, principally, sucrose.

3. Phloem sap flows from sugar sources (areas where sugar is made by photosynthesis or released from stored starch) to sugar sinks (areas where sugar molecules are required for either direct use, flower nectar, or storage for future use, taproots).

4. The pressure-flow mechanism is the most widely accepted model for the movement of phloem sap. At the sugar source, sugar is loaded into the phloem by active transport. Water follows by osmosis, raising the water pressure. At the sugar sink, sugar leaves the phloem, and water follows by osmosis, lowering the water pressure. Phloem sap flows from source to sink down a gradient of hydrostatic pressure. Excess water returns to the sugar source through the xylem (Acetate 32.5B).

5. This model is supported by tests using natural phloem tappers, aphids. Their mouthpieces (stylets) are severed while they feed on sap. The closer the aphid stylet is to a sugar source, the faster the sap flows out (Figure 32.5C).

III. Inorganic nutrients obtained from soil.

A. Plant health depends on a complete diet of essential inorganic nutrients (*Module 32.6*).

1. Plants do not need (and cannot use) a supply of organic nutrients because they make their own. All they need, in addition to raw materials for photosynthesis, is inorganic nutrients.

2. Inorganic nutrients are essential if a plant must obtain them to complete its life cycle. The most common symptoms of nutrient deficiency are stunted growth and leaf discoloration. Hydroponic culture can be used to determine which elements are essential (Master 32.6A).

3. Sixteen elements are essential in all plants. A few others are essential in only certain plant groups.

4. Nine of the essential nutrients are required in relatively large amounts and are known as macronutrients. Six are major atomic components of organic compounds: carbon, oxygen, hydrogen, nitrogen, sulfur, and phosphorus. Calcium, potassium, and magnesium are important in inorganic compounds.

5. Calcium is important in the formation of cells walls, in the functioning of certain proteins that help glue plant cells together in tissues, for the maintenance of cell membrane structure, and in regulating the selective permeability of cell membranes.

6. Potassium is an important part of several enzymes and plays a role in maintaining osmotic balance and in elongation during primary growth.

7. Magnesium is an important component of chlorophyll and is a cofactor for several enzymes.

Review: Cofactors are discussed in Module 5.7.

8. Seven essential nutrients are required in very small amounts and are known as micronutrients. Iron, chlorine, copper, manganese, zinc, molybdenum, and boron mainly function as components or cofactors of enzymes and are used over and over.

9. Growing plants in soil deficient in essential nutrients can produce plants of lower quality (nutritional value for humans, appearance, etc.) (Figure 32.6B).

B. You can diagnose some nutrient deficiencies in your house and garden plants (*Module 32.7*).

1. Symptoms of many forms of nutrient deficiency are often distinct enough to allow visual determination of the deficiency. Compare the following affected tomato plants with healthy plants (Figure 32.7A).

2. Nitrogen deficiency (in the form plants can use, NO_3^- or NH_4^+) exhibits stunted growth and yellow-green leaves (Figure 32.7B).

3. Phosphorus deficiency (in the form plants can use, $H_2PO_4^-$ or HPO_4^{2-}) exhibits green leaves but growth rate is greatly reduced and the new growth may be spindly and brittle (Figure 32.7C).

4. Potassium deficiency (in K^+ ions) exhibits yellow or dead brown spots or edges in localized areas on leaves. Stems and roots exhibit stunted growth (Figure 32.7D).

   NOTE: Commercial fertilizers give the values of these three nutrients, in percent weight in the fertilizer, of nitrogen, phosphoric acid (H_3PO_4), and potash (K_2O). A fertilizer labeled 4-8-4 is 4% N, 8% H_3PO_4, and 4% K_2O.

C. Soil in the life of plants: a closer look (*Module 32.8*).

1. The structure and nutrient content of soil are important characteristics in plant root absorption. Soil structure is categorized according to horizons. The microscopic details of structure affect the availability of nutrients.

2. The A horizon is the topsoil. It is subject to weathering and contains high levels of decomposing organic matter called humus. This horizon is usually intensely active with the decomposing activity of bacteria, fungi, protozoans, and small animals. Most plant roots branch out in the A horizon.

   NOTE: Much nutrient uptake takes place in the A horizon. It is also the region of most active mycorrhiza formation.

3. The B horizon contains fewer organisms and less organic matter, and is less subject to weathering.

   NOTE: The lower levels of the B horizon represent the deepest regions at which water is obtained by plant roots and mycorrhizae.

4. The C horizon is broken-down rock that has only been slightly weathered.

5. Water, dissolved oxygen, and nutrients are removed by root hairs in direct contact with water films on soil particles (Masters 32.8B, C).

6. Cation-ion exchange is the release of H^+ ions by root hairs to displace positive nutrient ions (such as Ca^{2+}, Mg^{2+}, and K^+) that naturally accumulate on negatively charged clay particles and thus free these nutrients for absorption by the root hairs.

D. Soil conservation is essential to human life (*Module 32.9*).

1. Irrigation makes possible the agricultural use of otherwise dry areas, but it has the disadvantages of overusing water and making the soil salty. Modern drip irrigation methods are the most efficient at avoiding theses problems.

2. Wind and water erosion can be prevented by minimal tillage farming, planting trees as windbreaks, and contour tillage. However, minimal tillage farming often relies on herbicides to destroy weeds that plowing

would otherwise remove, and herbicides may contribute to chemical pollution of the soil.

3. Most farmers use fertilizers containing nutrients that are either mined or produced by industrial processes. These fertilizers contain mixtures of N, P, and K that are rapidly released in soil and that may rapidly leach out to pollute groundwater. Organic fertilizers are of biological origin and have the advantages of releasing nutrients gradually.

   NOTE: Organic fertilizers also tend to help maintain beneficial soil structure, including humus.

   Preview: The environmental impact of agriculture (Modules 38.10 and 38.12).

E. Organic farmer Stephen Moore uses no commercial chemicals (*Module 32.10*).

1. Organic farming is farming without pesticides and inorganic fertilizers.

2. Thirty years of previous farming with inorganic chemicals had caused the land Stephen Moore inherited to be virtually infertile and lacking a healthy soil structure.

3. He uses organic fertilizers and green manures (turning under intergrown crops of grasses and legumes).

4. Although the operating expenses of organic farming are higher, some people are willing to pay more for safer, more nutritious food. In conventional farming, there are hidden costs, such as the loss of soil fertility and increased problems with pests from unbalanced soil conditions.

IV. Nutrients obtained from other biological sources.

A. Fungi help most plants absorb nutrients from the soil (*Module 32.11*).

1. *Review:* The characteristics of fungi, particularly hyphae, the means of heterotrophic nutrition, and the role of the mycorrhizae of citrus trees (Module 18.17 and Chapter 18, Opening Essay).

2. Mycorrhizae are structures formed by the roots of plants and fungi that invade these roots. Virtually all plants in naturally competitive situations have mycorrhizae. In the form of mycorrhiza shown in Figure 32.11, the fungal hyphae envelop the root with a covering, and some tips enter between the cortex cells of the root. In other mycorrhizae, fungal hyphae enter through root hairs or the root's epidermal surface into cells of the epidermis and the cortex.

3. The plant supplies the fungus with required carbohydrates, while the fungus supplies the plant with increased efficiency of nutrient (particularly phosphorus, but not nitrogen) and water uptake, and in some forms of mycorrhizae, protection of the root from root parasites.

   Preview: Symbiotic relationships, such as mutualism, are discussed in Module 36.5).

B. The plant kingdom includes parasites and carnivores (*Module 32.12*).

1. Dodder is a nonphotosynthetic plant that parasitizes other plant species using modified roots to tap the host plant's vascular tissue.

2. Mistletoes are photosynthetic parasites of trees that supplement their diets by tapping into the host's vascular tissue.

3. The sundew and Venus flytrap use ingenious insect traps formed from highly adapted leaves to trap, kill, and digest insects, and enhance their supplies of nitrogen. These carnivorous plants are found in bogs where nitrogen is limited because the acid conditions impede the decomposition of dead vegetation.

C. Most plants depend on bacteria to supply nitrogen (*Module 32.13*).

Preview: The nitrogen cycle (Module 36.16).

1. All plants and, indirectly, all animals depend on nitrogen supplies in soils in which plants grow. Nitrogen is available to plants only as NH_4^+ (ammonium ions) and NO_3^- (nitrate ions).
2. Although nitrogen is 80% of our atmosphere, this gas is not usable by plants but must be converted to organic form by bacteria. Nitrogen fixation is the conversion process by which bacteria (Module 17.17) convert gaseous nitrogen to ammonium (Master 32.13).
3. Ammonifying bacteria convert organic forms of nitrogen to NH_4^+. Little NH_4^+ is absorbed by plants because, being positive, it usually remains firmly bound to negative clay particles.
4. Nitrifying bacteria convert NH_4^+ to NO_3^-. NO_3^- is the form of nitrogen most often used by plants because it is negative and readily released from soils.

D. Many plants have built-in nitrogen-fixing bacteria (*Module 32.14*).

1. Legume roots house nitrogen-fixing bacteria of the genus *Rhizobium* in nodules made of plant cells (Figure 32.14A, B).
2. Some nonlegume plants, such as alders, have root nodules containing the nitrogen-fixing bacteria actinomycetes.
3. The plant provides the bacteria with carbohydrates and other organic compounds, while the bacteria provide ammonium.
4. Legume crops can be rotated with other crops or plowed into the soil prior to planting a second crop, both providing increased nitrogen to the soil.

V. **Other agricultural considerations.**

A. A major goal of agricultural research is to improve the protein content of crops (*Module 32.15*).

1. The majority of people in the world have vegetarian diets, but the plants they eat are low in protein. Many high-yielding modern crop plants require very high levels of nitrogen fertilization, which is difficult to obtain in developing countries.
2. One promising approach to improving the output of nitrogen-fixing bacteria is using molecular genetics techniques to remove the negative-feedback system that regulates the rate at which some of these bacteria fix nitrogen. This has been done in certain mutant strains, which may be incorporated into host plants in the future (Master 32.15B).

B. Genetic engineering could greatly increase crop yields (*Module 32.16*; Master 32.16).

1. *Review:* Recombinant DNA technology used in agriculture (Chapter 12, particularly Module 12.16).

2. A novel approach in genetically engineering plants is to use a .22-caliber gun to fire DNA-coated pellets into the cells. The pellets pass through the walls, into the cytoplasm where the DNA becomes integrated with that of the cell.

3. Using genes introduced by this approach and by plasmids, several transgenic plants have been produced, including potato plants that synthesize their own insecticide and tomato plants with fruit that is slow to spoil.

4. A not-yet-attained goal is to transfer nitrogen-fixing genes into nonlegume crop plants.

5. In this work, as in the work with transgenic animals, there are continuing questions about potential problems and risks.

CLASS ACTIVITIES

1. A simple device can be fabricated from two elongated balloons to demonstrate how the intake of water into and out of guard cells (with subsequent change of shape) opens and closes a stoma. Tie two balloons together at the far end and attach the open ends to the two "upper" ends of a glass or plastic "Y-tube." Blow the balloons up until they are about 8 to 10 inches long, lying parallel and touching. Apply a piece of masking tape along each of the elongated, touching surfaces. This represents the thicker, inner wall of each guard cell and, at this level of inflation, the whole device represents a closed stoma. Blow into the Y-tube to inflate the "guard cells" a bit, and a stoma-like opening will form within the inner, taped surfaces.

RESOURCES AND REFERENCES

Beardsley, T. "A Nitrogen Fix for Wheat." *Scientific American*, March 1991. Artificially induced root nodules on nonlegumes.

Death Trap. Seattle, WA: Videodiscovery, 1989. A videodisc that includes Oxford Scientific Films footage detailing many aspects of the physiology, ecology, distribution, and evolution of carnivorous plants. Time-lapse, close-up, and high-speed photography plus animation document all mechanisms. Videotape available from Carolina Biological Supply Company.

Gibbons, B. "Do We Treat Our Soil Like Dirt?" *National Geographic*, September 1984.

Heslop-Harrison, Yolande. "Carnivorous Plants." *Scientific American*, February 1978. The nutritional aspects of the evolution of insect traps among a variety of flowering plants.

Holmes, B. "Can Sustainable Farming Win the Battle of the Bottom Line?" *Science*, June 25, 1993. Environmental versus economic concerns

Kimball, Robert, and David Donoghue. *Botanical Gardens*. Scotts Valley, CA: Sunburst Wings for Learning, 1993. Macintosh or Apple II software simulation that allows experimentation with the growth of different plants under varying environmental conditions. Also featured is a genetics laboratory simulation for designing custom seeds.

Nap, Jan-Peter, and Ton Bisseling. "Developmental Biology of a Plant-Procaryote Symbiosis: The Legume Root Nodule." *Science*, November 16, 1990. The formation of this plant organ involves specific activation of genes in both plant and bacterium.

Reganold, J.P., R.I. Papendick, and J.F. Parr. "Sustainable Agriculture." *Scientific American*, June 1990.

Water and Plant Life. Princeton, NJ: Films for the Humanities and Sciences, 1987. A 28-minute videotape that covers the water cycle in plants and various types of adaptations to dry environment.

33

CONTROL SYSTEMS IN PLANTS

APPROACH

Experiments on phototropism provide a clear example of the sequential process of science: how the results of one experiment lead to others. To reinforce this principle, lead students through the work of scientists represented in Acetates 33.1C and D. The illustrations also can be used to discuss the role of controls in experimental procedures. In lecture, before students have read the chapter, show these illustrations and give some background. Ask students what conclusions can be deduced from each experiment, and which experimental illustrations show controls.

In prefacing this material, discuss the roles of plant hormones and other controlling processes in light of the basic needs of plants: to get more sun, not function at night, minimize water loss, maintain a connection to water and nutrients, etc.

Differences between the effects of plant hormones are subtle and generally involve a balance of one hormone against another that has an antagonistic effect. To avoid confusion on the part of students, state this clearly at the beginning of your treatment (see Module 33.2). Explain that all the necessary studies have not yet been done, that combination effects are particularly complex and difficult to study and have different effects in different plants. Therefore, an overall, simple picture of plant hormone actions cannot be presented.

CHAPTER OBJECTIVES

Describe the experiments leading to the discovery of the plant hormone auxin, showing how science proceeds in small, logical steps.

Name the source, target tissues, and overall roles played by each of the five main classes of plant hormones: auxins, cytokinins, gibberellins, abscisic acid, and ethylene. Emphasize that these hormones often function in combination, rather than alone.

Discuss the artificial application of these various hormones in agriculture.

Distinguish between the environmental cues responsible for phototropism, gravitropism, and thigmotropism.

Differentiate between long-day and short-day plants, explaining how the plants relate to seasonal changes in photoperiod, and the assumed mechanism of phytochrome in sensing the seasonal changes.

LECTURE OUTLINE

I. Introduction.

   A. The sensitive plant shows dramatic response to its environment (*Opening Essay*).

      1. Its leaves, small leaflets, and whole leaves fold up within a few seconds of being touched (Figure 33.0).

      2. The response is due to rapid loss from cells in specialized tissues at the base of leaflets, leaves, and leaf stalks. This water loss is induced by signals similar to action potentials (Module 28.5) that travel from plant cell to

plant cell. Once the water loss has been triggered, it takes about 10 minutes for the cells to regain turgor.

  3. Two possible functions have been hypothesized for this behavior: to reduce surface area (saving water) during windy times, and to make the plant less attractive to herbivores.

  NOTE: In light of other, better-documented experimental stories in this chapter, it is important to point out that there is little experimental evidence to support either of these hypotheses.

B. Plants, in general, show responses to their environments.

  1. Usually, the responses are not as dramatic as the case of the sensitive plant, but similar leaf folding occurs slowly as a kind of "sleep movement" in many plants. Charles Darwin hypothesized that sleep movements slowed the loss of heat in plants during the night.

  2. Most plants do not have the "action potential" mechanisms for communicating the presence of environmental stimuli. Instead, they rely on plant hormones to deliver the signals and on an innate sense of where the environment is, indicated by internal biological clocks.

C. Experiments on how plants turn toward light led to the discovery of a plant hormone (*Module 33.1*).

  1. Phototropism is an adaptive response of plant shoots and seedlings to grow toward a source of light. Examination of the cellular events during a phototropic response by a shoot tip shows that the cells on the darker side of the shoot have elongated faster (Master 33.1B).

  2. A series of classic experiments showed how a particular hormone controls the phototropic response. The series begins with experiments by Charles Darwin and his son, Francis, in the 1800s. They showed that grass seedlings bend toward light only if their tips are present and uncovered, suggesting that the tip was responsible for the sensing of light (Acetate 33.1C).

  NOTE: Because these plant parts are mentioned in Chapter 31, indicate that these classical experiments were performed in young seedlings that still had the sheath that covers the young shoots of monocots.

  3. In 1913, Danish botanist Peter Boysen-Jensen showed that a gelatin block placed between the plant tip and the lower plant would not inhibit the response, whereas a mica block would. This suggested that the response was mediated by a chemical that could diffuse through the gelatin but not through the mica.

  4. In 1926, Dutch botanist Fritz Went extended Boysen-Jensen's experiments to actually discover the hormone. He collected the hormone from severed shoot tips into agar blocks, and then applied the agar blocks in various positions to decapitated seedlings growing in the dark. Agar blocks placed off-center caused the seedlings to bend, away from the side the block was on. Went called the diffusible substance auxin (Acetate 33.1D).

  5. Follow-up experiments showed that shoot tips secrete auxin in equal amounts in light or dark, but that auxin diffuses from lighted sides of shoot tips to shaded sides. In the 1930s, biochemists identified the chemical structure of auxin: a small organic molecule made from the amino acid tryptophan.

II. Five classes of hormones regulate plant growth and development (*Module 33.2*).

   A. General overview (Table 33.2).

   1. Three of the hormone types (auxins, cytokinins, and gibberellins) are actually classes, including several chemicals with similar molecular structure and function. The other two types of plant hormones are abscisic acid (ABA) and ethylene.

   2. Plant hormones produce their effects in very small amounts and may do so by altering gene expression or by activating or inhibiting the action of enzymes or membranes.

   3. All five types influence growth, and four (all except abscisic acid) affect development.

   4. The effects of a hormone depend on plant species, the site of action, the developmental stage of the plant, the concentration of the hormone at the target site, and the balance of concentrations of different hormones.

   B. Auxins stimulate the elongation of cells in young shoots (*Module 33.3*).

   1. Auxins are a class of chemicals that promote the elongation of developing shoots.

   Review: The elongation of shoots is discussed in Module 31.7.

   2. The most important, naturally occurring auxin is indoleacetic acid (IAA). The major site of auxin synthesis is the apical meristem of the shoot tips.

   3. Relatively high concentrations of IAA stimulate shoot cell elongation and inhibit root cell elongation. However, higher levels of IAA stimulate the synthesis of ethylene. Ethylene counters the effects of IAA and is the likely cause of the inhibition of shoot growth at higher IAA levels (Acetate 33.3B).

   4. Relatively low concentrations of IAA have no effect on shoot cell elongation but stimulate root cell elongation (Acetate 33.3B).

   5. One hypothesis about how auxin elicits its effect on cells is that auxins may stimulate certain H^+ pumps in plasma membranes. The increase in H^+ outside the membrane but within the cell wall would stimulate enzymes that break bonds between cellulose molecules, thereby causing the walls to be more elastic and allowing cells to elongate in response to turgor pressure (Acetate 33.3C).

   6. Auxins can also trigger the development of vascular tissue and induce vascular cambium cell division and thus growth in stem diameter. Auxins produced in seeds promote the growth of fruit tissues.

   Review: Growth in diameter is discussed in Module 31.8.

   7. An agricultural use of auxin sprays is to promote fruit formation from seedless, unfertilized flowers (e.g., in tomatoes, cucumbers, and eggplants).

   C. Cytokinins stimulate cell division (*Module 33.4*).

   1. Natural cytokinins are produced in actively growing tissues in roots, embryos, and fruits and reach target tissues by flowing in xylem sap.

   2. The effects of cytokinins are often influenced by the concentrations of auxins; it is the ratio of the concentrations of these two types of hormones that has the effect.

3. In unpinched plants, the balance of auxin (diffusing down from the shoot tip) to cytokinin (flowing in xylem up from roots) affects where axillary buds will begin to extend laterally and may also help coordinate the growth of root and shoot systems.

4. When a terminal bud is removed, the elongating effects and inhibitory effects on lateral bud growth of auxin are removed, and the axillary bud-stimulating effects of cytokinins (still at the same concentration as when auxins dominate) activate lateral buds (Figure 33.4).

5. An agricultural use of cytokinin is to retard the aging of harvested flowers and fruits.

D. Gibberellins affect stem elongation and have numerous other effects (*Module 33.5*).

1. Gibberellins were first isolated from species of fungus of the genus *Gibberella* in Japan, where the fungus causes "foolish seedling disease" on rice plants. This disease is characterized by aberrant growth of rice shoots, causing weakening (Figure 33.5A).

2. Lower concentrations of gibberellins occur naturally in plants where the hormone functions as a growth regulator. Over 70 different gibberellins are known.

3. Gibberellins are made in tips of shoots and roots; their main effect is to enhance the action of auxin on cell elongation and fruit development; and they have a role in seed development. In seeds, they seem to be the link between environmental cues and the metabolic processes required for the renewal of embryo growth during seed germination.

4. An agricultural use of gibberellins is to enhance the size and spacing of Thompson seedless grapes (Figure 33.5B).

5. In some plants, gibberellins are antagonistic to abscisic acid.

E. Abscisic acid inhibits many plant processes (*Module 33.6*).

1. Abscisic acid (ABA) is produced in buds and produces effects that are adaptive for inducing or maintaining dormancy during unfavorable times.

2. ABA promotes dormancy in shoots by inhibiting cell division in buds and vascular cambium and by stimulating the formation of bud scales.

3. ABA seems to act as a growth inhibitor in seeds, particularly of annual plants, until a certain volume of water from rain washes the ABA out of the seeds. It may actually be the ratio of ABA to gibberellins that promotes germination.

4. ABA also acts as a stress hormone, accumulating in leaves during times of drought, and causing stomata to close, thus reducing transpiration.

F. Ethylene triggers fruit ripening and other aging processes (*Module 33.7*).

1. The role of this gas in fruit ripening was discovered when sheds warmed by kerosene stoves were used to ripen grapefruit. The gas was a by-product of the kerosene, which was absent from cleaner heaters.

2. Ethylene is produced in tissues of ripening fruits. It has a positive-feedback effect on ripening by inducing the breakdown of cell walls and color changes, and additional production of ethylene (Figure 33.7A).

3. To retard ripening and fruit spoilage, growers flush out stored fruit with CO_2 to remove the accumulating ethylene.

4. Ethylene probably also plays a role in the color changes in and falling of leaves during autumn. Pigment complexes are induced to change, and a breakable layer (abscission layer) of cells at the bases of leaves is induced to form when the shorter days and cooler temperatures of fall occur. Abscission is promoted by a change in the balance of ethylene to auxin (Figure 33.7B).

G. Plant hormones have many agricultural uses (*Module 33.8*).

1. *Review:* The roles of hormones mentioned in the last five modules. The additional uses below are also important.

2. Large doses of auxins are used to promote fruit drop.

3. Gibberellins are used for producing seedless fruit (Module 33.5). Large doses of gibberellins promote early (first-year rather than second-year) flowering and seed production in biennial plants such as carrots, beets, and cabbage.

4. 2,4-D is a synthetic auxin applied to eliminate dicot weeds in monocot grain crop plantings. Dicots are more sensitive to the toxic effects of this compound than monocots. One drawback to the use of 2,4-D is that a by-product of its production is dioxin, a compound known to be toxic to humans.

Review: The economics and ethics of the use of such hormones (Module 32.10).

Preview: Biological magnification and the pesticide DDT (Module 38.12).

III. Overall responses of plants to environmental stimuli.

A. Tropisms tune plant growth to the environment (*Module 33.9*).

1. The rapid responses of plants like *Mimosa* (Opening Essay) or the carnivorous Venus flytrap (Module 32.12) are exceptions to plant behavior. Most plants respond to environmental cues by slow growth responses known as tropisms. Phototropism, gravitropism, and thigmotropism are irreversible growth responses to light, gravity, and touch, respectively.

NOTE: Differentiate the term *tropism*, growth, from *taxis*, movement. Taxis is a common form of response seen in animals and motile protists.

2. *Review*: The mechanism for phototropism is the differential rate of cell elongation, largely under the control of auxin produced by shoot tips and differentially distributed in shoots by the effect of light and shade (Module 33.1).

3. Plants show positive gravitropism in roots and negative gravitropism in shoots. The cellular mechanism is unknown but may involve the effect of gravity on starch grain-containing organelles in cells (Figure 33.9A).

4. Tendrils (modified leaves) and stems of climbing plants exhibit thigmotropism. Contact on one side of these structures elicits cell growth on the opposite side. This causes the structure to coil and grasp the object that contacts the plant (Figure 33.9B).

B. Plants have internal clocks (*Module 33.10*).

1. *Review:* The opening and closing of stomata discussed in Module 32.4 is an example of an internal clock.

2. Like animals, plants display rhythmic activities, particularly those that exhibit circadian rhythms, cycles of about 24 hours. Plants will continue these activities if placed in completely dark environments or in orbiting satellites. The molecular bases of biological clocks are not known (Figure 33.10).

   Preview: The biological clocks of animals are discussed in Module 37.9.

3. Interestingly, most biological clocks are not precisely timed with the external environmental factors they keep the organism (plant or animal) tuned to. For example, the sleep movements of plants kept in dark exhibit 26-hour cycles. These imprecise timekeepers seem to be naturally reset each day by environmental cues.

4. Another interesting aspect of biological clocks is that their timing is not temperature-sensitive (this is an advantage), so the chemical bases of the timing must be quite complex because most chemical reactions are temperature-dependent.

C. Plants mark the seasons by measuring photoperiod (*Module 33.11*).

1. Biological clocks may also function in seasonal events such as flowering, seed germination, and the onset and ending of plant dormancy during seasonal periods of environmental stress.

2. The relative lengths of day and night (photoperiod, and, specifically, the length of the night) are the principal environmental cues to which plants relate (Acetate 33.11).

3. Short-day (actually, long-night) plants (e.g., chrysanthemums and poinsettias) flower in late summer or winter when light periods shorten. These plants respond to nighttime darkness of a critically long duration. A brief flash of light in the middle of this period will stop blossoming.

4. Long-day (actually, short-night) plants (e.g., spinach, lettuce, iris, and many cereal grains and other domesticated food plants) flower in late spring or early summer when light periods lengthen. These plants respond to nighttime darkness of a critically short duration. A brief flash of light in the middle of a longer period will induce blossoming.

D. Phytochrome is a light detector that may help set the biological clock (*Module 33.12*).

1. *Review:* The electromagnetic spectrum (Module 7.6).

2. Phytochrome is a colored protein that absorbs light. This protein has been found in all species of plants and algae examined to date, where it seems to help these photosynthetic organisms stay synchronized with seasonal changes in light conditions. It may also help trigger certain seasonal responses such as seed germination, flowering, and stomatal opening.

3. Phytochrome alternates between two forms that differ slightly in structure and wavelength of absorption, depending on which part of the spectrum has been most recently absorbed. Far-red light (characteristic of sunsets) is absorbed by P_{fr}, forming P_r, while red light (characteristic of sunrises) is absorbed by P_r, forming P_{fr}. Phytochrome is synthesized as P_r, and any P_{fr} that forms is converted to P_r in the dark. P_r is converted to P_{fr} at sunrise, and P_{fr} is converted to P_r at sunset. A plant can measure photoperiod by keeping track of the balance of P_r to P_{fr} (Master 33.12B).

4. This can be tested experimentally by exposing plants to brief flashes of light of red or far-red wavelengths. The wavelength of the last flash of light affects a plant's measurement of night length. In long-day plants (lower part of Figure 33.12), if the last flash is red light (signifying sunrise, and the shortened night), flowering is induced. In short-day plants (upper part of Figure 33.12), if the last flash is far-red light (signifying sunset, and the continued lengthening night), flowering is induced (Master 33.12A).

E. Talking About Science: Plant biochemist Eloy Rodriguez studies how animals use defensive chemicals made by plants (*Module 33.13*).

1. Dr. Rodriguez, a professor of plant biochemistry at Cornell University, is one of the world's leading authorities on defensive chemicals produced by plants. He is involved in a relatively new field of study called zoopharmacognosy, the study of how animals may medicate themselves with plants. Currently, Rodriguez's work focuses on the medicinal potential of rain forest plants.

Preview: The consequences of tropical rain forest destruction (Module 38.14).

2. There are numerous examples of defensive chemicals produced by plants that are now commonly used by humans. For example, aspirin was originally derived from willow trees and helps plants fight off infections; and chilies contain a naturally occurring antibiotic.

CLASS ACTIVITIES

1. The behavior of *Mimosa* and Venus flytraps can be demonstrated in class. These plants may be commercially available from greenhouses. If not, you may be able to find sundews, a carnivorous plant, in a bog.

RESOURCES AND REFERENCES

Evans, M.L., R. Moore, and K.-H. Hasenstein. "How Roots Respond to Gravity." *Scientific American*, December 1986.

Mores, P.B., and H.-H. Chua. "Light Switches and Plant Genes." *Scientific American*, April 1988. A link between environment and gene expression in plants.

Poethig, R. Scott. "Phase Change and the Regulation of Shoot Morphogenesis in Plants." *Science*, November 16, 1990. The unique physiological and morphological attributes of cells during different phases of development, and their genetic controls, are discussed in the context of whether control comes from the apical meristem or within the developing tissues themselves.

Sussman, M. "Shaking *Arabidopsis thaliana*." *Science*, May 1, 1992. The cell biology of a plant's response to touch.

Vogel, S. "When Leaves Save the Tree." *Natural History*, September 1993. How leaf responses help prevent wind damage to trees.

Wayne, R. "Excitability in Plant Cells." *American Scientist*, March-April 1993. Nervelike signals in plants.

Woodworth, Robert. *The Tropisms*. Concord, NH: Essayo, 1988. A 14-minute videotape that illustrates chemotropism, geotropism, phototropism, and thigmotropism using time-lapse and real-time photography of living plants. Available from Carolina Biological Supply Company.

34

THE BIOSPHERE: AN INTRODUCTION TO EARTH'S DIVERSE ENVIRONMENTS

APPROACH

Whether you treat ecology first, last, or somewhere in between, the crises in ecology, and specifically those concerning the global threats to biodiversity, are *the* critical biological issues of our time. The nature of the interactions between organisms and their environments is an important lesson that nonscience majors will take away from the course. Because it ties together all the other biological disciplines, teaching ecology at the end of a course in biology makes good sense.

Included in this chapter are all the basic definitions, descriptions of abiotic factors, and brief descriptions of each of the world's major biomes. You could briefly review the biomes at this point and include more detailed discussion later, just prior to covering human evolution and ecology in Chapter 38. Or, as the text does, cover the biomes now to prepare the student for details at smaller scales in chapters to come. In discussing each biome, be sure to cover the potential sensitivities to human disruption (usually mentioned in the last part of each description).

Rather than having students recall all the details for all the biomes, perhaps emphasize a single aspect (e.g., the defining abiotic parameters or the sensitivity to human disturbance) across all biomes, and discuss one aquatic and one terrestrial biome in some detail, perhaps focusing on those that are locally important. One goal might be to provide the details on the ecology of all local biomes and ecosystems in your particular geographic area.

Ecology and evolution (Units III and IV) are tied together. Ecology can be viewed as evolution occurring at a time scale closer to that in which humans operate. For example, the endosymbiotic origin of chloroplasts and mitochondria is viewed as the result of evolutionary change. However, if we had witnessed an intermediate stage of the process, we would have seen it as a type of symbiosis, traditionally studied in the ecology section of an introductory course.

CHAPTER OBJECTIVES

Define the limits of ecology, and discuss the approaches taken by biologists who focus at each of these levels: individual, population, community, ecosystem, and biosphere.

Identify the abiotic and biotic factors that control ecological relationships, stressing the role of natural selection in choosing the adaptations each organism has that allow it to function well in its environment.

Name the two types of aquatic ecosystems and nine types of terrestrial biomes, indicating for each the defining abiotic characteristics, the location of each on Earth, characteristic features of the organisms, and any particular sensitivity to human disturbance.

LECTURE OUTLINE

I. Introduction.

    A. The unit to come.

        1. Ecology is the scientific study of the interactions between organisms and their environments.

        2. A great number of humankind's gravest biological crises hinge on a firm understanding of ecological principles.

        3. This unit starts with an overview of Earth's different biological settings, details the principles and mechanisms of ecology, and ends with a discussion of human evolution and the impact of human societies on other organisms.

    B. A very special environment (*Opening Essay*).

        1. *Review:* Continental drift and the locations of crustal plates (Modules 16.3 and 16.4).

        2. Along mid-oceanic ridges between crustal plates, at depths of over a mile and in complete darkness, volcanic vents release hot, nutrient-rich gases (Figure 34.0A).

        3. Around these vents are groups of organisms, the most obvious of which are large, yard-long tube worms. A variety of other animals, including shrimps, crabs, clams, and a few fish, also live here (Figure 34.0B).

        4. Most of the bacteria in hydrothermal vents are chemoautotrophs (Module 17.10). Ultimately, all the other organisms depend on the food made by these bacteria because energy sources dependent on sunlight never make it in any appreciable quantity to these depths.

        5. This setting hints at many of the topics concerning populations, communities, and nutrient cycling that ecologists study.

    C. Ecologists study how organisms interact with their environments at several levels (*Module 34.1*).

    Review: The hierarchical organization of life (Module 1.1) and the scientific process (Modules 1.2 and 1.3).

        1. At the organism level, the focus is on how individual organisms interact with aspects of their immediate surroundings.

        2. At the population level, ecologists focus on functioning among all the members of an interbreeding group and its environment.

        3. Studies at the community level focus on all the organisms of all species and their interactions within one particular area.

        4. The ecosystem level of study adds the nonliving (abiotic) factors to the picture, in addition to relationships among the living (biotic) factors.

        5. Ecology can be enormously complex because it studies multidimensional problems. Ecological research still employs the scientific process but must often take into account the complexities produced by the multidimensional nature of ecological interactions. Studies can be done on idealized collections of organisms or environments assembled in artificial setups in the laboratory, or by a careful examination of natural systems.

II. **The biosphere.**

   A. The biosphere is the total of all of Earth's ecosystems (*Module 34.2*).

      1. The biosphere includes the atmosphere to an altitude of a few kilometers, the land to a soil depth of a few meters, all lakes and streams, and the ocean to a depth of several kilometers.

      2. The biosphere is isolated by space and totally self-contained except for energy inputs from the sun and heat loss to space (and an insignificant input of solid matter in the form of meteorites).

      NOTE: The Gaia hypothesis, introduced by British scientist James Lovelock and American biologist Lynn Margulis, suggests that all organisms, the atmosphere, oceans, and soils make up a single self-regulating system of global scale.

      3. The biosphere is patchy on many levels. At each level, the patchiness is a result of patchiness of habitats (places where organisms live).

   B. Environmental problems reveal the limits of the biosphere (*Module 34.3*).

      1. Technological solutions, particularly those using chemicals, to agricultural and other problems during the last 50 years have had environmental impacts.

      2. One of the first observers to realize the global dangers of the use of agricultural chemicals such as DDT was zoologist Rachel Carson, who wrote *Silent Spring* in 1962.

      Preview: Environmental problems are discussed in more detail in Chapter 38.

      3. Environmental degradation, famine, and species endangerment and extinction have been tied to many human activities, including land misuse, expanding human population, incursions on natural habitats, and the poisoning of soil and water by toxic wastes.

III. **Factors of the environment to which organisms relate.**

   A. Chemical and physical factors sculpt the biosphere (*Module 34.4*).

      1. Solar energy powers all living activities, except for a few ecosystems dependent on chemical energy, such as hydrothermal vents. In nearly all environments, the availability of light is a critical factor that affects the distribution of photosynthetic organisms and their dependents. In aquatic environments, most photosynthesis occurs near the surface.

      2. Water is essential to all life. In aquatic environments, water balance must be maintained in diverse ionic concentrations. In terrestrial environments, water must be conserved.

      3. Temperature affects rates of metabolism and the functioning of enzymes. Metabolism can occur only above 0°C and mostly below 50°C, although some organisms can survive freezing or live at temperatures around boiling.

      4. Soil structure, pH, and nutrient content are important factors determining where plants and other soil-dwelling organisms can live.

      5. Oxygen is usually not limiting for terrestrial organisms but may be so in aquatic habitats.

6. Fires and other catastrophes are infrequent and unpredictable in most ecosystems. In grasslands and drier forests, fire may play an important recurring role in modifying the physical, chemical, and biological parameters.

7. Wind brings nutrients to some organisms, but it can cause damage, increase evaporation, and lower effective temperatures.

B. Organisms are adapted to abiotic and biotic factors by natural selection (*Module 34.5*).

Review: Evolutionary principles are discussed in Chapter 14.

1. Species exist in a given area because they evolve there or disperse there. In either case, unique adaptations that fit the local environment allow each species to live there.

2. Each organism can usually tolerate environmental fluctuations only within the set of conditions to which it is adapted. As an example, the pronghorn antelope is a highly successful herbivore that evolved on the open plains of North America.

3. The abiotic factors of its environment include extreme daily and seasonal temperature fluctuations, aridity, and wind. The pronghorn is well insulated by a coat of hollow hairs and can obtain all the water it needs from the vegetation it eats.

4. Some biotic factors to which the pronghorn is adapted include its diet of coarse grasses and woody shrubs and its predators—wolves, coyotes, and cougars. Pronghorns have teeth adapted for biting and chewing these plants, and a ruminant-type digestive system that depends on the chemical digestion of cellulose by bacteria. Pronghorns can escape their predators by means of great speed, are camouflaged, have keen eyesight, and increase their environmental awareness by living in herds.

C. Regional climate influences the distribution of biological communities (*Module 34.6*).

1. Because of Earth's curvature, different latitudes receive different amounts of solar energy. This uneven heating drives winds and water currents. Seasonal differences in temperature result from Earth's inclination on its axis (Masters 34.6A, B).

2. Uneven surface heating affects wind patterns and rainfall. Air over equatorial areas rises, forming clouds and rain, but no wind. High-altitude air masses spread to the north and south, descend at about 30° latitude, then flow on the surface both north and south. The area around 30° includes the world's major deserts because these sinking air masses are very dry. The trade winds flow toward the equator, generally easterly because of the high spin of Earth's surface at the tropics. The winds flowing through the temperate zones to the poles trend westerly because of the slower spin of the surface there (Masters 34.6C, D).

3. Ocean currents are produced from a combination of the prevailing winds and Earth's rotational spin. The flow of warm and cold currents can have a major effect on regions with water of contrasting temperature (Master 34.6E).

4. Landforms also affect local climate by influencing wind speed and precipitation. For example, moist air flowing off the Pacific Ocean across the mountains of Pacific northwestern North America produces, first,

relatively warm, extremely wet climates and then, in the rain shadow of the Cascade Range, continental climates that are almost desertlike (Acetate 34.6F).

IV. Aquatic ecosystems.

   A. Oceans occupy most of Earth's surface (*Module 34.7*).

   1. Evaporation from the ocean surface loads the air with moisture that provides most of the world's precipitation. Photosynthesis of marine algae provides more than 50% of the biosphere's fixed carbon and the oxygen.

   2. Estuaries are flat, shallow areas where broad expanses of fresh water meet seawater. They are among the most productive of all environments on Earth.

   3. The intertidal zone characterizes all areas exposed to fluctuations in tidal height. Most pronounced are the intertidal zones of temperate coastlines with rocky shores, which may include tidepools (Figure 34.7B, Acetate 34.7C).

   4. Intertidal zones are just one type of wetland, regions with characteristics intermediate between aquatic and terrestrial ecosystems. Most wetlands have permanently or periodically saturated soils.

   NOTE: Most wetlands have specialized vegetation, can be highly productive, or play pivotal roles in the maintenance of aquatic ecosystems nearby.

   5. Estuaries and intertidal zones are among the most threatened ecosystems as a result of overharvesting, pollution, and the removal of habitat by human development of coastal areas. Laws in many countries now severely regulate activities in these areas in an attempt to restore and conserve these regions.

   6. The pelagic zone of the open ocean includes all the water and is a major habitat for phytoplankton and zooplankton and highly motile marine invertebrates, fishes, and mammals. It is divided between the photic zone, where photosynthesis can occur and which is usually rich in life forms, and the aphotic zone, which can depend only on input of detritus from above and is generally less rich in life forms.

   7. The seafloor is called the benthic zone and is characterized by communities of invertebrates and fishes. Depending on its depth, it may include photosynthetic organisms.

   8. The photic zone is that portion of the ocean that is penetrated by light and in which photosynthesis occurs.

   9. Underlying the photic zone is the aphotic zone, the most extensive part of the biosphere. The aphotic zone includes a diverse array of life whose sources of energy include organic remains that sink from the photic zone and the hydrothermal vents discussed earlier (Opening Essay).

   B. Freshwater communities occupy lakes, ponds, rivers, and streams (*Module 34.8*).

   1. Light and the scarcity of dissolved inorganic ions have a major effect on ponds, lakes, rivers, and streams.

   Review: Osmoregulation (Module 25.5).

2. All but the most shallow ponds and lakes have photic and aphotic zones. Because of the lack of currents, environments in the aphotic and benthic zones may be low in, or lack, oxygen.

3. Temperature stratification has important effects on oxygenation and on organismal distribution in lakes and ponds, particularly in temperate zones. Stratification can impede the mixing of surface and deep water, except during times when surface temperatures are changing rapidly, during spring and fall.

4. The presence and distribution of inorganic ions (particularly nitrogen and phosphorus) can limit organismal growth and distribution. Increased amounts of nutrients affected by inputs from sewage or agricultural runoff can lead to phytoplankton blooms. When these algae die and decompose, a pond or lake suffers from oxygen depletion.

   NOTE: Of particular danger is the increase in respiration by blooms overnight. This can deplete oxygen on a daily cycle and lead to the suffocation of many associated organisms, including fish.

5. Streams and rivers support different communities than ponds and lakes. Near sources, water is usually cold, clear, and low in nutrients, and currents are swift. Near outlets, water is usually warmer, murkier, and high in nutrients, and currents are slower. These differences support different species.

V. The major terrestrial ecosystems are called biomes (*Module 34.9*).

A. There are nine major types of such biological communities (Acetate 34.9).

   1. Biomes are usually named for their predominant vegetation but are characterized by distinct groups of organisms from all kingdoms.

   2. Each biome is a type, not a distinct assemblage.

   3. Biomes are distributed in broad patterns across planet Earth. Some occur in wide bands. Others are found in widely separated areas. The assemblages found in the separated examples of the same type may show convergence of traits in evolutionarily unrelated forms of plants and animals.

      Review: Convergent evolution (Module 16.11).

   4. Biomes do not abruptly change from one to another, but grade into each other.

      NOTE: Such gradients are referred to as (eco)clines.

   5. Fires are important in maintaining certain biomes, such as grasslands.

      Preview: Module 36.7 discusses the role of fire in ecosystems.

   6. Many natural biomes have been broken up by human activity. Some people recognize "urban biome" and "agricultural biome" as separate types.

B. Tropical forests cluster near the equator (*Module 34.10*; Master 34.10).

   1. Temperatures are warm, and days are uniformly 11–12 hours long year-round. Rainfall is variable and defines subtypes of this biome.

   2. Tropical thorn forests are common in equatorial lowlands, such as eastern Africa and northwestern India. They are characterized by scarce rainfall. The plants found here are thorny shrubs and trees.

3. Tropical deciduous forests dominate areas where there are distinct wet and dry seasons of about equal length, such as central West Africa, India, and Southeast Asia. Tropical rain forests occur where the dry season lasts no more than a few months. The deciduous trees and shrubs of these forests drop their leaves during the dry season and releaf during the wet season.

4. Tropical rain forest is the most complex of all biomes, with very high diversity (particularly of plant species—as many as 300 tree species per hectare) and complex structure. Larger animals are tree dwellers, and there is little organic soil because of high decomposition and recycling rates.

5. The impact of humans on the tropical rain forest is currently of great concern. The deforestation of these areas may cause large-scale changes in world climate, as well as the loss of numerous species.

Preview: The consequences of tropical rain forest destruction (Module 38.14).

C. Talking About Science: U.S. Forest Service scientist Ariel Lugo shares his views on tropical forest issues (*Module 34.11*).

1. Dr. Lugo is a tropical ecologist who works in Puerto Rico. He tempers his concerns about the rain forest organisms with an understanding of the social and economic needs of the people who live in and use these forests.

2. The main force causing deforestation at present is not the need for lumber, but the need for land to grow food. Unfortunately, in the tropics, once the forest is cut, the remaining land can be farmed only a short time before it is infertile. Increasing populations in tropical areas also add to the problem.

3. Lugo believes that it is critical that land use in these areas be sustainable, and that this depends on the cultural awareness of the people living there. Any deforestation that occurs needs to be planned, controlled, and integrated into the ecosystem mechanisms in the area to avoid losing species and degrading soil and water resources.

D. Savannas are grasslands with scattered trees (*Module 34.12; Master 34.12*).

1. Savannas cover wide areas of the tropics in South America, central and South Africa, and temperate North America.

2. In temperate North America, the grasslands of the west merge with the temperate forests of the east.

3. Savannas are simpler in structure and lower in diversity than tropical forests and are characteristic of areas of low but consistent rainfall.

4. Frequent fires and grazing animals inhibit the invasion by trees and maintain the wind-pollinated grasses and nonwoody insect-pollinated dicots.

Preview: Module 36.7 discusses the role of fire in ecosystems.

5. The savanna is home to many of the world's largest herbivores and their predators. Particularly famous are those of Africa (e.g., giraffes, zebras, antelope, baboons, lions, and cheetahs) and Australia (e.g., kangaroos). Savanna habitat in North America (home to bison, deer, black bear, coyotes, and wolves) has largely been replaced by farms.

6. Other animals characteristic of this region are burrowing animals (e.g., mice, moles, gophers, snakes, ground squirrels, worms, and arthropods), which must go underground to find shelter in a region with few trees.

E. Deserts are defined by their dryness (*Module 34.13*; Master 34.13).

1. Daytime temperatures reach 54°C, and nighttime temperatures drop below freezing. Rainfall is less than 30 cm per year, and evaporation is rapid. Rainfall often occurs at one short time (or not at all, in some years).

2. Desert areas are centered at about 30° north and south latitudes in regions of descending, dry air masses. Most famous are the Sahara (where evaporation exceeds rainfall), Arabian, and Kalahari deserts and the desert areas of North America in Mexico and the southwestern United States. Areas in rain shadows of mountains in more temperate areas may also be deserts.

3. The size of many deserts is increasing, by desertification, principally of savannas because of the pressures of overgrazing and dry-land farming.

4. Cycles of growth and reproduction are keyed to rainfall. Desert plants are adapted to conserve water and to produce great numbers of seeds that may remain dormant for several years before rain triggers germination. Desert animals are adapted to drought and temperature extremes. Many are burrowers and seed-eaters (e.g., ants, birds, and rodents). Lizards, snakes, and hawks are important predators.

F. Spiny shrubs dominate the chaparral (*Module 34.14*; Master 34.14).

1. The climate of this biome results from cool, offshore ocean currents that produce mild, rainy winters and long, hot, dry summers.

2. Important regions of chaparral include regions around the Mediterranean, and in coastal Chile, southwestern Africa, southwestern Australia, California, and northwestern Mexico.

3. Perennial shrubs and annual plants are adapted to periodic fires and require occasional fires to maintain their overall structure.

Preview: Module 36.7 discusses the role of fire in ecosystems.

4. Characteristic animals include browsers (e.g., deer), fruit-eating birds, and seed-eating rodents. Lizards and snakes are important predators.

G. Temperate grasslands include the North American prairie (*Module 34.15*; Master 34.15).

1. Grasslands are mostly treeless. This biome occurs in regions with relatively cold winter temperatures, seasonal drought, occasional fires, and grazing by large mammals.

Preview: Module 36.7 discusses the role of fire in ecosystems.

2. Grazers prevent the establishment of woody shrubs and trees.

3. Temperate grasslands include areas known as veldts in South Africa, pampas in Argentina and Uruguay, steppes in Asia, and prairies in central North America.

4. The area covered by grasslands increased dramatically following the last ice age. In North America, little native grassland remains because the area is used for grain production.

5. The amount of rainfall determines the height of the grassland vegetation. In addition to grazing mammals (e.g., bison and pronghorn of North America, gazelles and zebras of Africa, and wild horses and sheep of Asia), common animals include burrowing animals and carnivores such as badgers, skunks, and foxes.

H. Deciduous trees dominate temperate forests (*Module 34.16*; Master 34.16).

1. Temperate deciduous forests occur between latitudes of 35° and 50° in regions where there is enough precipitation to support trees. The seasonal temperature range can be great, ranging from –30°C to 30°C.

   Review: These forests are threatened by global warming (Module 7.14).

2. Temperate forests are in the eastern United States, most of central Europe, and parts of eastern Asia and Australia, and a small amount in southern South America. In many temperate zones, human occupation of this biome has drastically altered the region.

3. Worldwide, species of broad-leaved trees vary. In the northeastern U.S., this biome has maples, oaks, hickory, beech, and birch. The diversity of species may be high, but not nearly as high as in tropical rain forests. Leaf drop conserves water during cold, sometimes dry winters.

4. Animal life is diverse because of the diverse habitats and includes mammals (e.g., whitetail deer, bobcats, foxes, black bears, and mountain lions), insects, and spiders.

5. Unlike drier areas, temperate forests tend to recover after a disturbance.

I. Coniferous forests are often dominated by a few species of trees (*Module 34.17*; Master 34.17).

1. The main trees of coniferous forests are cone-bearers (e.g., spruce, pine, fir, and hemlock).

2. Prior to human disturbance, much of the southeastern United States was dominated by coniferous forest.

3. Fire is essential to this biome. Without periodic fires, deciduous trees will tend to replace the conifers.

   Preview: Module 36.7 discusses the role of fire in ecosystems.

4. The taiga forms a broad band across North America and Asia. In more southerly regions, it is located at higher elevation, below alpine areas. There is little or no taiga in the Southern Hemisphere

5. The region is characterized by harsh winters, short warm summers, and considerable precipitation, often as snow. Snow insulates the thin and acidic soil during the winter.

6. Taiga is usually characterized by a few species of conifer, whose shape and resiliency protect them from heavy snow loads. Scattered deciduous trees (birch, willow, aspen, and alder) are common, especially near openings.

7. Animals are adapted to cold winters, often remaining active in tunnels under the snow in winter (e.g., mice). Squirrels and birds feed on conifer seeds. Browsers include deer, moose, elk, snowshoe hares, beavers, and porcupines. Predators include bears, wolves, lynxes, and wolverines.

>   NOTE: The northwestern North American coastal forests are a special area of extremely high precipitation, and at lower elevations, relatively mild winter temperatures.

- J. Long, bitter-cold winters characterize the tundra (*Module 34.18*; Master 34.18).
    1. Tundra is found from the northern edge of the taiga to the northern limits of land and encircles the North Pole. Alpine tundras occur at high elevation, above the treeline and below permanent rock and ice, throughout all climatic regions.
    2. Vegetation is dominated by low-growing herbs and shrubs, and lichens and bryophytes. During short but warm summers, vascular plants grow and flower quickly.
    3. The arctic tundra is characterized by permafrost, a permanently frozen layer of ground beginning a few meters deep and ranging from several meters to over a kilometer in depth. In many tundra areas, there is little precipitation and then mostly as snow. But poor drainage due to permafrost and slow rates of evaporation retain the moisture and keep the soil saturated.
    4. Animals are adapted to the cold by thick fur, or by migratory behavior. Large herbivores include musk ox and caribou. The main smaller animal is the lemming. There are a few predators, such as the arctic fox and the snowy owl. Many species found in the tundra, especially the birds, use it as a summer breeding ground. Huge populations of blood-eating insects, such as mosquitoes, develop in mid-summer, and then experience an abrupt decline in population size.

    Preview: Population dynamics (Chapter 35).

CLASS ACTIVITIES

1. Feature dominant species from your own local biome. Ask students to characterize the local species first, and then show them specific examples, such as boughs or leaves from vegetation, skeletons (particularly skulls and teeth) of important vertebrates, and representatives from other kingdoms, either slides or actual specimens. For each example, discuss how the adaptations observed fit the abiotic and biotic factors that predominate in your area.

RESOURCES AND REFERENCES

Abrahamason, W.G., T.G. Whitham, and P.W. Prince. "Fads in Ecology." *BioScience*, May 1989. Changing ideas in a dynamic field.

Barlow, C., and T. Volk. "Gaia and Evolutionary Biology." *BioScience*, October 1992. A review of the current thoughts about the controversial idea that Earth itself functions as a self-regulating organismlike system.

"Beating the Heat." *Natural History*, August 1993. A collection of articles describing adaptations of animals from around the world to protect against overheating

Begon, M., J.L. Harper, and C.R. Townsend. *Ecology*, 2nd ed. New York: Blackwell Scientific, 1990.

Burman, A. "Saving Brazil's Savannas." *New Scientist*, March 2, 1991.

Childress, J.J., H. Flebeck, and G.N. Somero. "Symbiosis in the Deep Sea." *Scientific American*, May 1987. A description of the communities of organisms dependent on chemosynthetic bacteria mutualistic with tube worms in the deep sea.

"Ecology of Large Rivers." *BioScience*, March 1995. A special issue.

"Ecology from Space." *BioScience*, July/August 1986. A special issue that includes a number of articles featuring the roles of satellite imagery and data collection on the description and assessment of Earth's biomes and species.

Goulding, M. "Flooded Forests of the Amazon." *Scientific American*, March 1993. Parts of the vast rain forest are as much aquatic as terrestrial ecosystems. Unique adaptations allow creatures to thrive in these inundated woods.

Grassle, J.F. "Deep-Sea Benthic Biodiversity." *BioScience*, July/August 1991. The ocean bottom supports communities that may be as diverse as those of any habitat on Earth.

Hadley, N.F. "Desert Species and Adaptation." *American Scientist*, May/June 1972. Convergent evolution in the characteristics of plants and animals in arid environments.

Jackson, J.B.C. "Adaptation and Diversity of Reef Corals." *BioScience*, July/August 1991. Patterns result from species differences in resource use and life histories and from disturbances.

Kasting, J.F. "Earth, the Living Planet: How Life Regulates the Atmosphere." *The Planetary Report*, January/February 1990. A brief description of the mechanisms described for the Gaia hypothesis.

Kelly, D. "The Decadent Forest." *Audubon*, March 1986. The unique biological components and economic and sociological dilemmas involved in the old-growth coniferous forests of Pacific northwestern North America.

The Living Ocean. Washington, DC: National Geographic Society, 1988. A 25-minute videotape that explores the ocean's roles in the biosphere.

Monastersky, R. "The Cold Facts of Life." *Science News*, April 24, 1993. Adaptations to life in the antarctic as seen in several different organisms.

Nolan, R. *Coral Kingdom*. Scotts Valley, CA: Sunburst Wings for Learning, 1993. A CD-ROM using digitized audio and a huge database of coral reef organisms to document the ecosystem, organism adaptations, energy cycling, and human impact on coral reef environments. With Macintosh or IBM PC computer interfaces.

Odum, E.P. "Great Ideas in Ecology for the 1990s." *BioScience*, July/August 1992. This notable ecologist's twenty basic concepts for improving environmental literacy.

On Dry Land: The Desert Biome. Deerfield, IL: Coronet Film and Video, 1992. A videodisc with Macintosh or IBM PC computer interface that provides film footage and numerous stills on desert terrain and life forms.

Rain Forest. Washington, DC: National Geographic Society, 1983. A 59-minute videotape that details the organisms and interconnections in the tropical rain forest biome. Also available from Carolina Biological Supply Company.

Ray, G., and J. Grassle. "Marine Biology Diversity." *BioScience*, July/August 1991. Why we need a program for conserving marine communities.

Ricklefs, R.E. *Ecology*, 3rd ed. New York: Chiron Press, 1986.

Robinson, B.H. "Light in the Ocean's Midwaters" *Scientific American*, July 1995. Bioluminescence is a common adaptation in deep seas.

Stevens, J.E. "The Antarctic Pack-Ice Ecosystem." *BioScience*, March 1995.

Williams, T. "The Incineration of Yellowstone." *Audubon*, January 1989. A special report on the responses of organisms and humans to the fires in Yellowstone National Park during summer 1988.

Wuethrich, B. "Forests in the Clouds Face Stormy Future." *Science News*, July 10, 1993. A brief discussion of the high diversity in—and extreme sensitivity of—high-altitude cloud forests in tropical latitudes.

INTERNET RESOURCES FOR UNIT VII

CIA World Fact Book

http://www.odci.gov/cia/publications/pubs.html

Demographics, climate, geography, and so on, of just about every country in the world. The demographic information can be especially useful for Chapters 35 and 38.

EcoNet Home Page

http://www.igc.apc.org/econet

The International Wolf Center

http://www.wolf.org

New Jersey OnLine's Yucky Site: Worm World

http://www.nj.com/yucky/worm/index.html

Juvenile, but some useful information on worm ecology and anatomy.

The Rainforest Workshop

http://164.116.102.2/mms/rainforest_home_page.html

Terraquest

http://www.terraquest.com

Virtual field trips to the Galápagos and Antarctica.

U.S. Environmental Protection Agency

http://www.epa.gov

A particularly useful link for information concerning the environmental impact of humans.

Welcome to Cockroach World

http://www.nj.com/yucky/roaches/index.html

Juvenile, but some useful information on cockroach ecology and diversity.

World Population

http://sunsite.unc.edu

Follow "Alphabetic Index" link to "World Population Counter" link, a counter that keeps track of the size of the human population.

The World-Wide Web Virtual Library: Biodiversity, Ecology, and the Environment

http://conbio.bio.uci.edu/link

Many conservation-related links, including endangered species, habitats, legislation, pollution, exotic introductions, extinct species list, global warming, ozone, old-growth forest, value of biodiversity, etc.

The World-Wide Web Virtual Library: Environment

http://ecosys.drdr.virginia.edu/Environment.html

35

POPULATION DYNAMICS

APPROACH

Mechanisms describing demographics and population growth are critical for understanding ecological limits. Students may have more familiarity with the topics in this chapter because they are often the subject of courses in economics and anthropology. Human population growth puts invasive pressure on ecosystems, and population criteria of native species are responding to these pressures.

Ecologists work with mathematical models, which are introduced here, particularly in the modules on population growth. In this context, introduce the concept of a model as an idealized statement of a hypothesis. Models are particularly useful in ecological studies because they can simplify the numbers of often confusingly interrelated parameters. You may need to clarify the graphs and equations introduced in Module 35.3.

A growing aspect of population study is the population biology of plants and other clonal organisms. You might want to research this area and provide your own examples from nonanimal kingdoms.

CHAPTER OBJECTIVES

Define *population* in the sense used by population ecologists, and in contrast to the traditional biological definition of population.

Explain how density and dispersion describe populations and how each of these variables is measured.

Distinguish between exponential and logistic models of population growth, explaining the effects of existing population size and carrying capacity on growth rate.

Differentiate between density-dependent and density-independent factors that limit population growth.

Describe the interrelationships among predator, prey, and prey food that cause mixed populations to cycle.

Name the attributes of life histories, and distinguish between opportunistic and equilibrial life histories.

Discuss life tables and age structures, and examine them for human populations. Compare a stable population, such as Sweden's, with a population with an explosive growth rate, such as Mexico's.

Outline the history of the growth of the human population, including factors affecting that growth.

LECTURE OUTLINE

I. Introduction.

   A. Population studies on starlings (*Opening Essay*).

      1. Originally native to Europe and Asia, starlings are now abundant and destructive pests across North America.

2. Starlings were brought to New York City by a private group of citizens for arbitrary reasons at a time when importing foreign species was not regulated. Without native predators, parasites, or competitors, the starling populations rapidly spread across the United States (Master 35.0).

3. Such uncontrolled population growth has features in common with that of the human population on a global scale.

    NOTE: In general, this picture is also one seen commonly wherever nonnative species are introduced. Other examples include the spread of Dutch elm disease and American chestnut blight.

B. Population dynamics is concerned with the changes in population size and structure over time and the factors that regulate these changes.

II. Characteristics of populations.

A. Populations are defined in several ways (*Module 35.1*).

Review: Populations and their place in the hierarchical organization of life are first discussed in Module 1.1.

1. Biologists define populations as interbreeding groups of individuals of a particular species that are more or less isolated from other such groups. Some ecologists use a less rigid definition and consider a population to be a group of individuals of a species that use common resources and are regulated by the same natural phenomena.

2. Ecologists might restrict their definition to just the individuals in a very small but contained area, such as all the sea anemones of one species in one tidepool. Or they may expand their view to include all the individuals over the face of the Earth—for instance, the human population as it is exposed to the disease-causing HIV.

B. Density and dispersion patterns are important population variables (*Module 35.2*).

1. These variables can be compared between populations occupying different areas to contrast growth, stability, or other parameters.

2. Density is the number of individuals of a species per unit area or volume.

3. Density is usually measured by counting the number of individuals in a subsample and estimating the number (relative to a unit of volume or area) in the whole population. Another way to estimate density for animals is to use the mark-recapture method. A number of individuals from a population are trapped, counted, marked, and released. In a few weeks, traps are reset in the same population. The proportion of marked animals in the new sample gives an estimate of the total population size, assuming that the marked and unmarked animals are trapped with equal frequency.

    $$N = \frac{M_o \times T}{M_r}$$

    N = estimate for total population, M_o = number originally marked, T = total catch from the second trapping, M_r = total marked in second trapping.

4. Dispersion pattern refers to how the individuals in a population are spaced within their areas.

5. A clumped pattern shows local aggregations in patches, usually resulting from an unequal distribution of resources for plants and animals or from associations with mating and social behavior for animals.

NOTE: An example of a clumped distribution is the higher concentration of humans found in and near cities.

6. A uniform pattern shows an even distribution over an area and usually results from interactions among individuals, such as competition for resources by plants or territorial behavior of animals.

    Preview: Territoriality, which can be responsible for a uniform population distribution, is discussed in Module 37.17.

7. A random pattern shows a patternless, unpredictable distribution. Such patterns are rare but might be seen in, for example, clams distributed across a sandy ocean bottom, where the factors affecting them are numerous and complexly interrelated (Figures 35.2B, C).

8. Dispersion estimates are calculated from statistical (mathematical) descriptions of the spacing of individuals in a population.

III. Population growth.

A. Idealized models help us understand population growth (*Module 35.3*).

1. Mathematical equations of population growth provide useful starting points for studying populations and have stimulated many experiments and controversies. For the following two models, these abbreviations are used: G = growth rate, N = number of individuals in the population at the time the growth rate is studied, r = intrinsic rate of increase (estimated by subtracting the death rate from the birth rate), and K = carrying capacity (the maximum number of individuals of a particular species that can be supported by a particular environment).

2. Exponential growth model. This models the growth of a population under ideal conditions with unlimited resources. The rate of growth is exponential and depends on the number of individuals in the population:

$$G = rN$$

    The graph shows a J-shaped curve, representing population size increasing without limit. As N increases, so does G. This type of growth, if exhibited by a bacterium growing in an unlimited environment, would result in an inconceivably large number of bacteria in less than two days (Acetates 35.3A, C).

    NOTE: The graphs show different population sizes at different times. The equations relate growth rate to time. This may not pose a problem for nonscience majors, but generally an equation given with a graph represents that graph, so be sure to clarify this potential point of confusion. You may want to point out that the growth rate is represented by the slope of the curve in each graph.

3. In nature, population growth is rarely, if ever, best modeled by an exponential growth equation. There are always factors in the environment that limit population growth.

4. Logistic growth model. As with the exponential growth model, this is an idealized model of population growth. However, the logistic model takes limits to population growth into consideration. The rate of growth is exponential in the beginning, but limited by how close the population size (N) is to a critical size, the carrying capacity (K):

$$G = rN\,(K-N)/K$$

The graph shows a lazy-S-shaped (logistic) curve. At first population size increases slowly; N is very small relative to K, and $(K-N)/K$ is near 1; then the population size increases rapidly. As the population size approaches K, the growth rate slows down; as N approaches K, the value of $(K-N)/K$ approaches zero. This type of growth is typical of all organisms growing in limited environments. For example, actual data from a population of fur seals follow this model (Master 35.3B).

NOTE: The discussion of the model predicting different growth rates at different times, as noted above, can be clarified by again noting that growth rate is represented by the slope of the curves.

5. The growth rate of a population results from the combination of birth rate and death rate at any one time. Growth rate rises if birth rate rises or death rate falls. Growth rate falls if birth rate falls or death rate rises.

NOTE: Assuming there is no emigration or immigration (no gene flow).

B. Density-dependent and density-independent factors limit population growth (*Module 35.4*).

1. Density-dependent factors affect a greater percentage of individuals as the density of individuals (number per unit area) increases. Food supplies that become limited and the buildup of toxic wastes often depress growth rates by increasing the death rate, decreasing the birth rate, or both (Acetate 35.4A).

2. An example of the action of density-dependent factors is the leveling off of numbers of fruit flies (or any cultured organism) in a closed, laboratory environment.

3. Clear-cut cases of density-dependent factors operating in nature are sometimes hard to determine because of the many conflicting factors. Managed deer populations may show some density-dependent effects on growth (the tendency to produce twins).

4. Density-independent factors limit population size, no matter what the size, and are often abiotic factors such as fires, floods, storms, seasonal changes in temperature or moisture, or disruption by human activity (Master 35.4C).

5. An example of the action of density-independent factors is the exponential growth then sudden decline in mid-summer of populations of leaf-sucking aphids, due to drying conditions.

6. Over the long term, most populations are regulated by a mixture of density-dependent and density-independent factors, and the distinction between these two types may not be clear.

C. Some populations have "boom-and-bust" cycles (*Module 35.5*).

1. Populations of predator and prey often show periodic cycles, such as the 10-year cycle for the lynx and the snowshoe hare in the taiga of North America. Note that the cycles of both animals have about a 10-year period (Master 35.5).

NOTE: The numbers of prey are much larger than those for predator (note the two different Y axes), relative differences that hold true for these two different trophic levels across community types (see Module 36.11). Also notice that the hare population generally peaks before the lynx, and once the hare population falls, the lynx population follows. This leads into the following discussion.

2. For the lynx and many predators, the availability of prey often determines population changes.

3. For the hare, however, the force of predation may not be the determining factor. Hare populations cycle whether or not lynxes are present. Recent studies have shown that the 10-year cycle of the snowshoe hare population is due to both the effects of predation and fluctuations in the hare's food supply.

4. Other potential density-dependent factors suggested for population cycles (in addition to food supply and predation) include stress from crowding (e.g., lemmings).

   NOTE: Graphs relating parasites to their hosts often show similar cycles.

IV. Life histories and demography.

A. Evolution shapes life histories (*Module 35.6*).

1. Aspects of life history that influence growth rate include age of first reproduction, number of offspring, and amount of parental care. Each of these traits is shaped by natural selection.

2. An example is given of two different guppy populations evolving under two different regimes of predation. Killifish and cichlids both prey upon guppies. Killifish eat mainly small immature guppies. Cichlids eat mainly large, mostly mature, guppies. When preyed upon by cichlids, guppies tend to be smaller, mature earlier, and produce more offspring than they do in areas without cichlids. The heritability of these life history traits is demonstrated by the retention of these characteristics over the course of several generations in predator-free environments. Another experiment involved taking guppies from populations preyed upon by cichlids, removing them from that environment, and subjecting them to a pattern of predation in which small guppies were preyed upon. The results were that within 2 years the female guppies matured later and produced fewer larger offspring (Figure 35.6A).

   Review: These experiments are good examples of the scientific approach to problem solving (Modules 1.2 and 1.3) and the effects of directional selection (Module 14.20).

3. Opportunistic life histories are characterized by reproducing early in life, yielding many offspring, which generally exhibit high mortality rates, and populations showing exponential growth during favorable times. Such populations live in unpredictable environments and are controlled by density-independent factors. Examples include dandelions and molds (Figure 35.6B).

   NOTE: Such opportunistic species are said to be *r*-selected. An interesting piece of trivia: Most dandelions are parthenogenic and produce seed without fertilization, so their populations are of mixed clones.

   Preview: Opportunistic species are often the first recolonizers in secondary succession (Module 36.6).

4. Equilibrial life histories are characterized by reproducing later in life, yielding fewer offspring, which generally experience low mortality rates, and populations that are stable over time. Such populations are held near carrying capacity by density-dependent factors. Examples include many larger vertebrates and coconut palms (Figure 35.6C).

   NOTE: Such types are said to be *K*-selected.

B. Life tables track mortality and survivorship in populations (*Module 35.7*).
   1. Life tables relate these rates to size classes, usually grouped by decades for human life tables (Table 35.7).
   2. Data from life tables can be graphed to point out three basic survivorship curves (Acetate 35.7).
   3. Type I is characteristic of species (whales, elephants, and humans) that have low birth rates, low infant mortality, and life histories that fit the equilibrial model.

   NOTE: These are examples of *K*-selected species.

   4. Type II is characteristic of intermediate species (squirrels and *Hydra*).
   5. Type III is characteristic of species (oysters and sea lettuce) that have high birth rates, high infant mortality, and life histories that fit the opportunistic model.

   NOTE: These are examples of *r*-selected species.

C. A population's age structure indicates its future growth trend (*Module 35.8*).
   1. A typical population has three main age groups: prereproductive, reproductive, and postreproductive. In each diagram, all the bars add up to 100, the vertical "steps" represent age groups by 5-year intervals, and the left and right halves divide the population by gender (Acetate 35.8A).
   2. Tracing the U.S. population structure from 1960 to 1975 to 1990 shows how each age group moves up the graph as that age group gets older. The overall changes in shape of such distributions also help point out details about the population, such as the population "spurt" known as the "baby boom" of people born in the years following World War II.
   3. Human population age structures can also show differences in social conditions, as in the comparison of age structures of the populations of Sweden and Mexico. Sweden has a stable structure and relatively low birth rate, with older people continuing to represent a larger portion of the entire population than even the United States. Mexico's population structure shows the effects of a predominantly young population and an explosive population growth (Acetate 35.8B).

V. Human populations.

A. The human population has been growing very rapidly for centuries (*Module 35.9*).
   1. The worldwide human population grew nearly linearly during the very early years, up until it was about 500,000,000 in 1650. Between 1650 and 1850 (200 years), the population doubled to 1 billion. Between 1850 and 1930 (80 years), the population doubled to 2 billion. Between 1930 and 1975 (45 years), the population doubled to 4 billion. In 1993, the population was 5.6 billion. At present rates, there will be about 8 billion people on Earth by 2017; that is, about 1.5 times as many as there are today (Acetate 35.9B).
   2. Human population growth depends on birth and death rates. The exponential rate of growth has resulted in part from decreasing death rates, particularly infant mortality. Longer lives and decreased infant mortality have resulted from technological, not biological, attributes of humans—such as improved nutrition, health care, and sanitation. Recently in developed countries, birth rates have dropped, slowing the effects of the changes in death rates on overall growth.

3. The effect of decreasing mortality on population growth is compounded in most developing countries, which still have relatively high birth rates (Masters 35.9C, D).

4. Because no population can continue to grow forever in a limited environment, the human population *will* stop growing. The question is, at what carrying capacity or as a result of what unforeseen biotic or abiotic catastrophe? Fortunately, or hopefully (depending on one's view of human nature), humans can deliberately choose at what level to control the human population.

   NOTE: Regarding carrying capacity, debate must also concern the quality of life of humans and of the remaining species on Earth (Chapter 38).

B. Talking About Science: What does the future hold? Two opposing views (*Module 35.10*).

   1. Population biologist Paul Ehrlich regards human population growth as the single biggest problem facing humanity. Ehrlich is convinced that the economic mechanisms championed by Julian Simon and his followers are likely to dangerously deplete natural resources, and that ecological crises involving things like acid rain, global warming, and ozone-layer depletion dangerously imperil humans and the entire biosphere.

   2. Economist Julian Simon feels that humans will manage to avert catastrophe because of their innate ingenuity as problem solvers. He maintains that none of Ehrlich's doomsaying has come to pass and that humans will always find ways to avoid ecological calamities.

CLASS ACTIVITIES

1. Use the classroom population to demonstrate population density and dispersion and the procedure of estimating density based on a sample. Use various numbers of rows of students to represent the subsample, and calculate density each time, showing that the smaller the sample, the more variable the estimate. Patterns of dispersion may be obvious. The seats themselves are uniformly dispersed. Students may form aggregations (depending on how full the room is). You may want to ask students what aspects of their population (gender, distance from speaker, comfort of chairs, etc.) cause the patterns observed.

2. Modules 35.8 and 35.9 should provoke much discussion and debate: Discuss the economic impact of an aging population—fewer working individuals supporting a larger number of retired individuals. What, if any, is the carrying capacity of the planet for humans? What, if anything, should be done to limit human population growth? What are the moral and ethical issues involved? Ask students how they think these issues will affect their quality of life.

RESOURCES AND REFERENCES

Ackerman, L., et al. "The Successful Animal." *Science 86*, January/February 1986. Cultural and historical aspects of human population growth.

Baskin, Y. "Blue Behemoth Bounds Back." *BioScience*, October 1993. Population figures for the world's largest creature differ, but the endangered animal appears to be inching away from extinction.

Biology Explorer: Population Ecology. E. Arlington, MA: Logal Software, 1991. Part of a series of interactive, experimental simulations, this Macintosh computer program allows the user to make hypotheses, plan experiments, observe results, and draw conclusions about the various models of population growth.

Communities. Populations. Succession. Deerfield, IL: Coronet/MTI Film and Video, 1992. A 44-minute videodisc that distinguishes these facets of ecological study by providing living examples of the processes and organisms involved.

Daily, G.C., and P.R. Ehrlich. "Population, Sustainability, and Earth's Carrying Capacity." *BioScience*, November 1992.

Dasgupta, P.S. "Population, Poverty, and the Local Environment." *Scientific American*, February 1995. Child labor and the population explosion.

Ehrlich, P. *Paul Ehrlich's Earth Watch*. Pleasantville, NY: National Broadcasting Company, 1991. A 17-minute videotape that tours three of planet Earth's biggest environmental problems: the population explosion, extinction, and global warming. Available from Carolina Biological Supply Company.

Kates, R.W. "Sustaining Life on the Earth." *Scientific American*, October 1994. The cultural context of environmental problems.

Population Ecology. E. Arlington, MA: Logal Software, Inc. Computer simulations of population growth curves, competition, predation, equilibrium, adaptation, and systems management.

Reiners, W., W. Glanz, and S. Cornish. *Ecological Modeling*. Iowa City, IA: Conduit, University of Iowa, 1986. Computer software for the IBM PC that provides eight programs for teaching techniques for modeling population-level and ecosystem-level processes and systems. Available from Carolina Biological Supply Company.

Savonen, C. "One Salmon, Two Salmon . . . 10,000 Salmon: Counting the Fish in Alaska." *Oceans*, January/February 1985. A population density determination in action.

36

COMMUNITIES AND ECOSYSTEMS

APPROACH

This chapter lies at the heart of the ecology unit and introduces some of the most important biological concepts. Concepts involving the functioning of communities and ecosystems define the limits imposed by nature on ecosystem resilience, the structure and economics of human and other food chains, and the cycling and effects of pollutants.

Clarify the distinction between community and ecosystem. The term *community* can be used in the discussions of ecosystem (as happens in the latter part of the chapter) if one is dealing with interorganismal functions that do not involve the abiotic environment.

It may be challenging for some students to grasp the workings of, and implications of, ecological succession. Visiting or showing slides of local examples of community succession will help. Be sure to emphasize the roles of the organisms themselves in causing successional changes, and describe the changes as slow and incremental, not jumps from one phase to the next. Also, stress the changes in community function that occur during succession.

Popular attention to carnivores at high trophic levels has hidden some of the inner workings of food chains and energy pyramids. The efficiencies of energy transfer between low trophic levels are usually not detailed. The point made in Module 36.12, about the luxury of meat-eating for humans, should require little explanation. Although it may not convert any students to vegetarianism, it should provide understanding about the inefficiencies of energy transfer between trophic levels.

CHAPTER OBJECTIVES

Distinguish between the terms in the following pairs: *community* and *ecosystem*, *habitat* and *niche*, *coevolution* and *symbiosis*, and *energy flow* and *chemical cycle*.

List the properties that are used to compare different communities.

Discuss the forces that tie populations together into communities: competition, predation, and symbiosis. Give an example of each, and explain why it is difficult to fully assess these forces in natural communities.

Describe the process of community succession using a local example, and trace the changes in community parameters that occur during the process.

Define the trophic levels that occur in most ecosystems, and discuss how energy flow through trophic level structure results in an energy pyramid.

Outline the dominant pathways of the cyclic movement between organic matter and abiotic reservoirs of water, carbon, nitrogen, and phosphorus.

Explain how human disruption of ecosystems disturbs chemical nutrient cycles.

LECTURE OUTLINE

I. Introduction.

    A. A special food chain (*Opening Essay*).

1. Some insects go to extremes of interdependence. For example, chalcid wasps parasitize the eggs of ichneumon wasps, which have been laid in *Apanteles* wasp eggs, which have been laid in *Pieris* (cabbage butterfly) caterpillars (which eat cabbages) (chapter-opening photo; Figure 36.0).

2. One would be tempted to think such food chains could go on forever, as suggested by Jonathan Swift's poem about fleas on fleas. However, there are good reasons why this cannot be so.

B. This chapter focuses on two interrelated levels of ecology:

1. Community structure and function depend on the interactions among organisms.

2. Ecosystem structure and function depend on the interactions of the community with its abiotic environment.

II. Community structure and function.

A. A community is all the organisms inhabiting a particular place (*Module 36.1*).

1. Communities, like populations, are defined by sets of properties.

   Review: The hierarchical organization of life (Module 1.1).

2. Diversity refers to the variety of species present and has two components: richness (the number of different species) and relative abundance (the numbers of individuals of each species). A community with individuals divided equally among four species is very different from a community with the same species unequally represented.

3. The nature of the dominant organisms is an important property. This can involve either vegetation (in terrestrial communities, and then the structure of different plant "layers" or "clusters" is also important) or animals (more important in aquatic communities). An example is the dominant trees in a forest and the structure of a tree canopy over shrubs and herbs.

   NOTE: Another example is the dominance of heath-forming herbs, mosses, and lichens in alpine tundras and the clumps of different vegetation groups.

4. Community stability is the ability to resist change and return to the original species composition and structure following a disruption.

5. *Preview:* Another property of communities is their trophic structure, the nutritional relationships among all the components (Modules 36.9–36.11).

6. There are several forces that tie populations together in communities: competition, predation, and symbiosis.

B. Competition may occur when a shared resource is limited (*Module 36.2*).

1. *Review:* Population growth models (Module 35.3).

2. Interspecific competition between two species can inhibit the growth of both populations, sometimes to the point where one is eliminated.

3. In the 1930s, Russian ecologist G. F. Gause studied interspecific competition between two species of protozoans, *Paramecium aurelia* and *P. caudatum*. When grown in separate cultures with bacteria constantly added as food, each species' population followed a logistic growth curve. When grown together, *Paramecium aurelia* eliminated its competitor (Acetate 36.2A).

4. A population's niche is its role in its community, the total of a species' relationships to the biotic and abiotic factors of its habitat. The concept of niche is a theoretical construct that is difficult to assess in nature.

   NOTE: Compared to an organism's habitat (its address), its niche is like an organism's occupation. Although this analogy is not perfect, it clarifies the fact that *habitat* and *niche* refer to different characteristics of a species.

5. The ideas embodied in Gause's experiments have been termed the competitive exclusion principle. In modern wording, this principle states that populations of two species cannot coexist if their niches are identical.

6. However, it is often hard to determine whether interspecific competition has ever been involved between two species, and, if so, what has happened as a result to avoid the consequences of the competitive exclusion principle (extinction, migration, or natural selection to modify the species' niches).

7. Niche and interspecific competition are demonstrated by two species of suspension-feeding barnacles in the genera *Chthamalus* and *Balanus*. *Chthamalus* normally lives higher on intertidal rocks, but if *Balanus* is removed, *Chthamalus* will live lower. Ecologists conclude that *Chthamalus'* lower boundary is determined by competition, and the upper boundaries of both species are determined by how well a species resists drying out (Master 36.2B).

C. Predation leads to diverse adaptations in both predator and prey (*Module 36.3*).

   1. The predator is the eater and the prey is the eaten (including plants). No species is entirely free of predation, at least when young.

   2. Defense mechanisms against predators include size, fleeing, hiding, mimicry, and the production of defensive structures (spines, armor) or chemicals (in plants, strychnine, nicotine, and mescaline; in animals, skunk sprays and alkaloids in poison-arrow frogs). These mechanisms evolve by natural selection as predators and prey interact (Figures 36.3B, C).

      NOTE: In addition, predators (and parasites) often remove the weaker or infirm individuals of their prey, thereby helping improve the prey's genetic stock or, at least, the overall health of the prey population, and helping keep the prey population size below levels at which it may outstrip its food supply.

   3. Coevolution occurs when both predator and prey sequentially evolve reciprocal adaptations, each countering changes in the other. An example of coevolved features is the evolution of toxic chemicals in the leaves of *Passiflora* that protect the plant from consumption by most insects, and the coevolution of special enzymes in the butterfly *Heliconius*, which can eat *Passiflora*. Furthermore, some species of *Passiflora* produce sugar deposits on leaves that look like *Heliconius* eggs, thereby causing the butterfly to avoid laying eggs on them (Figure 36.3A).

   4. Mimicry protects the mimic (of a distasteful or dangerous species) from consumption. Batesian mimicry is shown by a harmless species (in lower numbers) mimicking a distasteful species. Such noxious species are often brightly colored. Müllerian mimicry is shown by two unpalatable species, each of which gains advantage from the other warning predators (Figures 36.3D, E).

D. Predation can maintain diversity in a community (*Module 36.4*).

Communities and Ecosystems

1. In laboratory situations, predators can sometimes consume all their prey and then die off from starvation. An example is the voracious protozoan *Didinium*, which will consume all of its *Paramecium* prey (Master 36.4A).

2. This rarely happens in nature because natural communities are complex, because predators are preyed upon, and because predators usually switch prey when the numbers of one prey species are too low to find easily.

3. Keystone predators help maintain community diversity by consuming, and reducing the density of, prey that are very strong competitors. This was shown in the 1960s in experiments by American ecologist Robert Paine. If the predatory sea star *Pisaster* is removed from a community, its main prey, *Mytilus* (a mussel), outcompetes many other shoreline organisms (e.g., barnacles, snails) (Figure 36.4B).

E. Symbiotic relationships help structure communities (*Module 36.5*).

1. A symbiotic relationship is an interaction between two or more species in which one or more species lives in or on another.

   NOTE: The distinction between each of the following categories of symbiosis is inexact. It is often difficult to determine the precise nutritional or other benefits provided by, or harm caused by, a symbiotic relationship. The categories are derived by ecologists to simplify discussion of some complex interrelationships.

2. Parasitism is much like predation, but the parasite is usually smaller than the host species. The parasite gains and the host loses from the relationship. Natural selection favors parasites that have adaptations to find and feed on hosts. Coevolution of host and parasite often takes place so that hosts have adaptive defenses against parasites. Such coevolution is seen in the rabbit (nonnative) in Australia and the myxoma virus, introduced to control the rabbit population. After 40 years of coevolution, rabbits are better able to resist infections, and the most virulent virus strains are absent, having died off with the rabbits they killed (Figure 36.5A).

   NOTE: The worst harm done by parasites is on nonnative hosts, which have not coevolved with the parasite. The fungal pathogens that cause Dutch elm disease and chestnut blight are two such parasites that have ravaged native tree populations in North America.

3. Commensalism involves close relationships between organisms in which one benefits and the other neither gains nor loses. An example is the relationship between certain insect-eating birds and grazing cattle. The cattle flush out insect prey for the birds. It is not clear what, if anything, the birds do for the cattle.

   NOTE: Another example is an owl roosting in an abandoned woodpecker hole. In commensal relationships, the neutral species may be neutral only in our present understanding, but may, in fact, gain or lose from the relationship.

4. Mutualism benefits both partners. Nitrogen-fixing bacteria gain a home, and their nodule-forming legumes gain nitrogen (Module 32.11). The acacia tree gains protection from plant pests by catering to its *Pseudomyrmex* inhabitants. These ants live in hollow thorns and eat sugar and protein-rich swellings as the tree grows. If the ants are removed, the trees die (Figure 36.5B).

NOTE: The various symbiotic relationships discussed in this module can be discussed in terms of + (benefit), – (harm), and 0 (no effect) relationships between the two populations involved. For example, parasitism is a + – relationship. The parasite benefits, and the host is harmed. Predation is also + –; commensalism is + 0; mutualism is + +; and competition is – –. Another possible type of symbiotic relationship is amensalism. Amensalism, a – 0 relationship, is very rare in nature. An example of amensalism may be bacteria killed by the mold *Penicillium*, the fungus from which the antibiotic penicillin was first derived. An example of mutualism involving humans is the presence of bacteria in the human GI tract: humans supply a nice place to live with plenty of food, and the vitamin K that the bacteria produce can meet up to 50% of the human's need.

F. Outside disturbances can radically alter community structure (*Module 36.6*).

1. Ecological succession is the transition in the species composition in a community over time, following the destruction of the original community by flood, fire, glacial retreat, or other natural or human disturbance.

   NOTE: In succession it is important to emphasize that it is the organisms themselves that influence the transition in both the biotic and abiotic factors.

2. Primary succession occurs in virtually lifeless areas that have no soil. An example is a newly formed volcanic island. Frequently, the first colonizers are autotrophic bacteria. Often, the first large photosynthesizers found on the barren ground are lichens and mosses. The decomposition of these organisms gradually forms soil. Once soil is present, the lichens and mosses are overgrown by other plants, such as grasses, shrubs, and trees from seeds that have been blown in or carried in by animals. This gradual process of succession to the community making up the biome can take hundreds or thousands of years.

   NOTE: One of the most thoroughly documented examples of primary succession is the recolonization of the island of Krakatau. In August 1883, following a series of eruptions, half of Krakatau disappeared and the other half was buried under lava and ash. Another example of primary succession would be the land exposed by a retreating glacier.

3. Secondary succession occurs when a previously existing community was destroyed but the soil was left intact. An example is a forest that was cleared for farmland and was later abandoned; this can be seen in areas of the eastern U.S. The earliest species to colonize are usually herbaceous plants with opportunistic life history patterns (Module 35.6). If the area remains undisturbed, woody shrubs may replace the herbaceous plants. The woody shrubs may, in turn, give way to forest trees.

4. Today, humans are the single greatest factor affecting community change and succession. Examples can be seen in overgrazing in parts of Africa resulting in desertification. Logging in the U.S. and Europe has reduced large tracts of forest to patchy woodlands. The grasslands of the midwestern U.S. have been converted to farmland. The tropical rain forest in the Southern Hemisphere is being cleared for lumber and pastureland.

   Preview: Human impact on the environment (Modules 38.8–38.16).

5. In some areas, succession proceeds to a self-perpetuating climax community. In other areas, the rate of natural disturbance is frequent

enough that a climax community type is never reached. This is true for grasslands, where periodic fire maintains the dominant grass vegetation, and for many forest types in the western United States, where fire also interrupts succession.

NOTE: When discussing community succession in your local area, show that as succession proceeds communities have more diversity (because of new habitats and new interconnections among the new organisms in them), greater overall biomass and lower productivity (because much of the energy input is used to maintain the biomass), and greater stability to catastrophic events (because of the greater diversity). These are important messages that underscore the fragility of agricultural ecosystems (which are like early successional communities) and provide another reason to keep representative old successional communities around (because the are innately more stable).

III. Ecosystem structure and function.

A. Talking About Science: Ecologist Frank Gilliam discusses the role of fire in ecosystems (*Module 36.7*).

Review: Modules 34.12, 34.14, 34.15, and 34.17 discuss biomes in which fires play a major role.

1. Dr. Gilliam is a professor of biological sciences at Marshall University. He has studied temperate deciduous forests, coastal pine forests, and tallgrass prairies. Rather than seeing fire solely as a destructive force, Gilliam is interested in the effects of natural fires (not caused by humans) on ecosystems.

2. In the pine forests of the southeastern U.S., acidic soils keeps the rate of decomposition low. Without human intervention, these forests tend to burn every 5 to 7 years. If the buildup of fuel is not excessive, these fires are not hot enough to kill large pine trees. What the fires burn is the organic material on the ground (such as pine needles) and low-growing plants. After a burn the number and variety of nonwoody plants present in these forests increase as a result of the fire having made more nutrients available to them.

3. In his studies of the tallgrass prairies of Kansas, Gilliam emphasizes that standing dead grass is an ideal fuel for fire. Barring human intervention, grasslands burn every 2 to 4 years, with some grasslands burning every year. He found that these fires are essential for maintenance of the grasslands. Without fires, the grasses are replaced by trees and shrubs.

B. Energy flow and chemical cycling are two of the fundamental processes of ecosystems (*Module 36.8*).

1. *Review:* The distinction between autotrophs and heterotrophs (Module 17.10).

2. Energy flow is the one-way passage of energy through the components of the ecosystem, usually starting with photosynthesis by autotrophs and proceeding through heterotrophs. Every use of chemical energy by an organism involves loss of heat to the surroundings (Acetate 36.8).

Review: An exception to the sun being the ultimate source of energy for an ecosystem, as for example hydrothermal vent ecosystems, is discussed in Chapter 34 (Opening Essay and Module 34.7). Also, review the laws of thermodynamics (Module 5.2).

3. Chemical cycling is the circular movement of elements among the biotic and abiotic parts of an ecosystem (or among components of many ecosystems).

C. Trophic structure determines ecosystem dynamics (*Module 36.9*).

1. The trophic structure of a community is the pattern of feeding relationships that determines the flow of energy and the routes of elements that are cycled.

2. The transfer of food from trophic level to trophic level is called a food chain. Food chains differ for each community type. Natural food chains are never single, unbranched chains (as implied by the diagrams).

3. In a terrestrial food chain, the trophic levels include producers (autotrophic plants), primary consumers (herbivores that consume the producers; e.g., grasshoppers), and secondary (e.g., a mouse eating an herbivorous insect), tertiary (e.g., a snake eating a mouse), and quaternary consumers (carnivores that consume the next lower consumer level; e.g., a hawk) (Acetate 36.9).

4. In the illustrated aquatic food chain, the trophic levels are the same, but the players are different.

5. Detritivores derive their energy from the breakdown of detritus, organismal waste and parts of dead organisms. Detritivores include, principally, fungi and bacteria, but also small animals and heterotrophic protists. The larger detritivores break apart the material physically and alter it chemically. The smaller detritivores function in decomposition, the breakdown of organic chemicals to inorganic chemicals.

IV. Food chains, food webs, and energy pyramids.

A. Food chains interconnect, forming food webs (*Module 36.10*).

1. Consumers usually eat more than one type of food. Each food type is consumed by more than one type of consumer. Detritivores feed on the dead remains of all (different detritivores are selective) (Acetate 36.10).

2. The arrows in a food web diagram outline an ecosystem's overall flow of chemical nutrients and energy, but the diagram still simplifies the relationships among organisms because the individual species at all the trophic levels are usually not represented.

B. Energy supply limits the length of food chains (*Module 36.11*).

1. Biomass is the amount of organic material in a collection of living organisms or their remains. The rate at which producers convert solar energy to chemical energy is called primary productivity.

2. Most energy Earth receives is absorbed, scattered, or reflected by the atmosphere. Of the energy reaching autotrophs, 1% is converted by photosynthesis into chemical energy. This amounts to 170 billion tons of organic matter per year for the entire biosphere.

3. Along a food chain, the energy available at each trophic level drops by an average of about 90% (this figure is actually on the low side; Module 36.12) because of the loss of heat and cellular respiration (about two-thirds of the total loss), and the inefficient use of food (i.e., waste output) (Acetate 36.11).

Review: Module 5.2 discusses the first and second laws of thermodynamics.

4. Applying this energy reduction between several trophic levels produces an energy pyramid, and explains why most food chains are limited to three to five levels. At the end of a food chain, little energy is available.

Preview: Since some toxins, such as DDT, persist in the food chain, they become increasingly concentrated at higher trophic levels. The result is biological magnification (Module 38.12; also see Module 34.3).

C. An energy pyramid explains why meat is a luxury for humans (*Module 36.12*).

1. Humans are omnivores. Among other foods, we eat fish (as tertiary or quaternary consumers), meat (as secondary or tertiary consumers), and plants (as primary consumers).

2. Humans have about ten times more energy available when we eat grain than when we eat grain-fed beef. In fact, the 10% transfer between trophic levels may be high for this system because, as endotherms, cattle expend much more energy in heat production (Master 36.12).

3. Meat production is expensive both economically and environmentally because it requires more land to be cultivated, more water for irrigation, and more chemical fertilizers and pesticides.

V. Chemical cycles.

A. Chemicals are recycled between organic matter and abiotic reservoirs (*Module 36.13*).

1. There are virtually no extraterrestrial sources of water or other nutrients life requires.

2. Chemicals cycle between abiotic reservoirs and an ecosystem's biotic components.

B. Water moves through the biosphere in a global cycle (*Module 36.14*).

1. The cycling of water is driven by heat from the sun through the processes of precipitation, evaporation, and transpiration (Acetate 36.14).

NOTE: In the diagram, the numbers indicate water in billion billion (10^{18}) grams per year (1 million billion kg).

2. The principal abiotic reservoir is the ocean. The atmospheric reservoir is mobile and gives the water cycle its global character. Over the ocean, evaporation exceeds precipitation, and there is a net movement of water vapor onto land. On land, precipitation exceeds evaporation and transpiration, and there is a net movement of liquid water into the ocean.

3. By removing forests and overirrigating, humans can have major impacts on the water cycle. Water vapor leaving land will decrease over forests. Over irrigated land, evaporation will increase and groundwater supplies will decrease.

C. The carbon cycle depends on photosynthesis and cellular respiration (*Module 36.15*).

1. *Review:* Equations for photosynthesis and cellular respiration (Figures 7.4A, B).

2. The principal abiotic reservoir is the atmosphere, which cycles globally. Within the biosphere, carbon moves along food chains between trophic levels. Carbon returns to the atmosphere as respired CO_2 (Acetate 36.15).

NOTE: Carbon cycle pathways are similar to food chain pathways because energy-storage molecules are all carbon-based compounds.

3. *Preview:* The burning of fossil fuels and wood has caused atmospheric levels of CO_2 to rise, and this may cause global warming (Module 38.13).

D. The nitrogen cycle relies heavily on bacteria (*Module 36.16*).

1. *Review:* Nitrogen fixation and other nitrogen-cycling processes involving bacteria (Modules 17.17, 32.13, and 32.14).

2. The principal abiotic reservoir is the atmosphere, which cycles globally. However, gaseous nitrogen (80% of the atmosphere) is only available directly to certain nitrogen-fixing prokaryotes. Within the biosphere, nitrogen is principally active as a part of proteins and nucleic acids (Acetate 36.16).

3. Plants can use nitrogen only in the form of NO_3^- (nitrate) or NH_4^+ (ammonium). Nitrogen-fixing bacteria convert atmospheric N_2 to NH_3 (ammonia), which then becomes NH_4^+. Nitrifying bacteria convert NH_4^+ into NO_3^-. NO_3^- is the main source of nitrogen for plants. Animals must eat plants or other animals to get usable nitrogen.

4. Fungi and bacteria decompose nitrogen-containing detritus to NH_4^+. Nitrogen is lost from the biotic cycling by the action of denitrifying bacteria converting soil NO_3^- into atmospheric N_2.

NOTE: Denitrification is anaerobic. Losses occur in agricultural areas where soil structure has broken down and become waterlogged and anaerobic.

5. Most nitrogen cycling by bacteria involves the inner cycle in Figure 36.16. Not shown in the figure is the NH_4^+ and NO_3^- made in the atmosphere, which reach the ground in precipitation and dust. Such sources are important in many tropical rain forests.

6. Sewage-treatment facilities and agricultural runoff release large amounts of biologically usable nitrogen into groundwater, streams, and lakes, resulting in eutrophication and the buildup of excess nitrogen compounds in water. Nitrates in drinking water are converted to nitrites, which can be toxic to humans.

NOTE: A considerable amount of nitrogen fixation (28% of the total) occurs in the combustion of fossil fuels and in synthetic fertilizers produced by industrial nitrogen fixation, pathways not shown in the figure.

E. The phosphorus cycle depends on the weathering of rock (*Module 36.17*).

1. The main abiotic reservoir of phosphorus is in rocks. The cycling time for this reservoir is extremely slow because it requires that phosphates precipitate out into sediments, reform into rocks, uplift, and be available again from rock weathering. Abiotic pools of inorganic phosphates in soil solution are often limiting in ecosystems (Acetate 36.17).

2. The biotic cycling of phosphate starts with plant uptake. The main biotic pool is in organic compounds such as phospholipids, ATP, and nucleic acids.

3. Phosphates from fertilizer made of crushed rocks from pesticides can be in excess in runoff, leading to eutrophication of streams and lakes.

Communities and Ecosystems

VI. The disruption of ecosystem function.

    A. Ecosystem alteration can upset chemical cycling (*Module 36.18*).

        1. Since 1963 the Hubbard Brook Experimental Forest, a deciduous forest in the White Mountains of New Hampshire, has been studied by a team of scientists. The Hubbard Brook Forest consists of several watersheds. Long-term studies such as this one are essential for gaining an understanding of the dynamics of an ecosystem.

        2. This study involves the monitoring of the water and nutrient dynamics that occur both under natural conditions and after human intrusion.

        3. From undisturbed forests, 60% of water entering as precipitation leaves the watershed in runoff, and 40% is lost by transpiration and evaporation. The flow of nutrients into and out of the watershed is nearly balanced, with small gains in some, particularly nitrates, in most years.

        4. In 1966 one of the watersheds was completely logged and sprayed with herbicide for 3 years to prevent the regrowth of plants. All the original material was left in place. Amounts of water and nutrients in runoff were monitored and compared to an unaltered watershed.

        NOTE: In the background of Figure 36.18B, experimental setups of other logging methods can be seen, particularly strip-cutting in the center.

        5. In the altered watershed, runoff increased 30–40%, apparently because there were no plants to absorb and transpire water from the soil. Further, there were huge losses of nutrients. For example, within 8 months of the cutting, nitrate losses were 60 times greater than in the undisturbed watershed (Master 36.18C).

        6. One of the conclusions of this study is that even before the study at Hubbard Brook began, none of the watersheds was free from human impact. For example, since the 1950s acid precipitation has dissolved and carried away most of the Ca^{2+} in the soil.

        Review: The effects of acid precipitation on the environment are discussed in Module 2.16, and the effects of pH on enzyme activity are discussed in Module 5.7.

    B. Altered ecosystems trigger changes in other ecosystems (*Module 36.19*).

        1. Increases in nutrients in runoff ultimately end up flowing into aquatic ecosystems, where they can lead to eutrophication.

        2. Eutrophication involves the increase of nutrients above natural levels in aquatic ecosystems. This causes the ecosystems to become more productive and results in changes in the kinds and relative numbers of organisms.

        3. Algal producers bloom in eutrophic conditions. This increases oxygen production during the day but greatly reduces oxygen levels at night when the algae respire. Increased production of organisms causes increased respiration by decomposers and the development of anaerobic conditions in lake-bottom sediments. Overall, all aerobic organisms may suffer and eventually die out.

        NOTE: Another problem is the buildup of acidity following acid rain, particularly in those lakes that do not have natural buffering (Module 2.16).

C. Megareserves are an attempt to reverse ecosystem disruption (*Module 36.20*).

    1. Megareserves are extensive regions of land that include one or more undisturbed areas. Within these areas, species and ecosystem processes involving them can be conserved.

    Review: Global warming will make managing reserves very difficult (Module 7.14).

    2. Countries exchange political and economic favors for establishing these areas. Costa Rica has become a world leader in establishing megareserves (Master 36.20).

    3. Local education about the ecological and economic benefits is needed, and buffers must be established around the margins of megareserves where social and economic climates are more compatible with the conservation of resources within the reserves.

CLASS ACTIVITIES

1. In most areas of the United States, successional changes in community structure occur and can be demonstrated. Demonstrate local changes by using slides of sites representing different time periods, or by visiting the sites with students and discussing the changes with them there. Have students measure or estimate the following at each site (or, with slides, during lecture): overall diversity of species; density and dispersion of the dominant vegetation; trophic level represented by each dominant organism; total biomass in the community; depth of the litter layers and organic soil; examples of parasitic, mutualistic, or commensal relationships; overall resistance of the community to catastrophic occurrences.

2. Construct a glassware-tubing device to demonstrate the flow of energy through three trophic levels in an ecosystem to show how little energy is available after the third level. Use Y-shaped tubes to divide the output of a corked flask into flow that goes to the next flask and flow that goes to the sink (energy in heat, respiration, and waste). Use a valve in the line to the next flask to adjust its flow to about 10% of the waste flow. Spinning-ball gauges at the start and between each flask will show the drop in available energy, and the output from the third flask will be a mere trickle. Charge up the entire system with water before lecture, and hook up the first flask with an "energy source" from the sink or other water supply.

3. See how many different examples of symbiotic relationships involving humans the class can think of.

RESOURCES AND REFERENCES

Beardsley, T. "Desert Dynamics." *Scientific American*, November 1992. An experiment tests the effect of kangaroo rats on desert vegetation.

Coutant, C.C. "Thermal Niches of Striped Bass." *Scientific American*, August 1986. Natural populations of this North American native spawn in fresh water and mature in coastal marine waters. The preferred water temperature changes with the age of the fish.

Culotta, E. "Biological Immigrants Under Fire." *Science*, December 6, 1991. How exotic species can disrupt natural communities.

Davies, N.B., and M. Brooke. "Coevolution of the Cuckoo and Its Hosts." *Scientific American*, January 1991. The cuckoo reproduces at the expense of other birds by laying eggs in their nests. This has produced an "evolutionary arms race" between parasite and host.

The Ecology of the Forest. Princeton, NJ: Films for the Humanities and Sciences, 1987. A 28-minute videotape that compares ecosystem structure and processes in temperate and tropical forests.

Enemies of the Oak. Oxford, England: Oxford Scientific Films, 1988. A 54-minute videotape that describes a 300-year-old English oak and its insect predators, including the oak's defenses and the balance between predator and prey. Available from Carolina Biological Supply Company.

Falkowski, P.G., Z. Dubinsky, L. Muscatine, and L. McCloskey. "Population Control in Symbiotic Corals." *BioScience*, October 1993. A discussion of how symbiotic associations of dinoflagellates remain stable in coral tissues.

Food Chains. Nutrient Cycles. Deerfield, IL: Coronet/MTI Film and Video, 1992. A 28-minute videodisc that covers the basics of ordinary and detritus food webs, keystone species, and cycles of nitrogen, carbon, oxygen, and phosphorus.

Gillis, A.M. "Sea Dwellers and Their Sidekicks." *BioScience*, October 1993. A discussion of an unusual mutualism between a squid and bacteria.

Gosz, J.R., R.T. Holmes, G.E. Likens, and F.H. Bormann. "The Flow of Energy in a Forest Ecosystem." *Scientific American*, March 1978. Details on trophic levels and an energy budget for a northeastern United States hardwood forest.

Greene, H.W., and R.W. McDiarmid. "Coral Snake Mimicry: Does It Occur?" *Science*, September 11, 1981. A detailed study to clarify the then-growing controversy surrounding the nature of skin patterns among coral snakes and look-alikes. The less harmless species are shown to exhibit Batesian mimicry of the most toxic species.

Gregory, S.V., F. J. Swanson, W.A. McKee, and K.W. Cummins. "An Ecosystem Perspective of Riparian Zones." *BioScience*, September 1991. A focus on the links between land and water in these sensitive wetland zones along rivers and streams.

Gressitt, J.L. "Symbiosis Runs Wild on the Backs of High-Living Weevils." *Smithsonian*, February 1977. Lichen-covered weevils from New Guinea are featured.

Intimate Strangers: Symbiosis. Burlington, NC: Oxford Scientific Films/Carolina Biological Supply Company, 1986. An 8-minute videotape that details the acacia–ant and the leaf-cutter ant–fungus mutualistic symbioses.

Lopez, G. *Community Dynamics*. Danbury, CT: Educational Materials and Equipment Company, 1987. An interactive computer tutorial and simulation of predatory-prey models that allows considerable manipulation of parameters. Available in Apple II and IBM PC versions.

Malecki, R.A., B. Blossey, S.D. Hight, D. Schroeder, L.T. Kok, and J.R. Coulson. "Biological Control of Purple Loosestrife." *BioScience*, November 1993. A case is made for using introduced insects as control agents of this European native, which has degraded many wetland habitats in North America.

Moore, J. "The Behavior of Parasitized Animals." *BioScience*, February 1995. Some parasites alter their host's behavior in self-serving ways.

Odum, H. T. *Systems Ecology: An Introduction*. New York: Wiley, 1984. A text that emphasizes important aspects of ecosystems.

Oliwenstein, L. "Royal Flush." *Discover*, January 1992. New debate about a classic case of mimicry, monarchs and viceroy butterflies.

Poulin, R., and A.S. Grutter. "Cleaning Symbiosis: Proximate and Adaptive explanations." *BioScience*, July-August 1996. The evolution of a type of mutualism.

Reiners, W., W. Glanz, and S. Cornish. *Predation Equilibria.* Iowa City, IA: Conduit, University of Iowa, 1986. A five-program package of computer programs that introduce predator-prey interactions. Available from Carolina Biological Supply Company.

Rennie, J. "Living Together." *Scientific American*, January 1992. Parasites and their hosts have devised many odd strategies in their endless game of adaptive one-upmanship. Sometimes they even cooperate. Details of several interesting parasitic relationships.

Rocky Mountain Beaver Pond. Washington, DC: National Geographic Society, 1987. A 44-minute videotape that describes the lives of beavers and the functions of this unique habitat.

A Swamp Ecosystem. Washington, DC: National Geographic Society, 1983. A 23-minute videotape that examines the animal and plant life in Okefenokee Swamp, including a discussion of succession in this ecosystem and the roles of fire and drought in maintaining the swamp's structure.

Tale of Plague. Deerfield, IL: Coronet/MTI Film and Video, 1990. A 30-minute videodisc that describes the biggest locust plague to hit Africa in 30 years, including details of the developmental stages of locusts, food chains, and human interactions.

37

BEHAVIORAL ADAPTATIONS TO THE ENVIRONMENT

APPROACH

Behavior fascinates all students. Very likely, they will have been introduced to several of the topics in this chapter in a psychology, sociology, or anthropology course. However, the adaptive explanations may have escaped them, and they should be one of your focal points in covering this material. As you detail the examples in the modules in this chapter, emphasize the evolutionary context of each behavioral trait.

The nature-versus-nurture controversy is of great interest because of the human implications. The distinction between the genetic and learned components of behavior is the other overall message to stress in this chapter.

In light of current interest in animal rights, there might be some concern with certain examples, such as imprinting a young goose on a ticking clock. Explain to students that knowledge about behavior in other species has been of great help in interpreting human behavior, particularly in a developmental context (child behavior) and in understanding the abnormal behaviors of criminals and the mentally ill.

CHAPTER OBJECTIVES

Differentiate between the interests of, and approaches taken by, behavioral biologists, behavioral ecologists, and sociobiologists.

Describe the early experiments of von Frisch, Lorenz, and Tinbergen, indicating in which contexts of behavioral biology each played a founding role.

Explain the relative roles of genes and environment in the following types of behavior: pattern recognition, fixed action patterns, habituation, imprinting, association, imitation, and innovation.

Describe the evolutionary context of examples of the following types of individual behavior: biological rhythms, kineses and orientation behaviors, migration, and feeding behavior.

Describe the evolutionary context of examples of the following types of social behavior: agonistic behavior, dominance hierarchy, territoriality, mating behavior, signaling, and altruism.

Outline the controversies surrounding sociobiology, particularly nature versus nurture.

LECTURE OUTLINE

I. Introduction.

    A. Behavior of the jaguar (*Opening Essay*).

        1. Jaguar activities have been studied in the world's only jaguar refuge, the Cockscomb Forest Jaguar Preserve, established in Belize in 1984.

        2. Jaguars are solitary hunters that avoid contact with other jaguars except during mating season. Male territories may overlap, but individuals

rarely occupy the disputed territory at the same time. Jaguars mark their boundaries by defecating in open areas and signal their presence to others by vocalizing grunts and growls. During breeding times, males and females pair off, and the males help with the cubs for several weeks after they are born.

3. The study of animal behavior is essential to understanding animal evolution and ecological interactions. Hunting techniques of jaguars are similar to those of many big cats, suggesting a common ancestry of these cats. Signaling and breeding behaviors are adaptive to preventing confrontations and improving the success of reproduction, respectively.

B. Behavioral biology is the study of how animals behave in their natural environments (*Module 37.1*).

1. Behavior refers to any externally observable muscular activity triggered by some stimulus.

2. Early workers in the field of behavioral biology were Karl von Frisch, Konrad Lorenz, and Niko Tinbergen, all Nobel laureates. Von Frisch pioneered the use of experimental methods in behavior, studying food hunting by honeybees. Lorenz, the "father of behavioral biology," compared behaviors in similar animals and found that the same stimulus often elicited different behavior. Tinbergen studied the relationship between innate (genetically programmed) behavior and learning.

3. An early experiment by Tinbergen on the nesting behavior of digger wasps shows one way natural behavior is studied. Female digger wasps build several nests, each containing one larva, and are able to find and return to each daily to care for it. Tinbergen hypothesized that the wasps use visible landmarks as cues. He arranged a distinctive landmark around a nest and later moved the landmark away from the nest. The wasps returned to the landmark. Further experiments showed that the wasps recognized the overall appearance of the landmark, not the individual parts (Master 37.1).

Review: This is another good example of the process of science (Module 1.2).

4. Behavioral biologists make a distinction between two levels of cause for behavioral activity. The proximate cause explains the behavior in terms of immediate activities (in Tinbergen's study, the pattern of the landmark). The ultimate cause is the evolutionary context of the behavior. (Tinbergen did not address this question, but presumably wasps gain reproductive advantage by performing the nest-finding behavior in the correct way.)

5. The search for ultimate causes is a subdiscipline of behavioral biology known as behavioral ecology, the dominant interest of students of behavior today.

II. The role of genes in creating behavior.

A. Programmed behavior and behavior modified by experience both have genetic underpinnings (*Module 37.2*).

1. The relationship between genes and behavior is complex. Most genetically programmed behaviors have some aspect that can be changed by experience. For digger wasps, the use of landmarks is programmed, but the exact landmark depends on the environmental context. How much behavior is governed by genes and how much by experience is a classic debate.

2. Another experiment showing the genetic basis of some behaviors involves African lovebirds. Hybrid females (between two species that have different nest-building behavior) exhibit behavior (some pointless) that combines aspects of both species. When this behavior fails to function to build the nest, the hybrid birds will learn more appropriate behavior but still carry out some of the movements of the inappropriate behavior (Master 37.2).

3. An example in humans is language. The ability to learn language is a result of the structure of a brain that develops under genetic guidance. However, the language that is learned is environmentally based.

Review: The cellular and chemical bases of learning (Module 28.20).

B. Innate behavior often appears as fixed action patterns (*Module 37.3*).

1. Fixed action patterns (FAPs) are a type of innate behavior (just as simple reflexes are innate; Module 28.14). FAPs are essentially unchangeable behavioral sequences that can only be performed as a whole and, once started, must be completed. The stimulus that triggers a FAP (i.e., is the proximate cause of the behavioral sequence) is called a sign stimulus.

2. For example, the graylag goose exhibits a FAP when she retrieves an egg that rolls out of her nest. She stands up, extends her neck, and sweeps her head from side to side, using her beak to bring the egg back. If an experimenter moves the egg aside while it is being brought back, the goose will continue the behavior (without the head swinging) until she sits down in the nest again, at which time she will notice that the egg is still outside (Master 37.3A).

3. The European cuckoo lays her eggs in other birds' nests, replacing a host bird's egg with one of her own. The host incubates the egg. The cuckoo egg often hatches first, and the young bird exhibits a series of FAPs. It first ejects the unhatched host eggs. Then, like all young birds, it begs for food from its foster parent by gaping and cheeping. The feeding FAP is performed over and over, even after the young cuckoo is larger than the foster adults (Master 37.3B).

4. Experiments with invertebrates show that a sign stimulus activates genetically programmed nervous pathways that induce appropriate movements, without the involvement of the brain.

5. FAPs are particularly important in animals with life spans too short to allow for learning and in newborn vertebrates, where, in the context of the ultimate cause, such behavior ensures that the young will obtain food from the parents (or directly from the environment) without any learning.

III. **The role of learning in creating behavior.**

A. Learning ranges from simple behavioral changes to complex problem solving (*Module 37.4*).

1. *Review:* Nervous system function in memory. Learned behavior requires that memory maintain the new pattern of behavior (Module 28.20).

2. Learning is a change in behavior resulting from experience. The different types of learning reflect the complexity of the changed behavior (Table 37.4).

3. Habituation happens when an animal learns not to respond to an unimportant stimulus. This is one of the simplest forms of learning and is common in all animals. It is highly adaptive because it prevents

animals from wasting energy on pointless activity. For example, if a hydra is repeatedly touched, it will eventually stop contracting (the normal response).

B. Imprinting is learning that involves both innate behavior and experience (*Module 37.5*).

1. Imprinting is learning to perform a response during a limited time period, the critical period. Its innate component is its actualization only during the critical period. Its learned component is the imprinting itself. Imprinting is particularly important in the formation of bonds between animals: young with parents, mating pairs, and members of social groups.

2. Konrad Lorenz's most famous study showed that graylag goslings would imprint on him as a mother figure if, during the first 2 days after hatching, the geese saw him and not the mother. The specific imprinting stimulus was found to be the movement of an object (normally the parent) away from the hatchlings (Figure 37.5A).

3. Salmon imprint on the complex of odors unique to the stream they hatch in so that, as adults, they can return to that same stream to spawn (Chapter 29 Opening Essay).

4. Each species of songbird has its own unique song. However, these songs exhibit both population (regional) and individual variation.

5. Song imprinting occurs in birds learning the mating or warning calls particular to each species. Each species of bird has an innate sound pattern it will sing without the imprinted component. During a critical period of 10–50 days after hatching, a bird imprints on the calls of its own species (Master 37.5B).

6. Birds raised in isolation who do not hear their species' song until after 50 days do not learn to sing normally; whereas isolated males played tapes of their species' song during the critical period do learn to sing normally.

7. Studies of the white-crowned sparrow demonstrate that there is a genetic component to their song. Isolated males who were exposed to the songs of species other than their own did not learn those songs; rather, they sang an abnormal song (lower strip of Master 37.5B).

C. Many animals learn by association and imitation (*Module 37.6*).

1. Association is learning that a particular stimulus, or behavioral response, is linked to a reward or punishment.

2. Classical conditioning is producing a behavioral response when a reward or punishment is associated with an arbitrary stimulus. Eventually, the response is elicited from the stimulus without the reward or punishment.

3. Natural associations occur by trial and error, where, over time, an animal associates the positive or negative results of a certain type of behavior with the reward or punishment it brings (Figures 37.6A, B).

4. Imitation is learning by observing and mimicking. But unlike imprinting, imitation can happen at any time, not just during a critical period. Many hunting behaviors are learned by imitation.

D. Innovation is problem solving without prior experience (*Module 37.7*).

1. Innovation is also known as reasoning. It implies the ability to think of possible solutions to a problem and analyze the possible outcomes of each. Innovation occurs mostly in primates.

2. Dogs solve problems by trial and error or by remembering a prior experience, not by innovation (as is seen in chimpanzee behavior) (Master 37.7A, Figure 37.7B).

3. It is difficult to differentiate clever innate (or learned without reasoning) behavior from innovation, and easy for humans to ascribe humanlike motivation and awareness to nonhuman behavior. This tendency is called anthropomorphism.

4. Historically, human intellect sets humans apart from other animals, but it is not known whether true consciousness (cognition) occurs in nonhuman animals (Jane Goodall, Module 37.16, has concluded that chimpanzees are cognitive). Some maintain that the subject lies outside of science because researchers cannot directly observe it. Others suggest that conscious thinking is part of the behavior of many animals, and that various levels of consciousness have developed over evolutionary time.

IV. **The ecological roles of behavior in individuals.**

A. An animal's behavior reflects its evolution (*Module 37.8*).

1. Many aspects of behavior are a result of the evolutionary process.

2. However, most complex behavior patterns involve an interplay between an organism's innate ability to do something and the organism's environment.

B. Biological rhythms synchronize behavior with the environment (*Module 37.9*).

1. *Review:* Biological (and circadian) rhythms in plants (Module 33.10).

2. Circadian (repeated daily) rhythms in animals are associated with sleep/wake cycles. Innate patterns may be slightly more or less than 24 hours but are adjusted by external cues, usually light/dark cycles.

3. For example, a flying squirrel kept in the dark will continue to show rhythmic patterns of activity for many days, but the activity cycles every 24 hours, 21 minutes. Thus, after 23 days of constant darkness, the experimental animal is nearly 8 hours out of synchronization with the time of day, and with the activity cycle of a control animal that is exposed to daily light/dark cycles (12 hours of light, 12 hours of dark) (Master 37.9A).

NOTE: Two other patterns can be seen in the experimental data that shed light on the maintenance of the rhythm. Extended activity gets greater, at least during the first couple of weeks. As the days proceed, the "memory" of the rhythm seems less sure, with more and more activity behavior outside the normal time period in which it would happen.

4. Humans exhibit circadian rhythms that usually follow approximately a 25-hour cycle. In one long-term test similar to the experiment with flying squirrels, a woman spent 131 days in isolation in a cave in New Mexico. Her sleep/wake cycle followed a 25-hour pattern, but other characteristics (blood pressure, heart rate) followed 48-hour to 7-day cycles. She lost weight, developed a calcium deficiency, and stopped menstruating. Normal body functions reappeared only after months of reconditioning

and hormone therapy. Extreme cycle changes are thought to be related to the emotional stress of isolation.

5. The proper functioning of human biological rhythms is associated with general feelings of well-being, work efficiency, and decision-making ability. In situations that require people to adjust their internal clocks (jet lag, odd work hours), some people can minimize the symptoms of being "out of sync" by adjusting when they eat, exercise, and sleep.

NOTE: It has also recently been shown that maintaining exposure to the light/dark regimen of the new schedule helps (for example, by not staying up late reading when one travels from west to east). See also Module 26.3 on the possible role of the pineal gland in seasonal cycles.

6. There is evidence that in mammals a region of the hypothalamus functions as the biological clock.

Review: The function of the hypothalamus is discussed in Module 28.20.

C. Kineses and orientation behavior place animals in favorable environments (*Module 37.10*).

1. Kinesis is a random movement response to a stimulus. The activity is repeated until the organism finds itself in a more favorable environment. Human body lice are more active in dry areas and less active in wet areas.

2. Orientation behavior involves directed movements toward (positive) or away from (negative) a stimulus. Such directed movement is known as a taxis. Numerous taxes are named from the stimuli that elicit the behavior (photo-, geo-, chemo-, rheo-). Thus, a positive rheotaxis is exhibited by fish swimming upstream. Taxes are more efficient behavioral responses than kineses.

NOTE: Even bacteria and protists exhibit kinetic and taxic behaviors. Distinguish between a taxis, movement, and a tropism, growth (as seen in plants; Module 33.9).

D. How do migrating animals navigate? (*Module 37.11*)

1. Seasonal migration is the regular movement from one place to another during a particular time of the year. Animals usually migrate to areas that are more suitable for feeding or reproducing.

2. Gray whales migrate north in summer to rich feeding grounds on the coast of Alaska, and south in winter to shallow lagoons off Baja California (Mexico), where they breed and produce offspring before the next northward migration. They have been shown to use coastal landmarks for orientation (Master 37.11A).

3. Monarch butterflies winter in certain forests in mountainous Cuba and Mexico; they live off stored food but are protected by uniformly cool temperatures. In spring, mated females begin a northward migration, lay eggs at regional destinations, and then die. Two or more generations continue the northward reinvasion of the monarchs into the United States and southern Canada. The last generation then moves back south. Their orientation must be innate because of the change in generations. They may use the sun as a compass.

4. Insect-eating birds winter in the tropics and summer at high latitudes. Many have been shown to navigate using the positions of stars and sun. Unless birds use the north star as a fixed landmark, such navigational skills require that the animals internally adjust their sensors to take into

account the daily changes in direction of stars and sun (Master 37.11C). There is evidence to suggest that when the sun or stars are not visible, birds can orient to the Earth's magnetic field.

E. Behavioral ecologists use cost/benefit analysis in studying feeding behavior (*Module 37.12*).

1. Some animals are feeding specialists (koalas eat only eucalyptus leaves). Some are generalists (gulls eat many foods). An animal is said to have a search image when it concentrates on feeding on an abundant food item to the exclusion of other food. (Figures 37.12A, B)

2. Foraging is the searching for, securing, and eating of food. Optimal foraging involves feeding behavior that provides maximum energy (or nutrient) gain for minimum energy and time expended by the behavior.

3. Optimal foraging depends on the animal's compromising among a number of trade-offs involved with choosing one feeding-behavior pattern over another. Different foods have different digestibilities. Further, prey differ in their relative sizes, population densities, and ease of capture; and these factors may change through time. Searching for one type of food might endanger an animal to more predation. Natural selection causes animal behavior to evolve in a direction that optimizes foraging over the entire range of possible environments and preys the animal could forage for. An animal may switch foraging behaviors, depending on conditions.

4. Jaguars concentrate on armadillos as prey because, though small, they are abundant and easy to catch. Alternative prey, the tapir, are not sought because, though large, they are fast, hide easily, and are less abundant (Figure 37.12D).

5. Kangaroo rats forage optimally on seeds. When a choice of seeds is available, they choose seeds with the most energy and, from a cache of seeds, will eat the richest ones first. Both activities minimize the time a kangaroo rat needs to stay aboveground, exposed to predators (Figure 37.12E).

V. **The behavior of animals relating to other animals.**

A. Social behavior is an important component of population biology (*Module 37.13*).

1. Many behaviors involve interactions between two or more individuals. Imprinting, migrating, and hunting behaviors have social elements.

2. Social behavior is broadly defined as any interaction between two or more animals, usually of the same species. Social behavior always involves communication. Much of social behavior is highly adaptive because it can affect the growth and regulation of populations.

3. Social behavior involved with breeding can affect fitness directly by determining which individuals get to mate.

Review: Fitness is defined in Module 14.18.

B. Rituals involving agonistic behavior often resolve confrontations between competitors (*Module 37.14*).

1. Agonistic behavior includes behavior that settles disputes among members of populations without harming the participants. It may involve tests of strength, posturing, or ritualized contests.

2. In most cases, agonistic behavior ends when one individual stops threatening, becomes submissive, and enters into some kind of surrender behavior. Further aggressive activity is inhibited.

3. Agonistic behavior is adaptive because the inhibition of aggressive activity minimizes the potential for damage that could adversely affect the reproductive success of both individuals.

C. Dominance hierarchies are maintained by agonistic behavior (*Module 37.15*).

1. The peck order established among hens in a hen yard is an example of a dominance hierarchy. The alpha (top-ranked) hen will not be pecked by any other hen, while the omega hen will be pecked by all other hens. The alpha hen gets first access to resources such as food, water, and nesting sites (Figure 37.15).

2. In a dominance hierarchy, the position of each animal is fixed for extended periods of time. By establishing the set pattern, the animals in the social group do not have to constantly waste energy testing each other's status and can devote their energies to other activities (e.g., finding food, watching for predators, locating a mate, and caring for young).

D. Talking About Science: Animal behaviorist Jane Goodall discusses dominance hierarchies and cognition in chimpanzees (*Module 37.16*).

1. Dr. Goodall has studied chimpanzees in their natural habitat in East Africa since the early 1960s. She promotes better understanding of primate behavior and pushes for better living conditions for animals kept in zoos or in medical research labs.

2. Chimpanzees are the closest living relative of humans. Dominance hierarchies are an important part of chimpanzee life. Adult males expend considerable effort improving or maintaining their social status, often by using agonistic charging displays. Higher-ranked males have preferred access to the best food, resting places, and mates, but other behaviors provide complex solutions to many social conflicts that do not always end with the dominant male having all the advantages (Figure 37.16B).

3. Females develop a different hierarchy. High status allows them to use the best food items. A high-ranking female's offspring start out in their social groups with higher social rankings.

4. Goodall's research has led her to conclude that chimpanzees are cognitive.

E. Territorial behavior parcels space and resources (*Module 37.17*).

1. A territory is an area, usually fixed in location, that individuals inhabit and defend from occupancy by other individuals of the same species. The sizes of territories depend on species, use, and the total size of the area out of which the territories are parceled. Not all animals are territorial.

Review: Population distribution patterns, some of which are a result of territoriality, are discussed in Module 35.2.

2. When space is at a premium, breeding birds maintain tiny nesting territories by agonistic behavior (Figure 37.17B).

3. Large hunters defend much larger territories for hunting and mating activities. They maintain—and spend considerable time proclaiming—their territorial boundaries, either by calling (birds, sea lions, squirrels) or marking (defecation by jaguars, scent marking by cheetahs and other cats).

4. Territorial animals gain exclusive use of the resources in their territories, gain familiarity with one area, and are better able to raise young without interference from other individuals.

F. Mating behavior often involves elaborate courtship rituals (*Module 37.18*).

1. Courtship rituals signal that certain individuals are not threats to others but are the correct species, sex, and physical condition.

2. Courtship rituals often contain actions that signal appeasement. For example, all of a loon pair's courting behavior consists of appeasement gestures. During their courtship, loons turn their heads away, dip beaks, and submerge heads and necks, and the male turns his head backward and down as a final invitation onto land to mate (Master 37.18A).

3. Many other animals go through group courtship rituals. Sage grouse males congregate in a lek, an area where they strut in front of watching females. All the females choose among all the males, usually mating with only a few of the most dominant individuals. If there is a positive correlation between a male's display and his evolutionary fitness (as behavioral ecologists think), such behavior will give a female the best chance at passing on her genes.

G. Complex social organization hinges on complex signaling (*Module 37.19*).

1. The signals used may be sounds, odors, visual displays, or touches. In general, the more complex the social organization of a species population, the more complex the signaling needed to maintain it.

2. Social insects exhibit some of the most complex social systems. A colony of honeybees includes over 50,000 individuals, mostly sterile workers that maintain the health of the hive by feeding, caring for young, and defensive behaviors. Feeding activities, in particular, involve complicated communication to pass on the locations of food sources.

3. Karl von Frisch first described the signaling system, hypothesizing that two types of behavior signaled two types of food. Signaling in the dark hive mostly involves physical contact and sound. Some modern researchers question von Frisch's model and are exploring alternatives.

4. "Round dances" indicate that food is near. "Waggle dances" indicate the distance and direction of the food. The waggle's speed indicates distance, and the angle relative to the hive's vertical position indicates direction relative to the horizontal position of the sun (straight up is equivalent to flying directly toward the sun). An internal clock must also be involved because the bees maintain their sense of where the food is even if several hours of sunless storm interrupt feeding (Figure 37.19B, Master 37.19C, Acetate 37.19).

H. Altruistic acts can be explained by evolution (*Module 37.20*).

1. Altruism is behavior that reduces an individual's fitness while increasing the fitness of a recipient. For example, naked mole-rats have a social structure resembling that of honeybees. The queen mates with three males (kings), while the rest of the colony consists of nonreproductive males and females who care for the queen and kings.

2. At first glance, altruism seems nonadaptive. However, it is thought to have evolved by either of two mechanisms, **kin selection or reciprocal altruism**.

3. Kin selection states that altruistic behavior would evolve in groups of related individuals because it increases the number of copies of a gene common to the whole group. For example, the great majority of the individuals in a naked mole-rat colony are closely related. The queen appears to be either a sibling, daughter, or mother of the kings; the nonreproductive individuals are either the queen's direct descendants or her siblings. Hence, by promoting the reproductive success of the queen and kings, the nonreproductive individuals are increasing the probability that copies of their own genes will be represented in the next generation.

NOTE: The example given of altruistic behavior in honeybee hives is a special case. In a sense, the whole hive is one individual with adaptive behavior among its parts. The hive is composed of one reproductive queen and thousands of sterile workers. The workers function like cells of other organisms, where the loss of one is not missed by the whole.

4. Reciprocal altruism is an altruistic act repaid at a later time by the beneficiary of the act (or social associates of the beneficiary). For instance, chimpanzees will sometimes save the life of a nonrelative. If a future "repaying in kind" always (or usually) follows, the altruistic behavior increases the fitness of an individual in the long run.

I. Talking About Science: Edward O. Wilson promoted the field of sociobiology and is a leading conservation activist (*Module 37.21*).

1. Behavioral ecologist Dr. E. O. Wilson's 1975 book, *Sociobiology: The New Synthesis*, promotes the idea that social behavior is genetically based and evolves along with other animal traits. Most of the book deals with nonhumans; however, two chapters discuss humans. This book rekindled the old debate about which most strongly influences human behavior: genes (nature) or learning (nurture).

2. In a more recent book, *The Diversity of Life*, Wilson discuss a "biodiversity ethic." This ethic holds that humans should never knowingly allow a species to go extinct if measures to save it can be implemented. He holds that biodiversity has an inherent value to humans apart from the physical welfare it provides.

3. In his autobiography, *Naturalist*, Wilson states his philosophy: "the human species is a product of biological evolution; that human beings arose in an arena of natural environments and biodiversity, and that therefore natural environments are a precious part of human heritage; and finally, that neither philosophy nor religion can ever make much sense unless they take the first two points into consideration."

J. Culture may reinforce innate human social behavior (*Module 37.22*).

1. Some argue that cultural change is so rapid that it now overwhelms any evolutionary influence on human behavior.

2. Alternatively, sociobiologists argue that human culture develops in such a way as to enhance behavioral activity derived through evolution.

3. The challenge in sociobiology is to uncover the ultimate causes of human cultural traits such as education.

Preview: Modules 38.8–38.10 discuss the three stages of human culture.

CLASS ACTIVITIES

1. Demonstrations of behavior are the clearest way to describe and explain the principles. Although social behaviors (other than those of students) are difficult to demonstrate in a classroom, individual types of behavior can sometimes be shown, particularly by animals that are familiar with their laboratory homes. In small classrooms or laboratory settings, reptiles and fish will show territorial, agonistic, and mating behaviors.

2. Many acquaria and zoos provide naturalistic environments in which a variety of normal (and abnormal) behavior can be observed. Ask zoo personnel when the best times are to observe behavior. Have small groups of students observe for one-half to one hour then describe a behavior and its function or distinguish between normal and abnormal behavior.

RESOURCES AND REFERENCES

Alcock, J. *Animal Behavior: An Evolutionary Approach*, 4th ed. Sunderland, MA: Sinauer Associates, 1989.

Ants and Evolution. Burlington, NC: Oxford Scientific Films/Carolina Biological Supply Company, 1986. A 28-minute videotape introduced by E. O. Wilson that describes his leaf-cutter ant–fungus colony and includes an interview on the subject of sociobiology.

The Birth of Bees. Burlington, NC: Oxford Scientific Films/Carolina Biological Supply Company, 1986. A 60-minute videotape that describes the evolution of cooperative social behavior in wasps and bees.

Borgia, G. "Sexual Selection in Bowerbirds." *Scientific American*, June 1986. The female bowerbird's mating choice depends on how well the male adorns its mating site.

Buck, J., and E. Buck. "Synchronous Fireflies." *Scientific American*, May 1976. Certain species of firefly in Asia and the Pacific islands flash in unison, unlike fireflies of temperate regions that flash unsynchronized. The evolutionary role of this behavior is discussed.

Davies, N., and M. Brooke. "Coevolution of the Cuckoo and Its Hosts." *Scientific American*, January 1991.

Diamond, J. "Sexual Deception." *Discover*, August 1989. Speculations on the evolutionary basis of seduction.

Gorilla. Washington, DC: National Geographic Society, 1981. A 59-minute videotape that describes the efforts of zoos to provide the proper physical and social habitat for mountain gorillas.

Gould, J.L., and P. Marler. "Learning by Instinct." *Scientific American*, October 1986. A review of the interrelationships between learning and innate behaviors in a number of animals.

Greenspan, R.J. "Understanding the Genetic Construction of Behavior." *Scientific American*, April 1995.

Gwinner, E. "Internal Rhythms in Bird Migration." *Scientific American*, April 1986. Migratory birds have a clock that tells them when to begin and end their flight. It is based on rhythms with a period of about a year.

Interactive Nova: Animal Pathfinders. New York: Scholastic Software, 1991. An interactive videodisc with Macintosh computer software that focuses on animal behavior, migration, and adaptation. The program links full-motion video and slides from the original Nova broadcast to a database of text and graphics HyperCards.

Introduction to Chimpanzee Behavior. Washington, DC: National Geographic Society, 1977. A 23-minute videotape featuring Dr. Jane Goodall observing wild chimpanzees as they feed, make tools, play, groom, court, and communicate.

Invertebrates: Conditioning or Learning? Washington, DC: National Geographic Society, 1975. A 15-minute videotape that describes experiments showing that some invertebrates can learn.

Jolly, A. "A New Science That Sees Animals as Conscious Beings." *Smithsonian*, March 1985.

Konrad Lorenz: Science of Animal Behavior. Washington, DC: National Geographic Society, 1975. A 14-minute videotape that focuses on the work of this pioneer ethologist.

Krebs, J.R., and N.B. Davis. *Behavioural Ecology*, 3rd ed. Cambridge, MA: Blackwell Scientific, 1991. An upper-division text.

Larsen, T. "Butterfly Mass Transit." *Natural History*, June 1993. The migratory behavior of a number of butterfly species on several continents.

Lorenz, K. *On Aggression*. New York: Harcourt Brace Jovanovich, 1966. A classic book on competitive social interactions.

Nova. *The Private Lives of Dolphins*. Princeton, NJ: Films for the Humanities and Sciences, 1993. A 60-minute videotape that investigates the sophisticated social system that has evolved in dolphins and the clues this knowledge provides about the origins and purpose of big brains and intelligence in these mammals.

Pfennig, D.W., and R.W. Sherman. "Kin Recognition." *Scientific American*, June 1995. The role of kinship in social evolution.

Pietsch, T.W., and D.B. Grobecker. "Frogfishes." *Scientific American*, June 1990. Masters of aggressive mimicry (displaying a tiny bait on a line in front of their mouths), these voracious carnivores can gulp prey faster than any other vertebrate predator.

Sense of Timing. Deerfield, IL: Coronet/MTI Film and Video, 1989. A 30-minute videodisc that examines the roles rhythms play in the lives of numerous invertebrate and vertebrate animals.

Sociobiology. New York: Time-Life Video, 1977. A 57-minute videotape of interviews expressing views for and against E. O. Wilson's ideas, including one with E. O. Wilson.

Wiley, R. Haven, Jr. "The Lek Mating System of the Sage Grouse." *Scientific American*, May 1978. An examination of the mating rituals of this unusual North American species.

Williams, T. C., and J.M. Williams. "An Oceanic Mass Migration of Land Birds." *Scientific American*, October 1978. Using radar to follow the journeys of songbirds and small shorebirds that migrate up and down the eastern coast of North and South America.

Wilson, E. O. *Sociobiology: The New Synthesis*. Cambridge, MA: Harvard University Press, 1975.

38

HUMAN EVOLUTION AND ITS ECOLOGICAL IMPACT

APPROACH

This is the time to gather together all the biological and ecological issues introduced in other chapters. Almost all the previous discussions are referred to here: acid rain, global warming, pollution, biodiversity, and threats from disease. You may have your own issues to add. Of all the themes introduced in this chapter, the loss of biodiversity and human population growth are the most serious: When considering human evolution and human impacts, they really are the ultimate themes. The most serious cause of loss of biodiversity is habitat destruction in the face of tremendous growth in human population and increased consumption of resources.

Beginning with the Opening Essay, the authors present some ironic intersections among the subjects in this chapter. Madagascar is home to the largest group of prosimians (the lemurs), the living group most similar to the earliest primates and now one of the most threatened of all animal groups.

As with the topic of evolution in general, and probably even more so, human evolution can be a touchy subject. Articles in *NCSE Reports* cover the major areas of controversy between scientists and creationists and can help you provide a sensitive and current approach to the subject of human evolution.

All the messages in this chapter are part of continual current debate, and you may want to spend extra time on them. Considerable resources are provided below to help you fill in details.

If you are using the text in sequence, it is time for a wrap-up of the course. In addition to the review incorporated in the chapter's content, you might want to retrace the dominant themes of life, introduced in Chapter 1. Particularly because of the serious tone of the last chapter, it is important to instill some sense of optimism in this part of your lecture. Emphasis on the positive relationships between humans and biology is one way: benefits to medicine, resources provided, understanding of how humans function biologically and culturally. Hopefully, students will have gained a realistic sense of the human position in ecosystems and the evolutionary time scale. Your final point could be that by taking a course like this, each student becomes a more thoughtful member of the human species.

CHAPTER OBJECTIVES

Name and describe the major groups of primates.

Trace the evolutionary development of hominids, discussing the changes in morphology, habit, and culture of the several species of *Australopithecus* and *Homo*.

Distinguish between the multiregional and monogenesis hypotheses of the origins of modern humans from their immediate ancestors.

Outline the phases in the cultural evolution of humans, and explain why this one characteristic of humans alone has brought about global environmental crises.

Explain the roles that technology, population growth, and species themselves play in the following threats to global environments: acid precipitation, ozone depletion, chemical toxicity, global warming, and loss of biodiversity.

State why the loss of biodiversity is such a serious problem.

LECTURE OUTLINE

I. Introduction.

  A. Focus on the human destruction of habitat (*Opening Essay*).

   1. *Review*: Islands as centers of specialization (Module 15.5).

   2. Bedo Jaosolo lived in Madagascar and was recognized as a skilled naturalist and guide, particularly around one of Madagascar's small nature reserves, Analamazoatra. Tragically, he was killed in 1989. His life and interests symbolize "the role to be played by a new generation of . . . biologists and naturalists, a small, crucial group of young people who care about nature."

   3. Madagascar is a large, isolated island off the eastern coast of Africa that was home to more than 200,000 species of plants and animals, 80% found nowhere else. Due largely to pressures from a population of 10 million people, Madagascar has lost 80% of its forests and about 50% of its native species. This country symbolizes the plight of the biosphere as a whole.

  B. Central issues of this chapter.

   1. Trace the evolutionary roots of humans, the not-entirely-worked-out evolutionary pathway of humans, and the pathway of cultural evolution.

   2. Review the major environmental problems humans face. The picture is not an optimistic one, but in order to fix these things, we need to understand them.

  C. Human beings are the most environmentally destructive animals to ever live (*Module 38.1*).

   1. *Review*: The human population explosion (Module 35.9).

   2. An underlying reason for the population explosion is human manipulation of natural forces (agriculture, control of disease, other biological resources). Other species are completely controlled by natural forces.

   3. Our manipulative abilities stem from our evolutionary heritage. Through the evolution of our intellect and social frameworks, we can develop and use complex tools and transmit learned information across generations.

   4. Manipulating the environment involves complex interrelationships; therefore, when we solve a problem in one place and time, we very likely produce another problem in another place and time. Many of our most serious ecological problems have global consequences.

   5. These problems are serious enough to threaten our own existence.

II. Human evolution.

  A. The human story begins with our primate heritage (*Module 38.2*).

1. *Review*: Geological timeline (Module 16.1), the comparison of human and baboon skeletons (Module 30.3), and the role of paedomorphosis in human evolution (Module 16.7).

2. The earliest primates were small, arboreal creatures similar to modern-day prosimians that lived at the end of the Age of Dinosaurs 65 million years ago (mya) (Module 16.5), and a group that expanded considerably during the early part of the Age of Mammals.

3. Most living primates are arboreal, and humans (never strictly arboreal) retain in their bodies many traits that evolved with our arboreal relatives. Primitive primate characteristics (as exhibited by the slender loris, a prosimian) include flexible shoulder and hip joints, maneuverable hands and feet, opposable thumbs and big toes, sensitivity to touch in the hands and feet, short snout, and eyes close together at the front (enhancing three-dimensional vision) (Figure 38.2A).

4. Modern-day primates include prosimians, monkeys, apes, and humans.

5. Prosimians include lorises, bushbabies, tarsiers, and lemurs. All live in tropical forests, and all are threatened by habitat destruction. Of 40 species of lemur to live on Madagascar, 18 are extinct because of human incursion into their habitat there.

6. The anthropoids include monkeys, apes, and humans. These differ from prosimians in having relatively larger brains and relying less on smell and more on eyesight. The earliest anthropoids were probably monkeylike primates that evolved from prosimian ancestors in Africa and Asia about 40 million years ago.

7. Monkeys differ from apes and humans in having a tail and equal-length forelimbs and hind limbs. They originated in Africa and migrated to other countries in Asia and the Americas. Old-world monkeys such as ground-dwelling baboons and many arboreal species never have prehensile tails and have narrow, close-set nostrils. New-world monkeys often have prehensile tails and have nostrils wide open and far apart (Figures 38.2B, C).

B. Apes are our closest relatives (*Module 38.3*).

1. Apes include the gibbon, orangutan, gorilla, and chimpanzee. All are tropical, lack tails, and have longer forelimbs than hind limbs.

    Review: In Module 16.13 alternative classifications of humans are discussed. Humans could very easily be placed in the same family as the apes.

2. Gibbons are the only apes that are entirely arboreal and monogamous for life. Nine species are found in Southeast Asia (Figure 38.3A).

3. The orangutan is a shy, solitary ape, mostly arboreal, living in forests in Borneo (Figure 38.3B).

4. The gorilla is the largest of all primates and spends most of its time on the ground in African rain forests (Figure 38.3C).

5. The chimpanzee (and very similar bonobo) inhabit rain forests in central Africa and African savannas. Many aspects of chimpanzee behavior resemble human behavior. They are fully capable of innovative behavior, can learn human sign language, and very likely have complex self-awareness. Recent biochemical evidence shows that chimpanzees and humans share over 97% of their DNA sequences (Figure 38.3D).

Review: In Module 37.16 Jane Goodall discusses dominance hierarchies and cognition in chimpanzees.

NOTE: Consider how different the classification of humans and chimpanzees might be were humans not doing the classifying, or how the classification scheme might differ if two nonhuman species exhibiting the same evolutionarily close relationship were being considered.

C. The human branch of the primate tree is only a few million years old (*Module 38.4*).

   1. Humans diverged evolutionarily from apelike ancestors, probably about 4–8 mya (Acetate 38.4).

   2. There are several difference species (and several genera, two of which are shown on the timeline) in the human lineage. Evolutionary connections between the species illustrated (and others not illustrated) are hotly debated. The figure shows when each species lived (as determined from the known fossil record) but does not join them in a phylogenetic tree.

   3. The first hominids were a diverse group called australopithecines. The different australopithecine species are distinguished on the basis of skull and tooth anatomy. The earliest and most primitive species may have been *Australopithecus ramidus*, which has been identified from fossilized fragments dated to 4.4 mya. The relationship between this species and *A. afarensis* (3–4 mya) is the subject of much debate. Three other species, *A. africanus*, *A. robustus*, and *A. boisei* lived at the same time as early members our own genus, *Homo*. The australopithecines and other species of *Homo* died out.

D. Upright posture evolved well before our enlarged brain (*Module 38.5*).

   1. Footprints that may be those of *A. afarensis* indicate that hominids have been bipedal for well over 4 million years. It is debatable whether the australopithecines were on the direct line to *Homo*. Some think they are a separate branch (that became extinct about 1.4 mya), and that the ancestor of *Homo* is older (Figure 38.5A).

   2. *A. afarensis* was a small-brained (and small) hominid that lived mostly on nuts, seeds, bird eggs, and scavenged kills from other predators. Spending most time on the ground, they probably became arboreal only to escape predators or scavenge (Figure 38.5B).

   3. Brain enlargement is first evident in fossils from East Africa dated at around 2.5 mya. Skulls that are intermediate in size between *Australopithecus* and *Homo sapiens*, some of which are found with associated stone tools, have been assigned to the species *Homo habilis* ("handy man"). This hominid species coexisted for almost 1 million years with later australopithecines on the African savanna, probably living in much the same way.

E. *Homo erectus* gave rise to *Homo sapiens* (*Module 38.6*).

   1. "Upright man" lived from 1.8 mya to 300,000 years ago. *H. erectus* was taller and had a larger brain and a more advanced culture than *H. habilis*. *H. erectus* lived in huts or caves, built fires, wore clothes, and made more complex tools.

   2. *H. erectus* spread out of Africa to Eurasia. With its broad geographic distribution, *H. erectus* became regionally diverse. One or more populations probably gave rise to our own species, *Homo sapiens*.

Review: Chapter 15 discusses geographic patterns of speciation.

F. When and where did modern humans arise? (*Module 38.7*).

   1. Fossil remains of the oldest representatives of *H. sapiens* date back 300,000 years, are restricted to Africa, and are likely the descendants of *H. erectus*. These earliest forms are sometimes separated as "archaic *Homo sapiens*" and were generally more heavily boned than modern humans, with thicker skulls and more pronounced brow ridges. We know most about one group of the earlier form, the Neanderthals. They lived in Europe, the Middle East, and parts of Asia from 130,000 to 35,000 years ago. They were short and stocky, were skilled toolmakers, and participated in burials and other rituals that required abstract thought (Figure 38.7).

   2. Fossil remains of the older representative modern *H. sapiens* (identical morphologically with present humans) are known from about 100,000 years ago from Africa (and nearly as old from Israel). They coexisted with the archaic stock until the latter disappeared about 35,000 years ago.

   3. The "multiregional hypothesis" of our origin states that humans arose from several archaic populations in Africa, Europe, and Asia, that modern races of humans stem from regional diversity that has been partly separate for as much as a million years, but that interbreeding among neighboring populations mixed most genes.

   4. The "monogenesis hypothesis" states that modern humans arose from a single population of the archaic stock that lived in Africa about 100,000 years ago. Neanderthals and all other populations of the archaic stock were evolutionary dead ends; they did not interbreed with the modern human line at all. This hypothesis is supported by analyses of the DNA in human (female) mitochondria.

III. Human cultural evolution.

A. Culture gives us enormous power to change our environment (*Module 38.8*).

   Review: Human population growth (Module 35.9), the impact of outside disturbances on community structure (Module 36.6), and the sociobiology of culture (Modules 37.21 and 37.22).

   1. Three milestones highlight human evolution: (a) erect stance, requiring skeletal remodeling; (b) brain enlargement, with prolonged postbirth development of the skull and its contents (Module 16.7); and (c) evolution of prolonged childhood, during which cultural information is passed between generations.

   2. Culture includes the accumulated knowledge, customs, beliefs, arts, crafts, and ideas that are passed between generations. Culture, more than any other human feature, and because it can change at a faster rate than biological evolution, has allowed our species to transcend the limitations imposed on other species.

B. Scavenging-gathering-hunting was the first stage of culture (*Module 38.9*).

   1. This way of life began with the earliest australopithecines and continued to be the way of life for the human lineage until about 100,000 years ago. Hunting became important only with the advent of sophisticated tools 50,000 years ago.

   2. Some modern cultures (e.g., the !Kung of southwestern Africa) still practice hunting-gathering (Figure 38.9).

3. The first major impact of human culture on the environment may have been to cause the extinction of some early predators, perhaps by hominids (as early as 1.5 mya) becoming skilled at stealing their killed prey. The use of thrown weapons and cooperative hunting definitely affected populations of prey species in Europe, Australia, and the New World.

4. With the development of more sophisticated hunting tools comes evidence of human-caused decimation of species such as the woolly rhinoceroses and giant deer of Europe.

5. Humans reach Australia about 50,000 years ago, an event which may be responsible for the extinction of the giant kangaroos.

6. Nomadic hunters migrated from Asia to North America about 11,000 years ago. Their actions are likely to have caused the extinction of many large animal species.

7. Other characteristics of this level of culture include the organization into communal groups that divided labor, the use of semipermanent homes, trading among populations, and (toward the end of the period) the growing of a few simple crops.

C. Agriculture was the second stage of culture (*Module 38.10*).

1. Agriculture developed in Eurasia and the Americas 10,000 to 15,000 years ago. Plots of rain forest were cut and burned, and crops were planted. Early Asian farmers domesticated and grew rice and millet.

    Review: The effects of slash and burn on tropical forests are discussed in Module 34.10.

    NOTE: Shifting (slash-and-burn) cultivation by modern peoples actually makes ecological sense in tropical forest areas, if it weren't for the incredible pressures placed on habitat by human populations that are too large. Small plots are cleared, farmed for a while, and then allowed to return to the native forest, during which time products from the successional stages are gathered. Scattered, small plots protect the land from the erosion that occurs when large areas are cleared. In addition, farmers take advantage of the high productivity of the land during the early stages of succession (Module 36.6), and later stages of the forest return what little fertility the soil can hold.

2. Increases in populations made possible by agriculture, and the use of more advanced plows and overgrazing, probably caused serious soil erosion in the Fertile Crescent, turning a rich forest area into a desert (Figure 38.10).

    Review: The runoff from the use of fertilizers can result in algal blooms (Module 17.14). Also, review soil conservation (Module 32.9) and organic farming (Module 32.10).

3. Agriculture changed forever the relationship between humans and the biosphere. Large areas of native vegetation have been converted to farming use. Agriculture allows the establishment of permanent settlements and cities, and it frees many members of cultures to specialize in other activities.

D. The machine age is the third stage of culture (*Module 38.11*).

Review: Human population growth (Module 35.9).

1. The change from small hand tools to large-scale machines also produced major effects on most human activities.

   NOTE: The first industries built up around the fabrication of metal tools for hunting and farming, long before the machine age.

2. Complex machines further reduced the need for agricultural workers. Energy consumption increased. Medical advances reduced deaths.

3. During all these cultural changes, humans have not changed genetically in any significant way. Rapid cultural change has changed environments that humans and other species live in. Nothing is new about this environmental change except the speed of the change, which now vastly outpaces the rate of biological evolution.

IV. Human impact on the biosphere.

A. Technology and the population explosion compound our impact on the biosphere (*Module 38.12*).

1. *Review:* Major prehistoric extinction events (Module 16.5).

2. Natural global crises that caused the major prehistoric extinction events have catastrophic effects on organisms and ecosystems, and they also have creative effects. For example, the extinction of the dinosaurs about 60 mya cleared the way for the rise of the mammals and birds.

3. Today, the impact of a single species, *Homo sapiens*, is causing a series of global crises. Cultural change has produced benefits but also puts a strain on many natural systems.

   Review: Farming and soil conservation (Module 32.9) and human population growth (Module 35.9).

4. The effects of cultural change are not produced evenly from all cultures, but are concentrated from the cultures of developed countries, particularly the United States. Although the United States has only 5% of the world's population, it uses far more than 5% of the world's resources. The United States uses more energy than the total populations of Central and South America, Africa, India, and China combined (Table 38.12).

5. *Review*: Acid precipitation (Module 2.16).

6. Such high resource use could deplete the stores of these resources. It can also result in other effects, such as increased acid precipitation, and the release of harmful chemicals, such as CFCs (chlorofluorocarbons used for refrigeration, which are now phased out) that deplete the ozone layer.

7. *Review*: Use of agricultural pesticides (Module 34.3).

8. The story of the pesticide DDT illustrates a particular danger of many toxic substances—biological magnification. The concentration of DDT increases in each succeeding trophic level in food chains. DDT is so concentrated in tertiary-consumer birds that it interferes with egg shell production in females. Although banned in the U.S., DDT is still used in many other countries. The DDT story also illustrates another point: Technological solutions that hold great promise may hold hidden dangers (Acetate 38.12B).

   Review: Energy flow through ecosystems (Modules 36.11).

   NOTE: This point makes a good case for a general conservative approach to applying any new technology.

B. Carbon dioxide and other gases added to the atmosphere may cause global warming (*Module 38.13*).
   1. *Review*: The introduction to global warming, the carbon cycle, and the equations for cellular respiration and photosynthesis (Modules 1.8, 5.21, 7.11, 7.13, 7.14 and 36.15).
   2. Carbon dioxide concentration in the atmosphere, as measured in Hawaii, has increased dramatically (over 10%) in the last 35 years and is projected to rise faster in the future (Masters 38.13A, C).
   3. CO_2 is one of the greenhouse gases. (CH_4 and NO are also greenhouse gases, from fossil-fuel burning, among other reasons.) These gases are good, in moderation, because they trap heat and allow Earth's atmosphere to be warmer than it would otherwise be (Acetate 38.13B).
   4. Global warming results from an excess amount of these gases. Increased temperatures could compound the problem by increasing the rates of respiration of soil bacteria. Some predictions show the mean global temperature rising by 2°–5°C over the next 50–100 years (Acetate 38.13C).
   5. This could cause sea levels to rise by 1 m, drastically affecting many important human settlements.
   6. The carbon cycle is complex, and it is difficult to be certain of the timing and extent of global warming. Some experts suggest that increases in greenhouse gases will affect cloud cover, or that marine algae will take up CO_2—both of which are situations that would tend to decrease temperature. Some researchers and politicians say we need more data.

   NOTE: A good analogy to use here is to ask the students what they would do if they had a tumor. Would they wait for it to metastasize to confirm that it was cancerous, or would they take precautions?

   7. To counter the trend, we need international cooperation, decreases in fossil-fuel consumption, reduced forest destruction, increased forest planting, and the general improvements that changes in consumer-based lifestyles in developed countries would bring.

   NOTE: Another major consequence of global warming is that tropical diseases such as malaria, schistosomiasis, filariasis, African sleeping sickness, yellow fever, and others are likely to increase their geographic range.

C. The consequences of tropical forest destruction shows that local activities can change the biosphere (*Module 38.14*).
   1. *Review*: The ecology of tropical rain forests (Module 34.10) and the role of tropical forest transpiration in the water cycle (Module 36.14). Coniferous forests are also threatened with destruction (Module 18.10).
   2. Extensive rain forests in Central and South American, Southeast Asia, and central Africa are being destroyed to support national economies. The cleared land is used for crops (often those, like coffee, consumed in developed countries) or grazing. An area roughly the size of Washington state is cleared annually.
   3. Although a local phenomenon, the loss of these forests has global consequences. Rain forests play important roles in the carbon and water cycles, releasing CO_2 and transpiring water. The loss of biodiversity inherent in these forests is also a major problem. These species represent a

storehouse of potentially useful human products (medicines, foods, and other useful tools).

Review: The water cycle (Module 36.14) and the carbon cycle (Module 36.15).

NOTE: With every outbreak of *Ebola,* the media has something to report about emerging viruses. With the destruction of forests, we are more likely to come in contact with such viruses (and bacteria).

4. Completely conserved tropical forest reserves (Module 36.20) and extractive reserves (which allow harvesting of natural products at natural rates of production) will help protect endangered species.

NOTE: But these small areas of land can have little effect on the global consequences to the carbon and water cycles.

Review: Eloy Rodriguez's studies of the potential health benefits of plant species (Module 33.13) that may disappear as a result of such destruction.

D. The rapid loss of species is a warning that the biosphere is in trouble (*Module 38.15*).

1. Habitat destruction is one of the main forces causing the extinction of species. For example, housing developments and other examples of human disturbance of the Florida Keys have decimated the populations of key deer, from 400 in the early 1970s (an increase over much lower numbers after hunting reduced the population to about 50 in the early 1900s) to about 240 today (Figure 38.15A). The deer are now dying on the roads faster than they can reproduce.

Review: Lichens can serve as environmental indicators since they are very sensitive to airborne pollutants (Module 18.19).

2. About 1.5 million species of organisms have been identified. Perhaps as many as 2.5 million are yet to be discovered. Because tropic forests contain about 80% of the world's species, their destruction will mean major losses of organisms. Forest destruction also disrupts migratory patterns, and the effects of climate change may also cause loss of species.

3. Overhunting and poaching also pose problems for populations of larger animals or of animals that provide unique products, like ivory (Figure 38.15B).

4. Competition, predation, and parasitism between nonnative and native species are also important forces leading to species loss.

Review: These symbiotic relationships are discussed in Module 36.5.

5. Some biologists predict that within 100 years, as much as half of Earth's current species will be extinct.

6. We should worry about species loss because it is an indication of the seriousness of global threats to ecosystem function, because we depend on other species for resources (food, clothing, shelter, oxygen, soil fertility), and because the species have their own right to be here.

E. Responsible environmental decisions require an understanding of the natural world (*Module 38.16*).

1. Considering present numbers, global extent, and impact, humans are one of the most successful species to ever inhabit Earth.

2. But in developed nations, we are using way more than our share of resources, and, as a result, are having a disproportionate effect on the world's human population.

3. It is unlikely that human nature will change suddenly. But we must become more aware of our surroundings and our impact on them.

CLASS ACTIVITIES

1. The free videotape *Scientists and the Alaska Oil Spill*, available from the Exxon Corporation, describes the various scientific approaches used in attempts to control and clean up oil spills. Emphasize the criteria that distinguish "good" science. Who did the research, and what were their qualifications? Who funded the research, and why? When was the research done, and has it been subjected to peer review?

2. To wrap up and review, display representative biological structures (either living or illustrations) and ask students which of the central themes of biology is best illustrated by each structure. Some examples are a TEM of a mitochondrion, a mushroom, a plant leaf, a bat, a pair of lizards, and an illustration of a subalpine forest.

3. This chapter provides the best opportunity to get students involved. Encourage involvement on a personal level (recycling) or on a larger level (e.g., volunteering in litter pickup). Ask the students what, if anything, needs to be done to preserve the environment. What are the relative merits of environmental and economic (both pro- and anticonservation) arguments?

4. Invite a representative of an environmental society, such as the Sierra Club, to speak to the class; invite speakers from contrasting groups on both sides of these issues.

RESOURCES AND REFERENCES

Biodiversity: The Videotape. Washington, DC: National Academy Press, 1988. A 45-minute videotape of an edited version of a teleconference involving some of the United States' main authorities in the crisis concerning loss of biodiversity. Includes a pictorial introduction to biodiversity.

Broeker, W. "Global Warming on Trial." *Natural History*, April 1992. How solid is the evidence that Earth is heating up?

Brown, L. *State of the World*. New York: Norton, 1996. A yearly publication of the Worldwatch Institute, assessing the state of the biosphere.

Bunkley-Williams, L., and E.H. Williams, Jr. "Global Assault on Coral Reefs." *Natural History*, April 1990. What's killing the great reefs of the world?

Can Polar Bears Tread Water? (The Changing Climate). Deerfield, IL: Coronet/MTI Film and Video, 1990. A 52-minute videodisc that investigates the factors influencing the greenhouse effect.

Children of Eve. Deerfield, IL: Coronet/MTI Film and Video, 1987. A 60-minute videotape that describes the (now debated) evidence for a single female ancestor of all modern humans.

"Choosing the Future." *Audubon*, March/April 1993. A set of brief articles documenting the extent of the world's ecological crises, with some revealing maps.

Daily, G.C., and P.R. Ehrlich. "Population, Sustainability, and Earth's Carrying Capacity." *BioScience*, November 1992. A framework for estimating population sizes and lifestyles that could be sustained without undermining future generations.

Franklin, J.F., C.S. Bledsoe, and J.T. Callahan. "Contributions of the Long-Term Ecological Research Program." *BioScience*, July/August 1990. An expanded network of scientists, sites, and programs can provide crucial comparative analyses of ecosystems.

"Frontiers in Biology: Ecology." *Science*, July 21, 1995. A special issue on ecology.

Gomez-Pompa, A., and A. Kaur. "Taming the Wilderness Myth." *BioScience*, April 1992. How Western beliefs affect environmental policy.

Hammond, A.L. *World Resources*, 1990–1991. New York: Oxford University Press, 1990. A revised version of this book, prepared by the staff of the World Resources Institute, appears every 2 years as an updated guide to the global environment. The book is prepared in collaboration with the United Nations Environment Program and the United Nations Development Program.

Houghton, R.A., and G.M. Woodwell. "Global Climatic Change." *Scientific American*, April 1989. Evidence suggests that the production of carbon dioxide and methane from human activities has already begun to change the climate, and that radical steps must be taken to halt any further change.

Interactive Nova: Race to Save the Planet. New York: Scholastic Software, 1991. An interactive videodisc with Macintosh computer software that focuses on global ecology and the effects of air and water pollution, waste disposal, and global warming. The program links full-motion video and slides from the original Nova broadcast to a database of text and graphics HyperCards.

Johanson, D. and M. Edey. *Lucy: The Beginnings of Humankind*. New York: Simon and Schuster, 1981. The story of the discovery of the first *A. afarensis* and its significance to the evolution of humans.

Man and the Biosphere. Princeton, NJ: Films for the Humanities and Sciences, 1991. A series of twelve 28-minute videotapes that looks at the ecological relationships in a variety of ecosystems, illustrating the problems imposed by human incursion into each. Based on UNESCO's *Man and the Biosphere* Program.

"Managing Planet Earth." *Scientific American*, September 1989. A special issue devoted to the environment.

May, R.M. "How Many Species Are There on Earth?" *Science*, September 16, 1988. A survey of the current best estimates of known and unknown biodiversity.

Milton, K. "Diet and Primate Evolution." *Scientific American*, August 1993. Many characteristics of modern primates, including our own species, derive from an early ancestor's practice of taking most of its food from the tropical canopy.

Monastersky, R. "The Deforestation Debate." *Science News*, July 10, 1993. Estimates vary widely over the extent of forest loss.

Nelson, M., T.L. Burgess, A. Alling, N. Alvarez-Romo, W. F. Dempster, R.L. Walford, and J.P. Allen. "Using a Closed Ecological System to Study Earth's Biosphere." *BioScience*, April 1993. Initial results from Biosphere 2.

Race to Save the Planet. Washington, DC: Corporation for Public Broadcasting, 1990. The Annenberg/CPB Project, a ten-part series of 60-minute videotapes that provides a comprehensive, global overview of major environmental questions ranging from population growth to soil erosion. The series presents scientific background but focuses on solutions, constructive ideas, and new approaches to dealing with these problems.

Ray, G.C., and J.F. Grassle. "Marine Biological Diversity." *BioScience*, July/August 1991. A scientific program to help conserve marine biological diversity is urgently required. One of three articles documenting the nature of marine biodiversity.

Redford, K.H. "The Empty Forest." *BioScience*, June 1992. Many large animals are already ecologically extinct in vast areas of neotropical forest where the vegetation still appears intact.

Rosenzweig, C., and G. Ropes. *Hothouse Planet*. Danbury, CT: Educational Materials and Equipment, 1990. This computer simulation focuses on the interrelated factors that may cause global warming. Produces projected changes and their effects for specific regions of North American. Apple II, Macintosh, and IBM PC formats. Available from Carolina Biological Supply Company.

Santer, B.D., et al. "A Search for Human Influences on the Thermal Structure of the Atmosphere." *Nature*, July 4, 1996. An international panel's evaluation of global warming.

The Search for Neanderthal. Princeton, NJ: Films for the Humanities and Sciences, 1988. A 28-minute videotape that discusses the ongoing controversy in the search for the origins of modern humans.

Sex and the Single Rhino. Princeton, NJ: Films for the Humanities and Sciences, 1992. A 60-minute videotape that examines the high-tech efforts to preserve Earth's animal diversity.

Shreeve, J. "Sunset on the Savanna." *Discover*, July 1996. Did our ancestors evolve mainly in forests or savannas?

Tattersall, I. *The Human Odyssey: Four Million Years of Human Evolution*. New York: Prentice-Hall General Reference, 1993. A readable and beautifully illustrated book based on the Hall of Human Biology and Evolution at the American Museum of Natural History.

Tattersall, I. "Madagascar's Lemurs." *Scientific American*, January 1993.

Tuttle, R.H. "The Pitted Pattern of Laetoli Feet." *Natural History*, March 1990. Who or what walked on the ancient African plain?

Washington, W.M. "Where's the Heat?" *Natural History*, March 1990. Greenhouse gases in the atmosphere are expected to raise temperatures, but shifts in ocean circulation may obscure the evidence of global warming.

Westman, W.E. "Managing for Biodiversity." *BioScience*, January 1990. A discussion of some unresolved scientific and policy questions.

"Where Did Modern Humans Originate?" *Scientific American*, April 1992. A debate about the origins of modern humans, including two articles documenting the single African origin (by Allan C. Wilson and Rebecca L. Cann) and the multiregional origin (by Alan G. Thorne and Milford H. Wolpoff).

Wilson, E.O. *The Diversity of Life*. Cambridge, MA: Harvard University Press, 1992. Why we should be concerned about the extinction of other species.

APPENDIX A

ORGANIZING A COURSE SYLLABUS

This section provides material to help you develop your syllabus. Most instructors will have general content and sequence in mind when they begin to plan a course syllabus using *Biology: Concepts and Connections*. The text is suitable, as is, for a one-year (perhaps, two-quarter) course, taught along a standard pedagogical sequence of chemistry/cells → evolution/diversity/systems → ecology. Those teaching the material in one semester or one quarter will wish to cut material, by either skipping chapters or skipping modules within chapters. Those who want to arrange the chapter topics around a different focus, pedagogical development, or seasonal need will have to rearrange the chapters accordingly.

Each of these options is considered below, with some general guidelines and methods you can use in designing your own course.

TIME

Shorter (one-quarter or one-semester) courses must eliminate some material, either whole chapters or modules from selected chapters. Depending on the pedagogical sequence you follow (see below), you may choose to greatly abbreviate or eliminate some areas, either chapters or whole units. For instance, many will want to cut much of the material from Units V and VI on animal and plant biology, following the chapters on evolution and diversity directly with those focusing on the environment. Others may want to emphasize the chapters on human biology and eliminate the chapters on diversity, plant structure and function, and many of the details on ecosystem structure and function.

In a one-quarter course, you may wish to include plant structure and function (some of the material from Chapters 31 and 32) following coverage of plant diversity (Chapter 18), and include three systems (digestive, immune, and nervous) following coverage of animal diversity (Chapter 19). Otherwise, follow the general progression of topics in the text.

DEPTH OF FOCUS

Many chapters (and most with conceptually difficult details of chemical processes, such as Chapter 6 on cellular respiration, Chapter 24 on the immune system, and Chapter 25 on controlling the internal environment) begin with several modules giving an overview, then move on to details, and, in some instances, review and go over a third level of detail. Human connections are made throughout the chapters but tend to be concentrated toward the ends of chapters. In a course with limited time, or one in which you want to minimize the chemical details, cover the background material at the beginning and skip to the human focus. The Approach sections of this guide include some additional thoughts on how to do this in the chapters where it should work.

PEDAGOGICAL SEQUENCE

Many instructors will have their own preferred sequence and specialties they want to present in a course. Below are three standard pedagogical sequences, each of which includes all the chapters in the text.

Chemistry and Cells First	Diversity First	Ecology First
Chapter 1
Chapters 2–7
Chapters 8–13
Chapters 14–16
Chapters 17–19
Chapters 20–30
Chapters 31–33
Chapters 34–38	Chapter 1
Chapters 2 (in part), 3 (in part)
Chapter 4
Chapter 8
Chapters 17–18
Chapters 31–33
Chapter 19
Chapters 20–30
Chapters 2–3, 5–7
Chapters 9–16
Chapters 34–38	Chapter 1
Chapters 2 (in part), 3 (in part)
Chapters 34–37, 38 (38.13–38.16)
Chapters 2–3 (remainder), 4–7
Chapters 8–13
Chapters 14–16
Chapters 17–19
Chapter 38 (38.1–38.12)
Chapters 20–30
Chapters 31–33

APPENDIX B

GENERAL RESOURCES AND REFERENCES

Included here are resources that provide useful background for more than one chapter in the text. If they present particularly good background for a given chapter, they are listed at the end of that chapter as well. If a resource is particularly good for a number of chapters in a unit, it will be indicated as such in the resource list for the first chapter in the unit. In addition, a few recent references are listed here that are not included in the Resources and References sections for the chapters.

Also included are general references with suggestions on how to organize and teach a course in biology to nonscience majors. Some separate journals that are particularly valuable sources of activities are also included.

Addresses for all sources that are not journal articles are in Appendix C.

Alberts, B., D. Bray, J. Lewis, M. Raff, K. Roberts, and J.D. Watson. *Molecular Biology of the Cell*, 3rd ed. New York: Garland, 1994. A comprehensive cell biology textbook for advanced students.

Ambron, Joanna. "Clustering: An Interactive Technique to Enhance Learning in Biology." *Journal of College Science Teaching*, November 1988. A technique of using free association to diagram course and lecture content, for both instructors and students. Can also be used in an activity to show the influence of right-brain and left-brain activity on creative thought.

The American Biology Teacher. Reston, VA: National Association of Biology Teachers. Published eight times a year, with articles on teaching and updating information content in biology. Usually includes articles featuring demonstrations to enhance student learning.

Becker, W.M., J.B. Reece, and M.F. Poenie. *The World of the Cell*, 3rd ed. Redwood City, CA: Benjamin Cummings, 1996. A student-oriented text.

Bierzychhudek, Paulette, and C. Gary Reiness. "Helping Nonmajors Find Out What's So Interesting About Biology." *BioScience*, February 1992.

Bio Sci II Videodisc. Seattle, WA: Videodiscovery, 1990. This is the best general multimedia resource for a great many visual aspects of biology instruction that is currently available. Includes stills, film clips, animations, demonstrations of technique, etc., with IBM PC and Macintosh interfaces.

BioShow. Redwood City, CA: Benjamin/Cummings, 1993. The videodisc to accompany the publisher's life sciences textbooks.

Carlson, Elof Axel. "A Model Introductory Course for the Life Sciences." *New Directions for Teaching and Learning*. No. 20. *Rejuvenating Introductory Courses*. San Francisco: Jossey-Bass, December 1984. Ideas on an innovative model for a course (centered on human genetics) that is meaningful to today's nonscience majors, in the context of the importance of biology in human affairs.

Cell Biology I Videodisc. Seattle, WA: Videodiscovery, 1987. An excellent source of film footage on a variety of topics centered on cell biology. Superb microcinematography from the Scientific Film Institute in Germany. Particularly good for subcellular structure and mitosis. Available with computer interfaces.

Cell Biology II Videodisc. Seattle, WA: Videodiscovery, 1994. Continued films on cell biology from the Scientific Film Institute. Particularly good for morphogenesis of protists and fungi, types of cell cleavage, gametogamy, and the cellular basis of animal development. Available with computer interfaces.

Cell Biology III Videodisc. Seattle, WA: Videodiscovery, 1994. Continued films on cell biology from the Scientific Film Institute. Particularly good for phototropisms, chemotropisms, and taxes, as well as tissue communications and parasitic infection processes. Available with computer interfaces.

Changeux, Jean-Pierre. "Chemical Signaling in the Brain." *Scientific American*, November 1993. Studies of acetylcholine receptors in the electric organs of fish have generated critical insights into how neurons in the human brain communicate with one another.

Cunningham, W.P. *Understanding Our Environment: An Introduction*. Dubuque, IA: W.C. Brown, 1994.

Darwin, C. *On the Origin of Species by Means of Natural Selection, or the Preservation of Favored Races in the Struggle for Life*. London: Murray, 1859.

"Disease Ecology." *BioScience*, February 1996. A special issue highlighting how disease organisms affect the global environment.

Ebert-Zawasky, Kathleen, and Gerald L. Abegg. "Integrated Computer Interfaced Videodisc Systems in Introductory College Biology." *ERIC Document ED 324240*. Paper presented at the National Association for Research in Science Teaching meeting, Atlanta, GA, April 1990. Includes a well-thought-out program of introduction and use of these resources.

Endler, J.A. *Natural Selection in the Wild*. Princeton, NJ: Princeton University Press, 1986. An excellent discussion of natural selection.

Flannery, Maura C. "Teaching Less Can Mean Better Teaching." *Journal of College Science Teaching*, May 1986. A brief point-of-view piece with a very pertinent method for biology teachers.

Gamlin, L. and G. Vines (eds). *The Evolution of Life*. New York: Oxford University press, 1986. A survey of life and life science.

Genetics. Cambridge, MA: Logal Software, Inc. Computer simulations of monohybrid and dihybrid crosses, sex linkage, incomplete dominance, codominance, multiple alleles, linkage and crossing over, population genetics, gene mapping, founder effects, environmental effects, and human genetic counseling.

Hartl, D.L. *Essential Genetics*. Boston: Jones and Bartlett, 1996. A good genetics text covering Mendelian and molecular genetics to evolution.

Huang, Samuel D., and Jane Aloi. "The Impact of Using Interactive Video in Teaching General Biology." *American Biology Teacher*, May 1991.

Journal of College Science Teaching. Washington, DC: National Science Teachers Association. Published six times a year, with articles relating to activities and alternative teaching methodologies to improve introductory science courses for science and nonscience majors.

Landman, Otto E. "Inheritance of Acquired Characteristics Revisited." *BioScience*, November 1993. Acquired heritable changes in traits, displayed in various experimental systems, can now be understood in terms of molecular genetics.

Lawson, Anton E., Steven W. Rissing, and Stanley H. Faeth. "An Inquiry Approach to Nonmajors Biology." *Journal of College Science Teaching*, May 1990. Uses some examples of how to deal with photosynthesis from an inquiry perspective in a class that starts with ecology and traces the historical development of biology.

Life Cycles Videodisc. Seattle, WA: Videodiscovery, 1985. A videodisc that includes motion picture and still-frame documentation on the reproductive activities of microbes, plants, and animals, emphasizing the mating and social behavior of a wide variety of animals. Available with computer interfaces.

Mange, E.L., and A.P. Mange. Basic Human Genetics. Sunderland, MA: Sinauer Associates, 1994. A basic text.

Marcus, Bernard A. "Community Colleges and the Nonmajors Biology Course." BioScience, October 1993. A brief review of the community colleges and students that explores ways to better suit the nonmajors course to its audience.

Margulis, L., and K.V. Schwartz. Five Kingdoms: An Illustrated Guide to the Phyla of Life on Earth, 2nd ed. New York: W.H. Freeman, 1987.

Marieb, E. Human Anatomy and Physiology, 3rd. ed. Redwood City, CA: Benjamin/Cummings, 1995. An excellent basic text.

Mathews, C.K., and K.E. van Holde. Biochemistry, 2nd ed. Redwood City, CA: Benjamin/Cummings, 1996. A general biochemistry text.

McKibben, Bill. "The End of Nature." The New Yorker, September 11, 1989. A moving, if somewhat pessimistic, discussion of the realms of natural history, their writers, and the inability of modern humans to come to grips with (or even comprehend) the changes in our natural environment.

Media Design Associates. BSCS Videodisc. A videodisc with transcribed BSCS Classic Inquiries presentations. Uses film footage from original experiments on a variety of biological processes, providing material for student observation, discussion, interaction, and data collecting. Topics covered are behavior, mimicry, mitosis, amoeboid locomotion, phototropism, peppered moth, and others.

Melear, Claudia T. "Profile of the Non-Major in College Biology by Learning Style." ERIC Document ED 325369. Paper presented at the Association for Psychological Type meeting, Atlanta, GA, May 1990.

Mellon, Margaret. "Altered Traits." Nucleus (Newsletter for the Union of Concerned Scientists), Fall 1993. A nice discussion of some of the potential risks of genetic engineering.

Miller, Judith E., and Ronald D. Cheetham. "Teaching Freshmen to Think—Active Learning in Introductory Biology." BioScience, May 1990. Describes ways to make introductory courses (for majors) more exciting.

Moore, J. A. Science as a Way of Knowing: The Foundation of Modern Biology. Cambridge, MA: Harvard University Press, 1993. A lively account by a famous developmental biologist.

Moore, R., W.D. Clarke, and K.R. Stern. Botany. Dubuque, IA: W.C. Brown, 1995. An excellent introduction to plant biology.

National Center for Science Education. The NCSE is the national clearinghouse for scientific information on the evolution-creation controversy, a source of information for students, members of the media, scholars, and anyone concerned about sectarian attacks on science education. Articles in the journal Creation/Evolution cover the major areas of controversy between scientists and creationists— science, education, law, religion, and culture. Articles, book reviews, debates, and correspondence keep readers up to date on the latest issues in the controversy and explain current evolutionary science to the nonspecialist.

The Nature of Sex. New York: WNET, 1993. Six 1-hour videotapes from the recent PBS series: (1) overview of sexual reproduction in nature, (2) effects of environmental forces, (3) courting behavior of males, (4) evolution of human sexuality, (5) protection of eggs and young, (6) life after sex.

OmegaWare. Courseware that provides a series of 52 Macintosh program modules in the Basic Biology Series, used in the freshman biology course at Colorado State University. The programs are tutorials with some experimental simulations, both including graphic displays, animation, and student interaction. Also available are sets of programs that access the *Bio Sci I* and *Bio Sci II* videodiscs from Videodiscovery.

Optical Data Corporation. *Mechanisms of Stability and Change*. Warren, NJ: Optical Data Corporation, 1991. A double-sided videodisc that focuses on the maintenance of biological stability/homeostasis from a variety of perspectives, including physiology, genetics, evolution, and ecology.

Optical Data Corporation. *Principles of Biology*. Warren, NJ: Optical Data Corporation, 1991. A videodisc with a variety of still and motion images for teaching biology.

Pääbo, Svante. "Ancient DNA." *Scientific American*, November 1993. Recent efforts to extract sequenceable DNA from the skin and bones of recently extinct species and from much older material.

Paul, Gregory S. "Giant Meteor Impacts and Great Eruptions: Dinosaur Killers?" *BioScience*, March 1989. An examination of evidence against the role of meteor impact on extinctions of dinosaurs and pterosaurs.

Pennisi, E. "Water: The Power, Promise, and Turmoil of North America's Fresh Water." *National Geographic*, 1993. A special issue examining use and abuse of water resources.

Reed, Christopher G. "What Makes a Good Teacher." *BioScience*, September 1989. How to give a good lecture, and the importance of showing concern for your students.

Reid, H. Bruce, Raghunath A. Virkar, and Robert W. Schuhmacher. "Outcomes Assessment as a Context for Evaluating the Biology Curriculum." *BioScience*, July/August 1992.

Ricklefs, R.E. *The Economy of Nature*, 3rd ed. New York: W.H. Freeman, 1993. An excellent basic ecology text.

Ridley, M. *Evolution*. Blackwell Scientific Publications: Boston, 1993. Covers Mendelian and molecular genetics and microevolution to macroevolution.

Scientific American Frontiers. Watertown, MA: Chedd-Angier Production Company. A series of five 1-hour television broadcasts on PBS that are shown one each month, from October to February. Each show consists of five or six (5- to 15-minute) segments on a variety of science subjects. The broadcasts may be copied with no restrictions on reshowing them. Free teacher guides, show dates, and information on past seasons' shows are available by calling (800) 523-5948.

Scientific American web page. http://www.sciam.com

Smith, J.M., and E. Szathmary. *The Major Transitions in Evolution*. New York: W.H. Freeman, 1995.

Suzuki, D., A. Griffiths, J. Miller, and R. Lewontin. *An Introduction to Genetic Analysis*, 4th ed. New York: Freeman, 1989. Includes Mendelian, molecular, and population genetics.

Taiz, L., and E. Zeigler. *Plant Physiology*. Redwood City, CA: Benjamin/Cummings, 1991.

Tyser, Robin W., and William J. Cerbin. "Critical Thinking Exercises for Introductory Biology Courses." *BioScience*, January 1991. Suggests a way to teach introductory biology students a model for evaluating information in science articles in the popular media.

Uno, Gordon E. "Teaching College and College-Bound Biology Students." *American Biology Teacher*, April 1988. An evaluation of some of the initial problems nonscience majors have in an introductory biology class.

WGBH. *The Secret of Life*. Princeton, NJ: Films for the Humanities and Sciences, 1993. Eight 1-hour videotapes from the recent PBS series, presenting the most up-to-date material on a range of molecular genetics topics suitable for most of Unit II and parts of Unit IV: basic molecular genetics, mutation and genetic diversity, aging, cancer, transgenic species, AIDS, gene therapy, and nature versus nurture.

APPENDIX C

ADDRESSES OF RESOURCE SUPPLIERS

Included are distributors or publishers of the computer software, CD-ROM discs, videodiscs, and videotapes listed in the Resources and References sections for all chapters. Also provided are the addresses of other sources, organizations, and distributors with audiovisual and other materials appropriate for introductory biology courses.

A.D.A.M. Software, Inc., 1899 Powers Ferry Road, Suite 460, Marietta, GA 30067. (800) 755-ADAM. http://www.aw.com/bc/sci/media

AIMS Media, 6901 Woodley Avenue, Van Nuys, CA 91406. (818) 785-4111.

American Educational Films, P.O. Box 8188, Nashville, TN 37207.

American Heart Association National Center, 7272 Greenville Avenue, Dallas, TX 75231-4596. A source for general information on nutrition and cardiovascular disease.

American Institute for Cancer Research, 803 West Broad Street, Falls Church, VA 22046. A source of materials for discussing changes in lifestyle as a way to reduce the risk of cancer.

American Medical Association, Department of Marketing Communications, 535 N. Dearborn Street, Chicago, IL 60610. (312) 645-5000.

American Nuclear Society, P.O. Box 97781, Chicago, IL 60678-7781. (800) 323-3044. An excellent source of information on radioactivity.

The Annenberg/CPB Collection: Corporation for Public Broadcasting, The Annenberg/CPB Project, 901 E. Street NW, Washington DC 20004-2037. (800) LEARNER.

Appleton-Century-Crofts, Publishers, P.O. Box 186, 150 White Plains Road, Tarrytown, NY 49201.

Arthur Mokin Productions, Inc., 2900 McBride Lane, Santa Rosa, CA 95401. (707) 542-4868.

Benchmark Films, 145 Scarborough Road, Briarcliff Manor, NY 10510. (914) 762-3838.

Berlet Films, 1646 W. Kimmel Road, Jackson, MI 49201.

BFA Educational Media, 468 Park Avenue South, New York, NY 10016. (800) 221-1274.

BioLearning, 420 Lexington Avenue, Suite 2735, New York, NY 10017.

William C. Brown, Publishers, 2460 Kerper Boulevard, P.O. Box 539, Dubuque, IA 52001. Product Information, (800) 351-7671; Educational Services Department, (800) 228-0459; Orders, (800) 338-5578. http://www.wcbp.com

Bullfrog Films, Oley, PA 19547. (215) 779-8226.

Cambridge Development Laboratory, Inc., 214 Third Avenue, Waltham, MA 02154. (800) 637-0047.

Carolina Biological Supply Company, 2700 York Road, Burlington, NC 27215. (800) 334-5551.

Churchill Films, 662 North Robertson Boulevard, Los Angeles, CA 90069-9990. (213) 657-5110.

Classroom Consortia Media, Inc., 57 Bay Street, Staten Island, NY 10301. (800) 237-1113.

Colour Images Unlimited, Inc., 8645 Kirk Avenue, Colorado Springs, CO 80908. (303) 495-4751.

Conduit, University of Iowa, Oakdale Campus, Iowa City, IA 52242.

Coronet/MTI Film and Video, 108 Wilmont Road, Deerfield, IL 60015. (800) 777-8100.

CRM/McGraw-Hill, 674 Via de la Valle, P.O. Box 641, Del Mar, CA 92014. (619) 453-5000.

Diversified Educational Enterprises, Inc., 725 Main Street, Lafayette, IN 47901.

Document Associates/The Cinema Guild, 1697 Broadway, Suite 802, New York, NY 10019. (212) 246-5522.

Edmund Scientific Company, 101 East Gloucester Pike, Barrington, NJ 08007-1380. (609) 573-6886.

Educational Dimensions Group, P.O. Box 126, Stamford, CT 06904.

Educational Images Ltd., P.O. Box 3456, West Side Station, Elmira, NY 14905. (607) 732-1090.

Educational Materials and Equipment Company (EME), P.O. Box 2805, Danbury, CT 06813-2805. (800) 848-2050.

Educational Media Corporation, 2036 Lemoyne Avenue, Los Angeles, CA 90026.

Education Express, 123 Skokie Valley Road, Suite 200, Highland Park, IL 60035. (800) 733-3396.

EduTech Software, 1927 Culver Road, Rochester, NY 14609. (716) 482-3151.

Encyclopaedia Britannica Educational Corporation, 425 North Michigan Avenue, Chicago, IL 60611. (800) 558-6968.

Exeter Software, 100 North Country Road, Building B., Setauket, NY 11733. (800) 842-5892.

Filmmakers Library, Inc., 133 E. 58th Street, Suite 703A, New York, NY 10022. (212) 355-6545.

Films for the Humanities and Sciences, P.O. Box 2053, Princeton, NJ 08543-2053 (800) 257-5126.

Films Incorporated, 733 Green Bay Road, Wilmette, IL 60091. (800) 323-4222.

Focus Media, Inc., 839 Stewart Avenue, P.O. Box 865, Garden City, NY 11530. (800) 645-8989.

Guidance Associates, Communications Park Box 3000, Mount Kisco, NY 10549-0900. (800) 431-1242.

HarperCollins Media, 10 East 53rd Street, New York, NY 10022. (800) 331-3761. http://harpercollins.com

Hawkhill Associates, Inc., 125 East Gilman Street, Madison, WI 53703. (800) 422-4295.

Health Sciences Center for Educational Resources, SB-56, University of Washington, Seattle, WA 98195. (206) 685-1186.

D. C. Heath and Company, Publishers, College Division, 125 Spring Street, Lexington, MA 02173. (800) 235-3565.

Human Relations Media, 175 Tompkins Avenue, Pleasantville, NY 10570. (800) 431-2050.

Intellimation, Intellimation Library for the Macintosh, Dept. 3SDA, 130 Cremona Drive, P.O. Box 1922, Santa Barbara, CA 93116-1922.

International Film Bureau, Inc., 322 South Michigan Avenue, Chicago, IL 60604. (312) 427-4545.

International Science Film Collection, AV Services, Special Services Building, Pennsylvania State University, University Park, PA 16502. (814) 865-6316.

J & S Software, 140 Reid Avenue, Port Washington, NY 11505.

Kalmia Science Series, Kalmia Company, Department B2, 21 West Circle, Concord, MA 01742. (617) 369-1829. Super-8 color film loops on a variety of biological topics.

Karol Media, 625 From Road, Paramus, NJ 07652. (201) 262-4170.

Key Line Educational Materials, P.O. Box 166, Cedarburg, WI 53012-0166. (414) 375-1999.

Landmark Films, 3450 Blade Run Drive, Falls Church, VA 22942.

Laser Learning Technologies, Inc., 120 Lakeside Avenue, Suite 240, Seattle, WA 98122-6552. (800) 722-3505.

Learning Corporation of America, 108 Wilmot Road, Deerfield, IL 60015. (800) 621-2131.

Life Science Associates, 1 Fenimore Road, Bayport, NY 11705. (516) 472-2111.

Logal Software, Inc., 125 Cambridge Park Drive, Cambridge, MA 02140. (800) LOGAL-USA.

Lucerne Media, 37 Ground Pine Road, Morris Plains, NJ 07950. (800) 341-2293.

Marty Stouffer Productions, P.O. Box 5057, Aspen, CO 81611.

Mastervision, 969 Park Avenue, New York, NY 10029. (212) 879-0448.

Media Design Associates, P.O. Box 3189, Boulder, CO 80307-3189. (303) 443-2800.

Media Guild, 11526 Sorrento Valley Road, Suite J, San Diego, CA 92121. (619) 755-9191.

Milner Fenwick, Inc., 2125 Greenspring Drive, Timonium, MD 21093. (800) 638-8652.

National Association of Biology Teachers, Membership Department, 11250 Roger Bacon Drive, #19, Reston, VA 22090.

National Center for Science Education, P.O. Box 9477, Berkeley, CA 94709-0477. (800) 290-6006. http://www.NatCenSciEd.org

National Film Board of Canada, 1251 Avenue of the Americas, 16th Floor, New York, NY 10020. (212) 586-5131.

National Geographic Society, Educational Services, Washington DC 20036. (800) 368-2728.

National Medical Audiovisual Center, 2111 Plaster Bridge Road, Atlanta, GA 30324.

National Science Teachers Association, 1742 Connecticut Avenue NW, Washington, DC 20009. (202) 328-5800. http://www.nsta.org

National Society of Genetic Counselors, Inc., 233 Canterbury Drive, Wallingford, PA 19086-6617. (215) 872-7608.

OmegaWare, P.O. Box 8024, Fort Collins, CO 80526. (303) 491-7858.

Optical Data Corporation, 30 Technology Drive, Warren, NJ 07060-0990. (800) 524-2481.

PBS Catalogs, 1320 Braddock Place, Alexandria, VA 22314. (800) 424-7963.

Penn Communications, 8425 Peach Street, P.O. Box 13010, Erie, PA 16514-1310. (814) 864-3278.

Perennial Education, Inc., 930 Pitner Avenue, Evanston, IL 60202. (800) 323-9084.

Perspective Films and Video, 65 E. South Water Street, Chicago, IL 60601. (312) 977-4000.

Phoenix Films, Inc., 468 Park Avenue South, New York, NY 10016. (800) 221-1274.

Plexus Communications Corporation, 15760 Ventura Boulevard, Suite 532, Encino, CA 91436. (800) 423-3061.

Projected Learning Programs, Inc., P.O. Box 3008, Paradise, CA 95967-3008. (800) 248-0757.

Pyramid Films, P.O. Box 1048, Santa Monica, CA 90406. (800) 421-2304.

QUEUE Software, Inc., 338 Commerce Drive, Fairfield, CT 06430. (800) 232-2224.

RMI Media Productions, 2807 W. 47th Street, Shawnee Mission, KS 66205. (913) 262-3974.

Scholastic Software, 730 Broadway, New York, NY 10003. (800) 541-5513.

Stanton Films, 2417 Artesia Boulevard, Redondo Beach, CA 90078.

Sterling Educational Films, 241 E. 34th Street, New York, NY 10016.

Sunburst Communications, Wings for Learning, 1600 Green Hills Road, P.O. Box 660002, Scotts Valley, CA 95067-0002. (800) 321-7511.

TimeBox Inc., P.O. Box 3060, Station D, Ottawa, Ontario, Canada K1P 6H6. (613) 236-8969 or (908) 782-5643.

Time/Life Video, Distribution Center, P.O. Box 644, Paramus, NJ. (201) 843-4545.

Trinity Software, P.O. Box 960, Campton, NH 03223. (603) 726-4641.

Tufts University Diet & Nutrition Letter, P.O. Box 57857, Boulder, CO, 80322-7857.

University of California, Extension Media Center, 2223 Fulton Street, Berkeley, CA 94720. (415) 642-0460.

Ventura Educational Systems, 3440 Brokenhill Street, Newbury Park, CA 91320.

Videodiscovery, 1700 Westlake Avenue N., Suite 600, Seattle, WA 98109-3012. (800) 548-3472.

WNET Video Distribution, P.O. Box 2284, South Burlington, VT 05407. (800) 336-1917.